Lecture Notes in Computer Science 5313

Vittorio Maniezzo Roberto Battiti
Jean-Paul Watson (Eds.)

Learning and Intelligent Optimization

Second International Conference, LION 2007 II
Trento, Italy, December 8-12, 2007
Selected Papers

Volume Editors

Vittorio Maniezzo
Università di Bologna
Scienze dell'Informazione
Bologna, Italy
E-mail: vittorio.maniezzo@unibo.it

Roberto Battiti
Università degli Studi di Trento
Trento, Italy
E-mail: roberto.battiti@unitn.it

Jean-Paul Watson
Sandia National Laboratories
Albuquerque, NM, USA
E-mail: jwatson@sandia.gov

Library of Congress Control Number: 2008941292

CR Subject Classification (1998): F.1, F.2, G.1.6, G.1-2, I.2.6, I.2.8

LNCS Sublibrary: SL 1 – Theoretical Computer Science and General Issues

ISSN 0302-9743
ISBN-10 3-540-92694-1 Springer Berlin Heidelberg New York
ISBN-13 978-3-540-92694-8 Springer Berlin Heidelberg New York

springer.com

Printed in Germany

Typesetting: Camera-ready by author, data conversion by Scientific Publishing Services, Chennai, India
Printed on acid-free paper SPIN: 12592319 06/3180 5 4 3 2 1 0

Preface

This volume collects the accepted papers presented at the Learning and Intelligent OptimizatioN conference (LION 2007 II) held December 8–12, 2007, in Trento, Italy.

The motivation for the meeting is related to the current explosion in the number and variety of heuristic algorithms for hard optimization problems, which raises numerous interesting and challenging issues. Practitioners are confronted with the burden of selecting the most appropriate method, in many cases through an expensive algorithm configuration and parameter-tuning process, and subject to a steep learning curve. Scientists seek theoretical insights and demand a sound experimental methodology for evaluating algorithms and assessing strengths and weaknesses. A necessary prerequisite for this effort is a clear separation between the algorithm and the experimenter, who, in too many cases, is "in the loop" as a crucial intelligent learning component. Both issues are related to designing and engineering ways of "learning" about the performance of different techniques, and ways of using memory about algorithm behavior in the past to improve performance in the future. Intelligent learning schemes for mining the knowledge obtained from different runs or during a single run can improve the algorithm development and design process and simplify the applications of high-performance optimization methods. Combinations of algorithms can further improve the robustness and performance of the individual components provided that sufficient knowledge of the relationship between problem instance characteristics and algorithm performance is obtained.

This meeting aimed at exploring the boundaries and uncharted territories between machine learning, artificial intelligence, mathematical programming and algorithms for hard optimization problems. The main purpose of the event was to bring together experts from these areas to discuss new ideas and methods, challenges and opportunities in various application areas, general trends and specific developments.

This second edition of LION received 48 submitted papers, with an approximate 50% acceptance rate, and 50 participants from 18 different countries. Eighteen papers were selected for inclusion in the conference proceedings following a rigorous review process.

The Conference Chair, Steering Committee and Local Chair, and the Technical Program Committee Chair wish to thank all the colleagues involved in the organization for their precious and professional contribution, including the Technical Program Committee listed here, the Steering Committee members Holger Hoos and Mauro Brunato, the Tutorial Chair David Woodruff, the IEEE Computational Intelligence Society Liaison Andrea Bonarini, the Publicity Chair Kenneth Sorensen, the local organization team Alessandro Villani, Roberto Cascella, Elisa Cilia, Paolo Campigotto, the Web Chair Franco Mascia and, last but not least, the Publication Liaison Thomas Stützle.

Technical Co-sponsorship was granted by the IEEE Computational Intelligence Society (local chapter) and by Associazione Italiana per l'Intelligenza Artificiale.

Financial support for this event was provided by our industrial sponsors, Eurotech Group S.p.A., which generously sponsored the two best papers awards, ESTECO, and Ars Logica IT Laboratories.

Vittorio Maniezzo
Roberto Battiti
Jean-Paul Watson

Organization

Conference Chair	Vittorio Maniezzo, Università di Bologna, Italy
Steering Committee and Local Chair	Roberto Battiti, University of Trento, Italy
Technical Program Committee Chair	Jean-Paul Watson, Sandia National Laboratories, USA

Technical Program Committee

Ethem Alpaydin	Bogazici University, Turkey
Roberto Battiti	University of Trento, Italy
Mauro Brunato	University of Trento, Italy
J. Christopher Beck	University of Toronto, Canada
Christian Blum	Universitat Politécnica de Catalunya, Spain
Immanuel Bomze	University of Vienna, Austria
Andrea Bonarini	Politecnico di Milano, Italy
Juergen Branke	University of Karlsruhe, Germany
Jehoshua (Shuki) Bruck	California Institute of Technology, Pasadena, CA, USA
Carlos Cotta	Universidad de Málaga, Spain
Valentin Cristea	Politehnica University of Bucharest, Romania
Marco Dorigo	Université Libre de Bruxelles, Belgium
Eugene Freuder	University College Cork, Ireland
Lee Giles	The Pennsylvania State University, USA
Michel Gendreau	Université de Montréal, Canada
Fred W. Glover	University of Colorado, USA
Marco Gori	University of Siena, Italy
Youssef Hamadi	Microsoft Research, Cambridge, UK
Geir Hasle	SINTEF Applied Mathematics, Norway
Holger Hoos	University of British Columbia, Canada
Bernardo Huberman	Hewlett - Packard, USA
Narendra Jussien	Ecole des Mines de Nantes, France
Zeynep Kiziltan	University of Bologna, Italy
Richard E. Korf	UCLA, USA
Michail G. Lagoudakis	Technical University of Crete, Greece
Vittorio Maniezzo	Università di Bologna, Italy
Elena Marchiori	Vrije Universiteit Amsterdam, The Nederlands
Lyle A. McGeoch	Amherst College, USA
Peter Merz	Technische Universität Kaiserslautern, Germany
Zbigniew Michalewicz	School of Computer Science, University of Adelaide, Australia

Nenad Mladenovic	School of Mathematics, Brunel University, West London, UK
Pablo Moscato	The University of Newcastle, Australia
Amiram Moshaiov	Tel-Aviv University, Israel
Raymond Ng	University of British Columbia, Canada
Panos Pardalos	University of Florida, USA
Marcello Pelillo	Università "Ca' Foscari" di Venezia, Italy
Vincenzo Piuri	Università di Milano, Italy
Christian Prins	University of Technology of Troyes, France
Franz Rendl	Institut für Mathematik, Universität Klagenfurt, Austria
Andrea Schaerf	University of Udine, Italy
Marc Schoenauer	INRIA, France
Yaroslav D. Sergeyev	Università della Calabria, Italy
Marc Sevaux	University of South-Brittany, France
Patrick Siarry	Université Paris XII Val De Marne, Paris, France
Thomas Stützle	Université Libre de Bruxelles, Belgium
Éric Taillard	University of Applied Sciences of Western Switzerland, Switzerland
Stefan Voss	Institute of Information Systems, University of Hamburg, Germany
Benjamin W. Wah	University of Illinois, Urbana-Champaign, USA
Jean-Paul Watson	Sandia National Laboratories, USA
David Woodruff	University of California, Davis, USA

Steering Committee	Roberto Battiti, Holger Hoos, Mauro Brunato
Tutorial Chair	David Woodruff
IEEE Computational Intelligence Society Liaison	Andrea Bonarini
Publicity Chair	Kenneth Sorensen
Publication Liaison	Thomas Stützle
Local Organization	Alessandro Villani, Roberto Cascella, Elisa Cilia, Paolo Campigotto
Organization Support	Stefano Poletti, Azienda Digitale
Web Chair	Franco Mascia

Industrial Sponsors

Eurotech Group S.p.A.

ESTECO

ARS LOGICA, IT Laboratories

Table of Contents

Nested Partitioning for the Minimum Energy Broadcast Problem*

Sameh Al-Shihabi[1], Peter Merz[2], and Steffen Wolf[2]

[1] Industrial Engineering Department, University of Jordan, Amman 11942, Jordan
s.shihabi@ju.edu.jo
[2] Distributed Algorithms Group, University of Kaiserslautern, Germany
{pmerz,wolf}@informatik.uni-kl.de

Abstract. The problem of finding the broadcast scheme with minimum power consumption in a wireless ad-hoc network is NP-hard. This work presents a new hybrid algorithm to solve this problem by combining Nested Partitioning with Local Search and Linear Programming. The algorithm is benchmarked by solving instances with 20 and 50 nodes where results are compared to either optimum or best results found by an IP solver. In these instances, the proposed algorithm was able to find optimal and near optimal solutions.

1 Introduction

Wireless ad-hoc networks have become very popular, as they are easily set up and do not need a wired backbone structure [1]. The nodes in such networks are usually battery powered, so the wireless ad-hoc network is a good choice for a first responders infrastructure, or even as the main communications infrastructure in regions where installing a wired infrastructure would be too expensive or time consuming.

Communication between the nodes in such ad-hoc networks can be performed by either a single hop, or by relaying the messages over intermediate nodes. To this end, each node is able to adjust its transmission power based on the distance to the receiver. Using omnidirectional antennas also brings the advantage of simple local broadcasts, as all nodes within the transmission range can receive the message without additional cost at the sender. This property of the wireless transmission is often referred to as the *wireless multicast advantage*. Because of the limited battery power of each node, it is crucial to find ways of communication that minimize the energy consumption.

One special kind of communication pattern is the one-to-all communication pattern (broadcast). Here, one source node needs to distribute information to all other nodes. Broadcast routing in wireless ad-hoc networks differs largely from routing in wired networks. In wireless settings such a broadcast can be achieved by simply adjusting the transmission power of the source to reach all

* Work was done while visiting the Distributed Algorithms Group at the University of Kaiserslautern.

V. Maniezzo, R. Battiti, and J.-P. Watson (Eds.): LION 2007 II, LNCS 5313, pp. 1–11, 2008.

nodes in the ad-hoc network in one hop. However, because of the physical laws of the power consumption over the distance, the total energy consumption can often be reduced by using intermediate nodes [2]. E. g., if the power consumption is proportional to the squared distance (this is the case when there are no obstacles), it is twice as expensive to send directly to the destination, instead of sending to a node half the way between sender and destination and have it relay the information to the final node.

In the case of a broadcast, we are looking for the broadcast tree that minimizes the total energy consumption. This problem is known as the Minimum Energy Broadcast problem (MEB) [3]. In this paper we present a new hybrid heuristic for the MEB. The steering component of the proposed algorithm is the Nested Partition (NP) algorithm [4]. NP is a global optimization algorithm that can be used for both stochastic [5] and deterministic problems [6,7]. The method works by successively partitioning regions expected to contain the best solution into smaller ones, where more concentrated sampling takes place, until a singleton is reached. The algorithm keeps a global view by aggregating the abandoned regions and sampling them. It backtracks to a larger region of the sample space if the abandoned regions are found to be better than the partitioned subregions. This behavior allows the algorithm to converge to the optimal solution with a positive probability.

This work presents a mixture of NP, Linear programming relaxation and a local search heuristic. The LP relaxation is used to find a lower bound for each subregion. If the global best solution is better than this bound, the corresponding subregion will not be sampled. The quality of the samples is further improved by a local search. It needs to be noted that the convergence of the algorithm to the optimum solution depends on the correct selection of the most promising subregion for further partitioning. This in turn depends on the quality of samples generated and number of samples taken from each subregion.

The paper is structured as follows. In the remainder of this section we give a formal definition of the MEB and summarize related work. In Section 2 we present our heuristic, and then give results of experiments carried out with this heuristic in Section 3. Section 4 summarizes our findings and gives an outline for future research.

1.1 Minimum Energy Broadcast

The Minimum Energy Broadcast Problem (MEB) is an NP-hard optimization problem [8,9]. It is also known under the name of Minimum Power Broadcast (MPB) or Minimum Energy Consumption Broadcast Subgraph (MECBS). The MEB can be defined as the problem of finding the broadcast tree $T = (V, E_T)$ (a directed spanning tree) rooted at a source node $s \in V$ in an ad-hoc wireless network $G = (V, E, d)$, that minimizes the necessary total transmission power $c(T)$ to reach all nodes of the network:

$$c(T) = \sum_{i \in V} \underbrace{\max_{(i,j) \in E_T} d(i,j)^\alpha}_{\text{transmission power of node } i}$$

Here, the distance function $d : E \rightarrow \mathbb{R}^+$ refers to the Euclidean distance and the constant α is the distance-power gradient which may vary from 1 to more than 6 depending on the environment [10]. Note that $(i, j) \in E_T$ does not imply $(j, i) \in E_T$, for T is a directed tree. Each node is required to send to the farthest child, all other children are then implicitly covered by this transmission. The leaves of the tree T do not send to other nodes and thus do not contribute to the total cost.

1.2 Related Work

The first major work on the MEB problem was the Broadcast Incremental Power algorithm (BIP) by Wieselthier *et al.* [3]. This heuristic builds the broadcast tree in a way that resembles Prim's algorithm for building Minimum Spanning Trees (MST). While Prim's algorithm does find the optimal MST, BIP does not necessarily provide an optimal solution for the MEB. The MST itself can also be used as a heuristic solution for the MEB, but the BIP explicitly exploits the wireless multicast advantage and thus produces solutions with lower cost than the corresponding MST solutions. The approximation ratio of MST is later shown to be 6 for the case of $\alpha \geq 2$ [10], while for $\alpha < 2$ the MST does not provide a constant approximation ratio [8]. The approximation ratio of BIP for $\alpha = 2$ is shown to be between $13/3$ and 6 [11].

The BIP heuristic can be further improved by a local search, e. g. r-shrink [12]. Here, the transmission power for one node is reduced by r steps, cutting off r nodes. These nodes will be assigned to other nodes, which increases the latter nodes' transmission power. If the total cost is not reduced, this change is rejected, otherwise it is accepted and the local search is repeated. Experiments showed that the initial BIP solutions could be improved considerably. Another improving heuristic called Embedded Wireless Multicast Advantage (EWMA) is presented in [9]. Here, the transmission power of a node is increased, such that other nodes can be switched off completely. This can be thought of as the opposite of the r-shrink heuristic. In [13], an Iterated Local Search heuristic is presented. It is based on an edge exchange neighborhood perturbation and the Largest Expanding Sweep Search local search (LESS, [14]), an improved local search based on EWMA. Unfortunately, there are no established standard test instances, so comparisons between the heuristics is difficult.

Several Mixed Integer Programming formulations (MIP) have been presented to compute optimal solutions [15,16]. While both approaches are based on a network flow model, the MIP from [16] uses an incremental mechanism over the transmission power variables, and is claimed to give better linear relaxations.

Nested Partitioning algorithms (NP) have been used to solve a number of problems such as the Traveling Salesman Problem [17], Product Design [6] and Scheduling parallel machines with flexible resources [7]. Local search heuristics can easily be implemented in NP such as using 2-exchange or 3-exchange local search in case of the TSP [17]. NP can also be used in combination with a number of meta-heuristic algorithms such as Genetic Algorithms [6] and Max-Min Ant Systems [18].

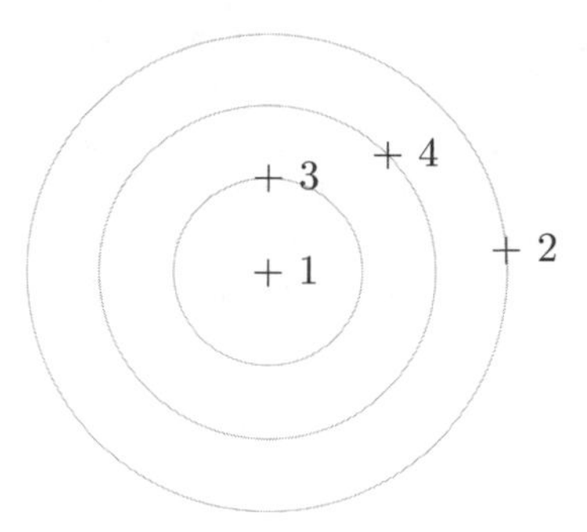

Fig. 1. An example network of four nodes in the plane. The different transmission ranges for source node 1 are shown. For $\alpha = 2$ the areas of the circles also represent the necessary transmission power.

2 Algorithm

The general structure for an NP algorithm is as follows. Starting with the whole feasible solution space Θ at $d = 1$ where d stands for depth, the NP algorithm partitions the feasible region into $M(d)$ disjoint subregions where samples are generated from each subregion. This process continues by partitioning the most promising subregion $\sigma(d) \in M(d)$ at each depth. To guarantee convergence, the abandoned subregions at $d > 1$ are aggregated in a subregion commonly called the surrounding region that is also sampled so the whole feasible solution space Θ is covered but with different sampling intensities. If the surrounding region is found to be more promising than the subregions forming $\sigma(d-1)$, the algorithm backtracks to a larger region. A simple example having four nodes, as shown in Fig. 1, is used to explain the NP algorithm for the MEB problem. Here, node 1 is the source node, and the remaining nodes ordered by distance from the source are 3, 4 and 2. The proposed algorithm consists of the following steps:

1. Partitioning
2. Finding Lower Bounds
3. Sampling
4. Calculating the Promising Index
5. Backtracking

2.1 Partitioning

In this work, we use a generic partitioning scheme [4]. The partitioning for our example is shown in Fig. 2. At depth 1, the feasible region Θ is divided into three subregions according to the different nodes the source node transmits to. Assuming that subregion 2 where the solution has the $1 \rightarrow 4$ arc is found to be the best subregion; at depth 2 the algorithm divides this subregion again into three subregions: one subregion having the arc $3 \rightarrow 4$ as part of the solution, a second having $3 \rightarrow 2$ and a third subregion where node 3 does not transmit

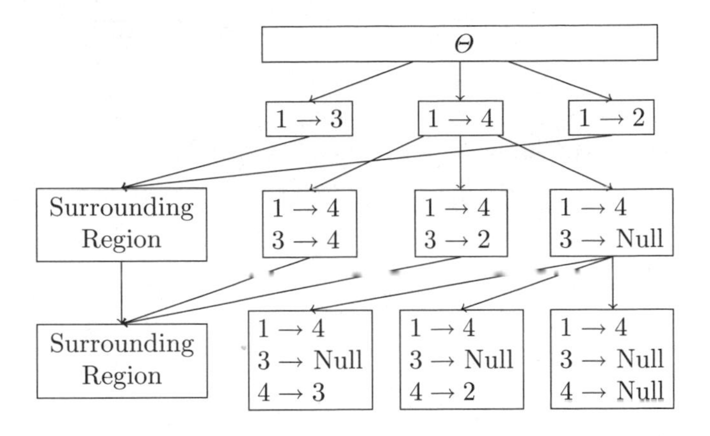

Fig. 2. A simple example for the partitioning steps. The feasible solution space Θ is divided in smaller regions by iteratively adding arcs to the model. Only the most promising region is expanded, while all remaining regions form the surrounding region.

to any other node (denoted by $3 \rightarrow$ Null). The arc $1 \rightarrow 4$ is imposed on all of these subregions. The surrounding region at this depth is $\Theta \setminus \{1 \rightarrow 4\}$. It needs to be noted, that subregion $3 \rightarrow 4$ is not sampled since it is guaranteed not to be optimal due to the extra redundant cost of $3 \rightarrow 4$. To finish the example, assuming that subregion $3 \rightarrow$ Null is the best subregion, then this subregion is partitioned into the subregions $4 \rightarrow 3$, $4 \rightarrow 2$, and $4 \rightarrow$ Null. The abandoned subregions are again aggregated and sampled. This process of partitioning the regions is continued until a singleton, i. e. a region with a single solution, is reached.

2.2 Finding Lower Bounds

Lower bounds for each subregion are found using the LP relaxation of the MIP formulation from [16], where the integrality constraints are relaxed by having $0 \leq y_{ij} \leq 1$. In this model, y_{ij} is a binary variable denoting that node i has transmission power to reach node j. To generate the three subregions at $d = 1$, a new constraint is added such that $y_{13} = 1$, $y_{14} = 1$, and $y_{12} = 1$ for each respective subregion. The lower bounds are calculated using an LP solver, and subregions having higher lower bounds than the global best solution S^* are not sampled. This reduces the number of subregions to be studied. The solution S^* found at any stage of the algorithm is passed to the appropriate subregion in the next depth or iteration of the algorithm.

2.3 Sampling

For subregions resulting from partitioning the most promising subregion, the sampling step begins by forcing the arcs found up to depth $d - 1$ to be part

```
while not all nodes are covered do
  let sum = 0
  for i = 1...n do
    for j = 1...n do
      if node i is covered and j is not then
        let P_ij = 1 / (Extra cost to reach j from i)
        let sum = sum + P_ij
      else
        let P_ij = 0
      end if
    end for
  end for
  generate random number u ∈ [0, 1]
  if u < q then
    select the arc(i, j) that has the highest P_ij, i.e. the lowest extra cost
  else
    randomly select arc(i, j) with probability P_ij/sum
  end if
end while
```

Fig. 3. Pseudo code for the weighted sampling step

of the solution. For each subregion, the arc that distinguishes it from the other subregions is also chosen to be part of the solution. The rest of the solution is generated using a weighted sampling step [4]. The pseudo code for this algorithm is shown in Fig. 3. Here, parameter $q \in [0, 1]$ controls the degree of diversification. With probability q the arc that increases the total energy consumption the least is chosen in each step. All other arcs are chosen according to their individual probability, which is again higher for arcs with less additional cost. For $q = 1$ this algorithm matches the BIP construction heuristic [3].

Sampling the surrounding region is done in a similar way but without imposing any arc on the generated samples. A local search algorithm (r-shrink with $r = 1$ as in [12]) is then applied to all samples taking into account not to change any of the arcs forced on the subregions. After the local search, the samples of the surrounding region are checked again, and all samples that follow exactly the chosen arcs up to depth $d - 1$ are disregarded since they effectively left the surrounding region.

2.4 Calculating the Promising Index

The promising index of the algorithm is the best sample found at each subregion. Again it needs to be noted that the solution S^* is inherited along the different depths. The most promising subregion is the one having the lowest feasible energy broadcasting. This region is then partitioned as explained earlier.

2.5 Backtracking

In case the surrounding region is found to be superior to the other subregions, backtracking takes place. In this work, a simple backtracking scheme is used where full backtracking takes place and the algorithm starts again with the whole feasible region. Other backtracking schemes are possible, such as backtracking step by step until the best region is not the surrounding region.

3 Results

A number of experiments have been conducted to check the quality of solutions obtained using the suggested hybrid algorithm. Two sets of problems have been generated where n nodes are randomly located in an area of $1000 \times 1000\,\mathrm{m}^2$. The Euclidean distance was used and the distance-power gradient was set to $\alpha = 2$. The first set has $n = 20$, while the second has $n = 50$. Each set contains 30 instances. The sets are available at `http://dag.informatik.uni-kl.de/research/meb/`.

We used the commercial LP solver ILOG CPlex 10.1 to obtain optimal solutions for the problem instances. Since the MEB is NP-hard, CPlex was only able to provide optimal solutions for the 20 nodes problems and about half of the 50 nodes problems. For the remaining problems, we stopped CPlex after 24 hours.

In the experiments, we set the number of samples that are generated for each subregion that passed the bounds test to 100. The value of q implemented in the samples generation algorithm is chosen as 0.5. These settings have proven to be a good choice in preliminary experiments. Again, CPlex 10.1 was used to calculate the lower bounds at each step of the algorithm. Each experiment was repeated 30 times and average values were used for the following discussion. Calculation times refer to the CPU time on a 3 GHz Pentium D running Linux 2.6; the algorithm was implemented in C++.

The results for the first ten of the 20 nodes problems are shown in Table 1, the NP algorithm shows the same behavior when applied to the other problems of this set. The NP algorithm was able to find the optimum in almost every run. Also, the non-optimal solutions are very close to the optimum, with average excess of less than one percent. Comparing the results of the NP against the results obtained by applying r-shrink to the BIP solution shows how much can be gained by using NP. In some of the instances, BIP+r-shrink already finds the optimal solution, but in general the BIP+r-shrink solution is 5 % to 33 % more expensive than the optimum. The NP results are also better than the results of the Iterated Local Search [13] (ILS), where an average excess of 1.1 % over the optimum is given for a similar setting ($n = 20$ placed in an area of $1000 \times 1000\,\mathrm{m}^2$, $\alpha = 2$, average over 1000 instances). However, a direct comparison is not possible, because different problem instances were used.

Table 2 shows the results for the 50 nodes problems. Optima are known only for about half of these instances. E. g., for problems p50.08 and p50.14, the NP algorithm found the proven optimum in 27 and 22 out of 30 runs. Also, the average excess in the remaining cases is quite low. Problem p50.02 is the hardest

Table 1. Results for the 20 nodes problems. The NP results are compared to the optimal solution and the solution found by BIP+r-shrink. Only the first ten instances are shown, the results for the remaining instances are similar.

Instance	Optimum	NP	BIP + r-shrink	Excess over Optimum	#Optimum found	CPU time
p20.0	407 250.81	407 250.81	467 919.92	0 %	30/30	0.30 s
p20.1	446 905.52	446 905.52	446 905.52	0 %	30/30	0.36 s
p20.2	335 102.42	335 102.42	335 102.42	0 %	30/30	0.41 s
p20.3	488 344.90	489 149.48	511 740.22	0.16 %	27/30	0.46 s
p20.4	516 117.75	516 117.75	615 991.07	0 %	30/30	0.43 s
p20.5	300 869.14	300 869.14	394 315.34	0 %	30/30	0.35 s
p20.6	250 553.15	250 553.15	332 998.09	0 %	30/30	0.18 s
p20.7	347 454.08	347 454.08	372 636.22	0 %	30/30	0.31 s
p20.8	390 795.34	390 795.34	390 795.34	0 %	30/30	0.46 s
p20.9	447 659.11	447 665.81	514 005.04	0 %	30/30	0.41 s
avg				0.06 %	28.7/30	0.33 s

problem in this set for the NP algorithm. Here, the optimum is not found, and the average excess is 10.11 %.

For the instances of the 50 nodes problems where no optimal solution is known we use the best solution found by CPlex within 24 hours as a comparison. We have also observed that instances which are harder to solve for CPlex are also harder for the NP algorithm. However, in some cases the NP algorithm still finds the best known solutions (e. g. p50.06). For all other instances, the NP algorithm finds solutions that are close to the best known solutions. The average excess over all instances of this set is 4.03 %.

The NP algorithm is again competitive as comparisons to other heuristics show. The BIP+r-shrink solutions are between 3 % and 37 % percent more expensive than the NP solutions. Comparison to ILS [13] for the 50 nodes problems can only be achieved through the average gap to the BIP solution (without r-shrink), which is about 19 % for ILS but 21 % for the NP. Here, a higher gap means a better solution, since the BIP is worse than both NP and ILS. This again shows the need for standard test instances.

In Fig. 4, we compare the optimal and the worst solution found by the NP algorithm for problem p50.19. As can be seen, both solutions show many similarities. Source node 3 sends to a majority of the nodes in both cases. A minor difference is in the left part, where node 37 sends to node 38. The main difference is in the bottom part, where the path from node 39 was not found and a path from node 29 is used instead.

The CPU times for the different problem sizes match the expected average time complexity of $\mathcal{O}(n^4)$. In each depth $1 \ldots n$ of the NP, n subregions are sampled, where the weighted sampling step uses $\mathcal{O}(n^2)$ time. However, when backtracking takes place more often, the time complexity will increase.

Table 2. Results for the 50 nodes problems. The NP results are compared to the optimal solution or the best known solution found by CPlex, and the solution found by BIP+r-shrink.

Instance	Optimum	NP	BIP + r-shrink	Excess over Best/Opt	#Best/Opt found	CPU time
p50.00	399 074.64	423 894.57	440 640.28	6.22 %	0/30	11.4 s
p50.01	≤ 373 565.15	387 316.68	475 102.25	3.68 %	0/30	7.1 s
p50.02	393 641.09	433 450.66	480 988.66	10.11 %	0/30	10.3 s
p50.03	316 801.09	337 165.59	386 205.59	6.43 %	0/30	6.1 s
p50.04	≤ 325 774.22	342 784.91	381 304.56	5.22 %	0/30	7.5 s
p50.05	382 235.90	394 791.97	422 809.25	3.28 %	1/30	10.9 s
p50.06	≤ 384 438.46	389 013.13	456 813.09	1.19 %	5/30	10.2 s
p50.07	≤ 401 836.85	428 741.55	461 307.09	6.70 %	0/30	8.9 s
p50.08	334 418.45	334 749.92	384 384.31	0.10 %	27/30	4.6 s
p50.09	≤ 346 732.05	378 630.95	399 725.84	9.20 %	0/30	12.9 s
p50.10	416 783.45	425 682.71	474 002.59	2.14 %	0/30	8.9 s
p50.11	≤ 369 869.41	385 915.15	411 906.13	4.34 %	0/30	7.7 s
p50.12	≤ 392 326.01	404 820.55	433 126.59	3.18 %	0/30	13.5 s
p50.13	≤ 400 563.83	427 105.91	485 333.75	6.63 %	0/30	11.2 s
p50.14	388 714.91	389 006.86	532 971.81	0.08 %	22/30	6.6 s
p50.15	371 694.65	373 179.70	427 741.66	0.40 %	0/30	8.1 s
p50.16	≤ 414 587.42	436 493.53	439 920.69	5.28 %	0/30	15.1 s
p50.17	355 937.07	363 652.32	387 976.03	2.17 %	1/30	11.9 s
p50.18	376 617.33	399 078.25	405 057.03	5.96 %	0/30	11.3 s
p50.19	335 059.72	342 670.93	451 377.38	2.27 %	5/30	9.8 s
p50.20	414 768.96	427 780.96	462 060.14	3.14 %	0/30	10.4 s
p50.21	≤ 361 354.27	371 950.59	443 070.76	2.93 %	2/30	10.4 s
p50.22	329 043.51	329 043.51	413 037.47	0 %	30/30	7.2 s
p50.23	383 321.04	407 803.26	425 242.32	6.39 %	0/30	11.1 s
p50.24	404 855.92	427 875.82	452 893.25	5.69 %	0/30	10.0 s
p50.25	363 200.32	363 200.32	471 153.44	0 %	30/30	3.2 s
p50.26	406 631.51	445 632.46	458 168.22	9.59 %	0/30	11.5 s
p50.27	451 059.62	469 912.83	525 401.41	4.18 %	0/30	9.5 s
p50.28	≤ 415 832.44	434 466.55	451 758.28	4.48 %	0/30	11.6 s
p50.29	380 492.77	380 492.77	452 424.68	0 %	30/30	5.9 s
avg				4.03 %	5.1/30	9.5 s

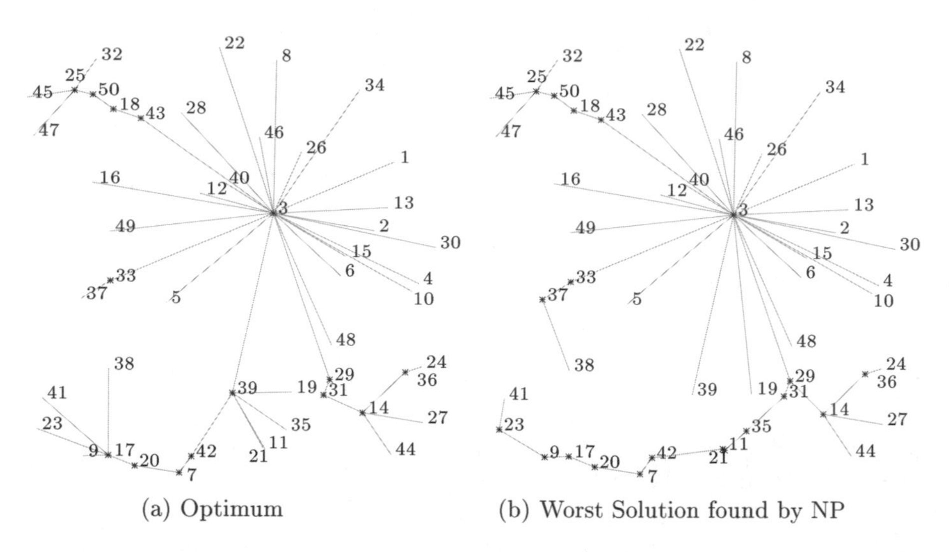

(a) Optimum (b) Worst Solution found by NP

Fig. 4. Optimal and sub-optimal solutions for problem p50.19. The source is node 3, and transmitting nodes are highlighted.

4 Conclusion

We have presented a hybrid heuristic for the Minimum Energy Broadcast problem. The major part of this work is the Nested Partitioning, where a generic partitioning scheme was implemented. LP relaxation was used as a probing operation to check which subregions are worth sampling. The r-shrink local search was also adjusted to fit the constraints imposed on the samples generated from each subregion. The algorithm has shown to find optimal or near-optimal solutions in short time.

Comparisons to other heuristics have shown that the proposed heuristic is competitive, but also that there is a need for standard test instances. We hope the instances provided in this paper can be used as such test instances.

A number of extensions are currently under study by the authors. The LP relaxations' solutions can offer additional information that can be used to generate better samples. Trying different local search heuristics in addition to improving the quality of the surrounding region samples is also investigated. Advances in exact solution techniques such as improving the bounds by using cutting planes or other relaxation techniques can also be implemented.

Acknowledgements

We would like to thank the DFG for sponsoring Sameh Al-Shihabi's visit to the Distributed Algorithms Group at the University of Kaiserslautern during the summer of 2007. This work was also partially supported by the Rhineland-Palatinate cluster of excellence DASMOD.

References

1. Haas, Z.J., Tabrizi, S.: On some challenges and design choices in ad-hoc communications. In: IEEE MILCOM 1998, Bedford, USA, pp. 187–192 (1998)
2. Rappaport, T.S.: Wireless Communications: Principles and Practices. Prentice Hall, Englewood Cliffs (1996)
3. Wieselthier, J.E., Nguyen, G.D., Ephremides, A.: On the construction of energy-efficient broadcast and multicast trees in wireless networks. In: Proceedings of the 19th IEEE INFOCOM 2000., pp. 585–594 (2000)
4. Shi, L., Ólafsson, S.: Nested partitions method for global optimization. Operations Research 48(3), 390–407 (2000)
5. Shi, L., Ólafsson, S.: Nested partitions method for stochastic optimization. Methodology and Computing in Applied Probability 2(3), 271–291 (2000)
6. Shi, L., Ólafsson, S., Chen, Q.: An optimization framework for product design. Management Science 47(12), 1681–1692 (2001)
7. Ólafsson, S., Shi, L.: A method for scheduling in parallel manufacturing systems with flexible resources. IIE Transactions 32(2), 135–146 (2000)
8. Clementi, A.E.F., Crescenzi, P., Penna, P., Rossi, G., Vocca, P.: On the complexity of computing minimum energy consumption broadcast subgraphs. In: Ferreira, A., Reichel, H. (eds.) STACS 2001. LNCS, vol. 2010, pp. 121–131. Springer, Heidelberg (2001)
9. Čagalj, M., Hubaux, J.P., Enz, C.: Minimum-Energy Broadcast in All-Wireless Networks: NP-Completeness and Distribution Issues. In: MobiCom 2002: Proceedings of the 8th Annual International Conference on Mobile Computing and Networking, pp. 172–182. ACM Press, New York (2002)
10. Ambühl, C.: An optimal bound for the MST algorithm to compute energy efficient broadcast trees in wireless networks. In: Caires, L., Italiano, G.F., Monteiro, L., Palamidessi, C., Yung, M. (eds.) ICALP 2005. LNCS, vol. 3580, pp. 1139–1150. Springer, Heidelberg (2005)
11. Wan, P.J., Călinescu, G., Li, X.Y., Frieder, O.: Minimum-energy broadcasting in static ad hoc wireless networks. Wireless Networks 8(6), 607–617 (2002)
12. Das, A.K., Marks, R.J., El-Sharkawi, M., Arabshahi, P., Gray, A.: r-shrink: A heuristic for improving minimum power broadcast trees in wireless networks. In: Global Telecommunications Conference, GLOBECOM 2003, pp. 523–527. IEEE, Los Alamitos (2003)
13. Kang, I., Poovendran, R.: Iterated local optimization for minimum energy broadcast. In: 3rd International Symposium on Modeling and Optimization in Mobile, Ad-Hoc and Wireless Networks (WiOpt), pp. 332–341. IEEE Computer Society, Los Alamitos (2005)
14. Kang, I., Poovendran, R.: Broadcast with heterogeneous node capability. In: Global Telecommunications Conference, GLOBECOM 2004, pp. 4114–4119. IEEE, Los Alamitos (2004)
15. Das, A.K., Marks, R.J., El-Sharkawi, M., Arabshahi, P., Gray, A.: Minimum power broadcast trees for wireless networks: Integer programming formulations. In: Proceedings of the 22nd IEEE INFOCOM 2003, pp. 1001–1010 (2003)
16. Montemanni, R., Gambardella, L.M., Das, A.: The minimum power broadcast problem in wireless networks: a simulated annealing approach. Wireless Communications and Networking Conference (WCNC) 4, 2057–2062 (2005)
17. Shi, L., Ólafsson, S., Sun, N.: New parallel randomized algorithms for the traveling salesman problem. Computers and Operations Research 26(4), 371–394 (1999)
18. Al-Shihabi, S.: Ants for sampling in the nested partition algorithm. In: Hybrid Metaheuristics, pp. 11–18 (2004)

An Adaptive Memory-Based Approach Based on Partial Enumeration

Enrico Bartolini[1], Vittorio Maniezzo[1], and Aristide Mingozzi[2]

[1] Department of Computer Science, University of Bologna, Italy
[2] Department of Mathematics, University of Bologna, Italy

Abstract. We propose an iterative memory-based algorithm for solving a class of combinatorial optimization problems. The algorithm generates a sequence of gradually improving solutions by exploiting at each iteration the knowledge gained in previous iterations. At each iteration, the algorithm builds an enumerative tree and stores at each tree level a set of promising partial solutions that will be used to drive the tree exploration in the following iteration.

We tested the effectiveness of the proposed method on an hard combinatorial optimization problem arising in the design of telecommunication networks, the Non Bifurcated Network Design Problem, and we report computational results on a set of test problems simulating real life instances.

1 Introduction

There is a vast class of combinatorial optimization problems exhibiting a regular substructure that can be decomposed into n smaller (and possibly easier) subproblems which are linked together by a set of coupling constraints. These problems can often be modeled by defining, for each subproblem k, a set $\mathcal{S}_k$ containing all the feasible solutions for subproblem k and by reformulating the coupling constraints so that the resulting problem consists in choosing, from each set $\mathcal{S}_k$, $k = 1, ..., n$, a single item $s^k_{i_k} \in \mathcal{S}_k$ in such a way that the selected items $X = \{s^1_{i_1}, ..., \bar{s}^n_{i_n}\}$ satisfy all the constraints. With each item $s^k_i \in \mathcal{S}_k$ is associated a cost c^k_i and the cost $c(X)$ of solution X is a function of the selected items. The objective is to find a solution X of minimum cost. As an example, consider the Multiple Choice Knapsack problem (MCKP). In the MCKP are given n item sets $\mathcal{S}_k$, $k = 1, ..., n$ and a bin of size W. With each item $s^k_i \in \mathcal{S}_k$ is associated a weight w^k_i. It is required to select exactly one item from each set so that the sum of the item weights does not exceed W and the sum of the item costs is minimized.

In this paper we describe an iterative heuristic algorithm, called F&B, that tries to avoid being trapped in local minima by adopting a memory-based look ahead strategy that exploits the knowledge gained in its past search history. Algorithm F&B iterates a partial exploration of the solution space by generating a sequence of enumerative trees of two types, called *forward* and *backward* trees.

V. Maniezzo, R. Battiti, and J.-P. Watson (Eds.): LION 2007 II, LNCS 5313, pp. 12–24, 2008.

Each node at level h of the trees represents a partial solution X' containing h items. At each iteration t, the algorithm generates a forward tree, if t is odd, or a backward tree if t is even. In generating a tree, each partial solution X' is extended to a feasible solution using the partial solutions generated at the previous iteration and the cost of the resulting solution is used to guess the quality of the best complete solution that can be obtain from X'.

An obvious non exact way for solving combinatorial optimization problems is using metaheuristcs, i.e., "master strategies that guide and modify other heuristics to produce solutions beyond those that are normally generated in a quest for local optimality" [6]. A vast body of literature exists describing metaheuristic approaches for solving combinatorial optimization problems, and surely there is no need to recap it here. The best known metaheuristics, like tabu search, VNS or genetic algorithms, are quite different from our proposal, however some other ones, such as the Pilot Method (C.Duin, S.Voss, 1999 [3]), the Filter&Fan method (F.Glover, 1998 [5]; P.Greistorfer, C.Rego, 2006 [7]) and the ANTS metaheuristic (Maniezzo et al., 1999 [10], 2002 [11]) have some similarities with the algorithm F&B described in this paper.

The Pilot Method consists in a partial enumeration strategy where the possible expansions of each partial solution are evaluated by means of a pilot heuristic [3]. The Filter&Fan method starts with a feasible solution S and builds an enumerative tree where branches correspond to submoves in the neighborhood space of S and each node corresponds to a solution obtained as a result of the submove sequence associated with the root-node path. The initial candidate list of moves is filtered at each level by evaluating each move in the list with respect to all the solutions at that level. The best moves at each level are included in the candidate list of the next level and the corresponding solutions are the nodes of the successive level. The ANTS metaheuristic is a particular instance of the ACO class, where ants are defined as computational agents which let iteratively grow a partial solution into a complete one. At each step ants compute a set of feasible expansions of the associated partial solution and choose one of these expansions according to bounds and previous search history. Variant BE-ANT [11] is particularely close to F&B, but is nondeterministic and lacks the forward - backward construction interleave.

The proposed algorithm F&B, that uses an interleaved sequence of forward and backward trees to evaluate the completion of partial solutions, is an alternative to the look-ahead strategies described above. F&B clearly is a metaheuristic, according to the definition of Glover and Laguna [6] given above. The subordinate heuristics is the partial enumeration method tht is embedded as a subroutine, while the master strategy dictates the forward - backward periodicity at the heart of search intensification. The remainder of this paper is organized as follows. In section 2 we give a detailed description of the algorithm, in section 3 we introduce the Non Bifurcated Network Design Problem (NBP) and we describe how the proposed algorithm can be tailored to solve this problem. Finally, in section 4 we report computational results on a set of large NBP instances simulating real life problems and we compare F&B against the commercial package CPLEX 10.1 and two other heuristics. Concluding remarks follow in section 5.

2 Algorithm F&B

A forward tree is an n-level tree where each level $h = 1, ..., n$ is associated with the item set $\mathcal{S}_h$ and each node at level h corresponds to a partial solution containing one item of each set $\mathcal{S}_1, \mathcal{S}_2, ..., \mathcal{S}_h$. Conversely, in a backward tree each level h is associated with the set $\mathcal{S}_{n-h+1}$, so that a node at level h, represents a partial solution containing one item of each set $\mathcal{S}_n, \mathcal{S}_{n-1}, ..., \mathcal{S}_{n-h+1}$. Associated with each level h of a tree built at iteration t there is a list, called $\mathcal{L}ist^t(h)$, containing Δ nodes generated at level h, where Δ is an a priori defined parameter. Once the tree at iteration t has been completely expanded, the nodes in the lists $\mathcal{L}ist^t(h)$, $h = 1, ..., n$, represent the algorithm memory of past iterations $1, ..., t$ that will be used to guide the tree exploration in the following iteration $t + 1$. In order to make the exposition simpler, in the following no distinction is made between a node and the corresponding partial solution.

2.1 Evaluation of Partial Solutions

The key idea is to evaluate the completion cost of each partial solution generated at level h of the tree at iteration t, using the partial solutions stored in $\mathcal{L}ist^{t-1}(n-h)$ at iteration $t-1$. Suppose we are building the forward tree associated with an odd iteration t. Let $\mathcal{T}(h)$, be the set of all partial solutions generated at level h and consider two partial solutions $X \in \mathcal{T}(h)$ and $\overline{X} \in \mathcal{L}ist^{t-1}(n-h)$. Notice that, since t is odd, X contains one item of the sets $\mathcal{S}_1, \mathcal{S}_2, ..., \mathcal{S}_h$ while $\overline{X}$ contains one item of each set $\mathcal{S}_n, \mathcal{S}_{n-1}, ..., \mathcal{S}_{h+1}$. These two solutions can be combined to obtain a (not necessarily feasible) complete solution $X \cup \overline{X}$ of cost $c\left(X \cup \overline{X}\right)$. Clearly, if the resulting solution $X \cup \overline{X}$ satisfies all the coupling constraints, the associated cost represents a valid upper bound. Then, at each iteration t algorithm F&B builds the associated tree and computes, for each node $X \in \mathcal{T}(h)$, a label $\beta(X)$ which is computed as follows:

$$\beta(X) = \min_{\overline{X} \in \mathcal{L}ist^{t-1}(n-h)} \left\{ c\left(X \cup \overline{X}\right) + \alpha\left(X \cup \overline{X}\right) \right\}, \tag{1}$$

where $\alpha(X \cup \overline{X})$ is a (strongly problem specific) function whose value is related to the degree of infeasibility of $X \cup \overline{X}$ and that is equal to 0 if $X \cup \overline{X}$ is a feasible solution.

When building the first forward tree at iteration $t = 1$ we assume that the lists $\mathcal{L}ist^0(h) = \emptyset$, $h = 1, ..., n$. Therefore, at iteration 1, expression (1) gives $\beta(X) = c(X)$, where $c(X)$ is the cost of the partial solution (X).

2.2 Description of Algorithm F&B

Let Δ be an a priori defined parameter that controls the number of nodes expanded at each level of both forward and backward trees. To expand level h of a tree at iteration t, the algorithm computes the value $\beta(X)$ for each node $X \in \mathcal{T}(h)$ and builds the set $\mathcal{L}ist^t(h) \subseteq \mathcal{T}(h)$ containing the Δ nodes in $\mathcal{T}(h)$ having the smallest label value $\beta(X)$. For every $X \in \mathcal{L}ist^t(h)$ such that $\beta(X)$ represents the cost of a feasible solution, we update $z_{best} = \min\{z_{best}, \beta(X)\}$, where z_{best} represents the cost of the best solution achieved by F&B and is initialized equal to ∞ at the beginning of the algorithm. Each node X included in $\mathcal{L}ist^t(h)$ is expanded to create a new node $X \cup \{s\}$ for each item s of the set S_{h+1} associated with level $h+1$. Notice that a feasibility test is required to eliminate any new node $X \cup \{s\}$ that violates some constraint.

Algorithm F&B terminates after $MAXT$ iterations (where $MAXT$ is an a priori defined parameter) or after two consecutive iterations where the value of z_{best} does not improve.

Figure 1 shows an example of algorithm F&B at iteration $t+1$ (even) expanding $\Delta = 2$ nodes per level. The label value for each node $X \in \mathcal{T}(4)$ at level 4 of the backward tree associated with iteration $t+1$ is computed using the partial solutions stored in $\mathcal{L}ist^t(6)$ of the forward tree computed at the previous iteration t. The Δ nodes having the smallest label value are then included in $\mathcal{L}ist^{t+1}(4)$ and further expanded.

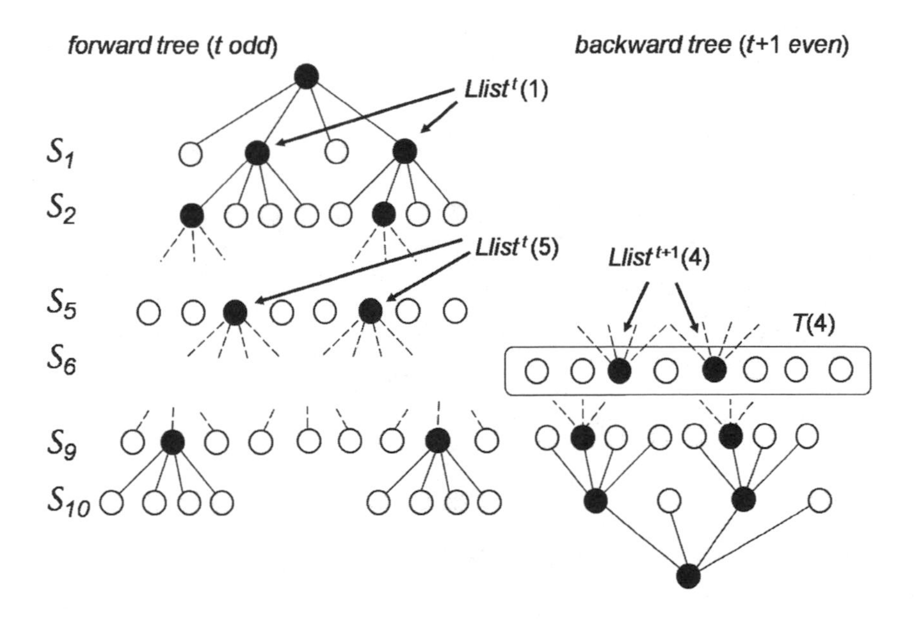

Fig. 1. Example of algorithm F&B

Algorithm 1. F&B

```
initialize t = 1, zbest = ∞ and flag = 0;
while t ⩽ MAXT do
    flag = flag + 1;
    [build the tree associated with iteration t]
    let List^t(0) = {{∅}};
    foreach level h = 1, ..., n do
        set T(h) = {∅};
        [generate the node set T(h) ]
        if t is odd then
            set k = h;
        else
            set k = n − h + 1;
        end
        foreach node X ∈ List^t(h − 1) do
            foreach item s ∈ S_k do
                let X' = X ∪ {s};
                if X' does not violate any constraint then
                    set T(h) = T(h) ∪ X';
                    if h = n and c(X') < zbest then
                        set zbest = c(X') and flag = 0;
                    end
                end
            end
        end
        [ extract the subset List^t(h) ∈ T(h)]
        foreach node X ∈ T(h) do
            compute β(X) according to expression (1) and let X̄ be the
            partial solution of List^{t−1}(n − h) producing the minimum in
            expression (1);
            if X ∪ X̄ is a feasible solution and c(X ∪ X̄) < z_best then
                update z_best = c(X ∪ X̄) and set flag = 0;
            end
        end
        if |T(h)| ⩽ Δ then
            set List^t(h) = T(h);
        else
            let List^t(h) be the set containing the Δ partial
            solutions of T(h) having the smallest label β(·);
        end
    end
    if flag = 2 then
        the upper bound z_best has not been improved in the last two
        consecutive iterations: stop;
    end
end
```

3 An Application of Algorithm F&B

In this section we describe an application of algorithm F&B for solving an optimization problem arising in the design of telecommunication networks, called Non Bifurcated Network Loading Problem (NBP), (see Barahona, 1996 [1]).

3.1 The Non-bifurcated Network Loading Problem

The NBP asks to connect a set of cities by a given network installing integer multiples of a fixed base capacity on its edges in order to route a set of commodities. Each commodity consists of a flow that must be sent from an origin node to a destination node through a single path along the network. The total capacity installed between two nodes allows traffic on both directions and must be greater than or equal to the total flow on each direction. A fixed cost must be paid to install a base capacity on each link of the network and each commodity must pay a routing cost to pass through any link. The objective is to install sufficient capacity so that all the commodities can be routed, while minimizing the sum of capacity and routing costs.

The NBP plays a fundamental role in the design of telecommunications networks running asynchronous transfer mode (ATM) protocol and production-distribution with single sourcing and express package delivery and has been studied in many variants with respect to network layout, capacity usage and commodity routing options. The majority of methods proposed in the literature for solving network design problems are restricted to the special case where a single base capacity can be installed on each arc and where the flow of each commodity can be split among different paths. For this variant see for example Ghamlouche, Crainic and Gendreau (2003) [4] who propose a tabu search using cycle-based neighborhood structures that take into account the impact on the total cost of potential modifications to the flow distribution of several commodities simultaneously. The only heuristic method we found for the NBP is due to Barahona (1996) [1] who solves the corresponding relaxation where commodity flows can be split among different paths using a branch-and-cut algorithm and then uses a heuristic procedure to obtain a solution where each commodity is assigned to a single path.

3.2 Solving the NBP Using Algorithm F&B

Let $G = (V, E)$ be a connected and undirected graph associated with the network where V is the set of nodes representing the cities and E is the set of edges representing the links. We denote by (i_e, j_e) the endpoints of edge $e \in E$. Let $\overrightarrow{G} = (V, A)$ be a directed graph associated with G where A is the set of arcs obtained from E by replacing every edge $e \in E$ with two arcs in opposite directions, i.e. $A = \{(i_e, j_e), (j_e, i_e) : e \in E\}$. The mapping $e(i, j)$ gives the edge of E corresponding to arc $(i, j) \in A$. It is given a set R of n commodities where each commodity $k = 1, ..., n$ specifies a flow d_k that must be sent through a single path from an origin node $s_k \in V$ to a destination node $t_k \in V$. Let r_{ij}^k be the cost for routing commodity k through arc $(i,j) \in A$. On each edge $e \in E$ can be installed integer multiples of a base capacity $u_e \in Z^+$ at unit cost g_e. The total capacity installed on edge e represents an upper bound for the flow on edge e in each direction, so that the required capacity on each edge e is determined by the maximum flow on the corresponding arcs (i_e, j_e) and (j_e, i_e).

The NBP can be modeled using, for each commodity $k = 1, ..., n$, a set $\mathcal{S}_k$ containing all the simple paths in $\overrightarrow{G}$ from the origin node $s_k \in V$ to the destination node $t_k \in V$. Then, an NBP solution is represented by an ordered list of n paths, one path from each set $\mathcal{S}_k$, $k = 1, ..., n$. Consider an enumerative tree where the nodes at level h represent a set $\mathcal{T}(h)$ of partial solutions generated at level h involving commodities $1, 2, ..., h$. Each partial solution $X \in \mathcal{T}(h)$ is represented by an ordered list of h paths, i.e. $X = (s_{j_1}^1, ..., s_{j_k}^k, ..., s_{j_h}^h)$ where $s_{j_k}^k$ is the $j_k - th$ element of the path set $\mathcal{S}_k$, $k = 1, ..., h$. For each path $s_{j_k}^k \in \mathcal{S}_k$ of commodity k let $[f_{ij}(s_{j_k}^k)]$ be a $(0-1)$ matrix where $f_{ij}(s_{j_k}^k)$is equal to one if path $s_{j_k}^k$ uses arc $(i,j) \in A$, zero otherwise. For each partial solution $X = (s_{j_1}^1, ..., s_{j_k}^k, ..., s_{j_h}^h)$ let $q_{ij}(X)$ and $y_e(X)$ be, respectively, the total flow on each arc $(i,j) \in A$ and the minimum number of base capacities on each edge $e \in E$ required by solution X. We have:

$$q_{ij}(X) = \sum_{k=1}^{h} d_k f_{ij}(s_{j_k}^k), \quad \forall (i,j) \in A \tag{2}$$

and

$$y_e(X) = \max \left[\left\lceil \frac{q_{i_e j_e}(X)}{u_e} \right\rceil, \left\lceil \frac{q_{j_e i_e}((X)}{u_e} \right\rceil \right] \forall e \in E. \tag{3}$$

The cost $c(X)$ of partial solution X is given by:

$$c(X) = \sum_{e \in E} g_e y_e(X) + \sum_{k=1}^{h} \sum_{(i,j) \in A} r_{ij}^k f_{ij}(s_{j_k}^k). \tag{4}$$

The problem is then to select a single path from each path set $\mathcal{S}_k$, $k = 1, ..., n$, to obtain a complete solution $X = (s_{j_1}^1, ..., s_{j_k}^k, ..., s_{j_n}^n)$ of minimum cost $c(X)$.

Notice that, since there is no restriction on the maximum number of base capacities u_e that can be installed on each edge $e \in E$, at each iteration t of algorithm F&B any two partial solutions $X \in T(h)$ and $\overline{X} \in \mathcal{L}ist^{t-1}(n-h)$, always provide a feasible solution $X \cup \overline{X}$ of cost $c(X \cup \overline{X})$. This means that the label $\beta(X)$ of any node X, computed by means of expression (1), always represents a valid upper bound on the NBP. In order to use algorithm F&B for solving the NBP it is necessary to compute the path sets $\mathcal{S}_k$, $k = 1, ..., n$. Since these sets are typically exponential in size we limit $\mathcal{S}_k$, for each commodity $k = 1, ..., n$, to contain the largest subset of the least cost paths in $\overrightarrow{G}$, from s_k to t_k, such that $|\mathcal{S}_k| \leqslant \rho$, where ρ is an a-priori defined parameter. The paths included in each set $\mathcal{S}_k$ are computed by associating with each arc $(i,j) \in A$ a cost $\bar{g}_{ij}^k = g_{e(i,j)} \left\lceil \frac{d_k}{u_{e(i,j)}} \right\rceil + r_{ij}^k$. Notice that $\left\lceil \frac{d_k}{u_{e(i,j)}} \right\rceil$ represents the minimum number of links required to route commodity k through edge $e \in E$.

4 Computational Experiments

Algorithm F&B has been coded in ANSI C and experimentally compared on two classes of NBP instances with the the integer programming solver CPLEX 10.1, using an integer programming formulation of the NBP with two heuristic algorithms, called PEM and TPH, described in Bartolini and Mingozzi (2006) [2]. Algorithm TPH is a two phase heuristic that generates an initial feasible NBP solution in phase 1 and iteratively improves it in the second phase using a local search procedure. Algorithm PEM is a partial enumeration method that uses a modified version of heuristic TPH to compute an upper bound for each node of the enumerative tree. Algorith PEM can be viewed as a variation of the Pilot Method described in C.Duin and S.Voss, (1999) [3].

Barahona (1996) [1] proposed a set of test instances corresponding to practical problems arising in the design of telecommunication networks. Since the test instances described in Barahona (1996) are not publicly available, we randomly generated a set of test instances sharing the same network structure (i.e., the underlaying network graphs correspond to complete graphs). In computing the base capacity size and installation costs we used the information provided in Kousik, Ghosh and Murthy, (1993) [8]. Moreover, since the model described in Barahona (1996) does not take into account the commodity routing costs we set all routing costs equal to 0. All the instances generated correspond to complete undirected graphs with 30 nodes and one commodity for each edge. The instances are partitioned in two classes, A and B, with respect to the method used for computing the base capacity installation costs $\{g_e\}$. For each class we randomly generated 5 instances as follows:

- the node set V is randomly generated in a square $[3000 \times 3000]$;
- the edge capacity u_e is set equal to 56 units for each edge $e \in E$, imitating the link capacity of DS0 channels (see Kousik, Ghosh and Murthy, (1993) and Magnanti, Mirchandani and Vachani, (1995) [9]);
- there is a commodity k_e associated with each edge $e \in E$, having as origin and destination nodes the endpoints of edge e (i.e. $s_{k_e} = i_e$ and $t_{k_e} = j_e$);
- the commodity demands are integers chosen from the set $\{8, 16, 24\}$ with probability 70%, 20% and 10% respectively;
- all routing costs $\{r_{ij}^k\}$ are set equal to 0.

In computing the installation costs $\{g_e\}$ for class A instances, we used the costs for the annual leasing of DS0 channels as reported in Kousik, Ghosh and Murthy, (1993). Let $ed(i,j)$ be the Euclidean distance between nodes $i, j \in V$, then the costs $\{g_e\}$ are computed as follows:

$$g_e = \begin{cases} 232 + 7.74 \cdot ed(i_e, j_e) & \text{if } ed(i_e, j_e) \leqslant 50, \\ 435 + 3.68 \cdot ed(i_e, j_e) & \text{if } 50 < ed(i_e, j_e) \leqslant 100, \\ 571 + 2.32 \cdot ed(i_e, j_e) & \text{if } 100 < ed(i_e, j_e) \leqslant 500, \\ 1081.4 + 1.30 \cdot ed(i_e, j_e) & \text{if } ed(i_e, j_e) > 500. \end{cases} \quad (5)$$

For class B instances the costs $\{g_e\}$ are computed by setting:

$$g_e = 572 + ed(i_e, j_e). \quad (6)$$

We made several experiments to identify good parameter settings for our algorithm F&B. As a result we decided to use the following set of parameters.

- parameter ρ was set equal to 100,
- parameter $MAXT$ was set equal to 50,
- parameter Δ ranges between 1 and 20, as no dominating value could be found, as detailed below.

In our experiments we found that algorithm F&B is particularly sensitive to the value of parameter Δ that controls how many nodes are expanded at each level of a tree. We found that in most cases the best results were obtained when using small values of Δ, i.e. $\Delta \leqslant 20$. We noticed that by using larger values of Δ the algorithm was able to obtain better results in the first iterations but it was then unable to improve them significantly later on, given the time bound for the runs and the increasing slowness of each iteration for increasing Δ values. As an example, figure 2 plots the best upper bound achieved by F&B on a class B instance (problem p2-b), in one hour of CPU time, when Δ ranges between 1 and 100 and shows that the best solutions are achieved when expanding a small number of nodes at each level. This behavior was common for all the NBP instances, showing that the computing time is better invested in performing more iterations (thus refining the algorithm's knowledge of the solution space) rather than trying a wider exploration. This could be specially true at earlier iterations, when the algorithm memory is largely ineffective in guiding the exploration toward good solutions. Figure 3 shows the percentage distance of the obtained solution costs, for increasing values of Δ, with respect to the best known values, averaged over all class A instances. The general structure is the same as that of figure 2, even though smoothed by the averaging.

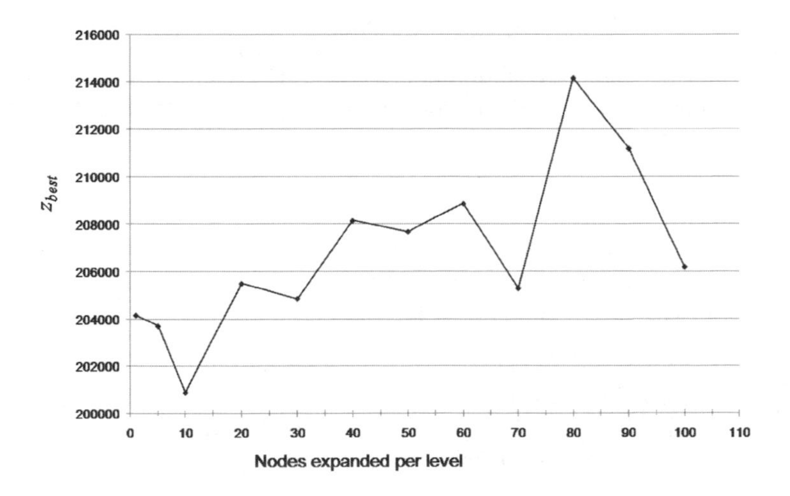

Fig. 2. Problem p2-b: best results achieved for increasing values of the parameter Δ

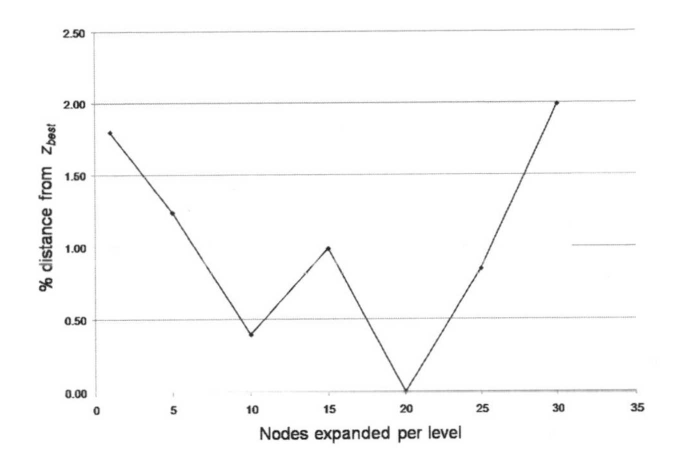

Fig. 3. Class A instances: percentage distance from the best result achieved for increasing values of the parameter Δ

In the following tables 1 and 2 we report computational results obtained on NBP instances of class A and B. We compare the results achieved within 1 hour time limit by the integer programming solver CPLEX10.1 , by algorithm F&B using different values of the parameter Δ and by the heuristic algorithms PEM and TPH. In CPLEX we activated both the Node heuristic and the RINS heuristics, moreover, the final solution obtained by CPLEX within the imposed time limit was further improved by allowing CPLEX to run the Solution Polishing heuristic for an extra hour of computing time. All the computational tests were run using an Intel Pentium D 3.2GHz. equipped with 3Gb of RAM.

Tables 1 and 2 report for each algorithm the following columns:

- z_{UB}: cost of the best solution achieved by the algorithm within the time limit,
- Gap: percentage distance between the cost of the best known solution (z_{best}) and the cost of the solution found by the algorithm (i.e. Gap $= 100 \times (z_{UB} - z_{best})/z_{best}$),
- Time: total computing time in seconds.

From the results reported in tables 1 and 2 it is clear that it is not possible to choose a value of Δ which gives the best results on all the instances, even though, on average, the best results are obtained using $\Delta = 10$, for class B instances, and $\Delta = 20$, for class A instances. Note moreover, that the value of the parameter Δ also impacts significantly on the total computing time so that smaller values of Δ can provide a better trade off between solution quality and computing time.

Tables 1 and 2 show that algorithm F&B outperforms all other algorithms on the NBP instances under consideration and achieves, within the 3600 sec. time limit, much better solutions than CPLEX does in 7200 seconds. Moreover, algorithm TPH produces in a few seconds better solutions than CPLEX using two hours of computing time.

Table 1. Results on class A instances

Prob.	CPLEX [1]		F&B ($\Delta = 1$)			F&B ($\Delta = 5$)			F&B ($\Delta = 10$)			F&B ($\Delta = 20$)			PEM [2]		TPH		
	zub	gap	zub	gap	time	zub	gap	time	zub	gap	time	zub	gap	time	zub	gap	zub	gap	time
p1-a	320281	16.56	279340	1.66	14	275705	0.34	213	**274781**	0.00	1263	275381	0.22	3197	293058	6.65	294034	7.01	31
p2-a	337136	17.40	291269	1.43	16	296623	3.29	234	**287163**	0.00	893	290144	1.04	2531	303799	5.79	314051	9.36	32
p3-a	309701	10.21	295853	5.28	17	286697	2.02	399	285442	1.58	992	**281015**	0.00	2571	300875	7.07	306463	9.06	31
p4-a	331173	17.02	285773	0.98	17	**283007**	0.00	198	284547	0.54	754	284684	0.59	2695	298088	5.33	307136	8.53	27
p5-a	327313	15.20	288574	1.57	14	290839	2.37	245	289003	1.72	735	**284118**	0.00	3568	302164	6.35	312829	10.11	32
Avg.	325121	15.28	288162	2.18	16	286574	1.60	258	284187	0.77	927	**283068**	0.37	2912	299597	6.24	306903	8.81	31

(1) Reaches the time limit of 3600 sec. in all problems and uses an additional.
3600 sec. for running the CPLEX Solution Polishing heuristic.
(2) Reaches the time limit of 3600 sec. in all problems.

Table 2. Results on class B instances

Prob.	CPLEX [1]		F&B ($\Delta = 1$)			F&B ($\Delta = 5$)			F&B ($\Delta = 10$)			F&B ($\Delta = 20$)			PEM [2]		TPH		
	zub	gap	zub	gap	time	zub	gap	time	zub	gap	time	zub	gap	time	zub	gap	zub	gap	time
p1-b	225785	17.08	197747	2.54	16	200125	3.78	424	**192842**	0.00	1886	197249	2.29	3600	207824	7.77	210122	8.96	30
p2-b	232121	15.55	204162	1.63	15	203715	1.41	281	**200890**	0.00	605	205503	2.30	3600	218781	8.91	219064	9.05	33
p3-b	216794	14.58	191162	1.03	18	190516	0.69	273	**189214**	0.00	1333	191609	1.27	3600	196450	3.82	205648	8.69	29
p4-b	197212	12.44	175931	0.31	18	179121	2.13	268	**175389**	0.00	1026	178965	2.04	3600	188224	7.32	191218	9.03	32
p5-b	220353	20.32	185760	1.43	21	184460	0.72	463	183312	0.10	976	**183138**	0.00	3600	194955	6.45	195481	6.74	33
Avg.	218453	15.99	190952	1.39	18	191587	1.74	342	**188329**	0.02	1165	191293	1.58	3600	201247	6.85	204307	8.49	31

(1) Reaches the time limit of 3600 sec. in all problems and uses an additional.
3600 sec. for running the CPLEX Solution Polishing heuristic.
(2) Reaches the time limit of 3600 sec. in all problems.

5 Conclusions

This paper presents a new metaheuristic approach, named F&B for Forward - Backward partial enumeration. The method is a general metaheuristic for combinatorial optimization problems, and can be directly applied to all problems that can be decomposed into smaller subproblems which are linked together by a set of coupling constraints. The essential trait of this new method is the guidance of a partial enumeration search by means of alternate forward and backward visits of the enumerative tree associated with the solution space, where each visit makes use of the results of the previous one for estimating partial solution completion costs. This corresponds to a memory-based look ahead strategy that exploits the knowledge gained in previous iterations for escaping from local minima.

We tested the effectiveness of the proposed method solving a hard combinatorial optimization problem arising in the design of telecommunication networks, called Non Bifurcated Network Design Problem, and we report computational results on a set of test problems simulating real life instances. On all instances the new method proved able to outperform competitive approaches, be they alternative heuristics from the literature or an advanced usage of a commercial MIP solver.

References

1. Barahona, F.: Network design using cut inequalities. SIAM Journal on Optimization 6(3), 823–837 (1996)
2. Bartolini, E., Mingozzi, A.: Algorithms for the non-bifurcated capacitated network design problem. Report, Computer Science, Univ. of Bologna (2006)
3. Duin, C., Voß, S.: The pilot method: A strategy for heurisic repetition with application problem in graphs. Networks 34, 181–191 (1999)
4. Ghamlouche, I., Crainic, T.G., Gendreau, M.: Cycle-based neighbourhoods for fixed-charge capacitated multicommodity network design. Operations Rsearch 51(4), 655–667 (2003)
5. Glover, F.: A template for scatter search and path relinking. In: Hao, J.-K., Lutton, E., Ronald, E., Schoenauer, M., Snyers, D. (eds.) AE 1997. LNCS, vol. 1363, pp. 3–51. Springer, Heidelberg (1998)
6. Glover, F., Laguna, M.: Tabu Search. Kluwer, Boston (1997)
7. Greistorfer, P., Rego, C.: a simple filter-and-fan approach to the facility location problem. Computers & Operations Research 33, 2590–2601 (2006)
8. Kousik, I., Ghosh, D., Murthy, I.: A heuristic procedure for leasing channels in telecommunications networks. The Journal of The Operational Research Society 44(7), 659–672 (1993)
9. Magnanti, T.L., Mirchandani, P., Vachani, R.: Modeling and solving the two-facility capacitated network loading problem. Operations Research 43, 142–157 (1995)
10. Maniezzo, V.: Exact and approximate nondeterministic tree-search procedures for the quadratic assignment problem. INFORMS J. on Computing 11(4), 358–369 (1999)
11. Maniezzo, V., Milandri, M.: An ant-based framework for very strongly constrained problems. In: ANTS 2002: Proceedings of the Third International Workshop on Ant Algorithms, pp. 222–227. Springer, Heidelberg (2002)

Learning While Optimizing an Unknown Fitness Surface*

Roberto Battiti, Mauro Brunato, and Paolo Campigotto

DISI - Dipartimento di Ingegneria e Scienza dell'Informazione,
Università di Trento, Italy
battiti@disi.unitn.it

Abstract. This paper is about Reinforcement Learning (RL) applied to online parameter tuning in Stochastic Local Search (SLS) methods. In particular a novel application of RL is considered in the Reactive Tabu Search (RTS) method, where the appropriate amount of diversification in prohibition-based (Tabu) local search is adapted in a fast online manner to the characteristics of a task and of the local configuration. We model the parameter-tuning policy as a Markov Decision Process where the states summarize relevant information about the recent history of the search, and we determine a near-optimal policy by using the Least Squares Policy Iteration (LSPI) method. Preliminary experiments on Maximum Satisfiability (MAX-SAT) instances show very promising results indicating that the learnt policy is competitive with previously proposed reactive strategies.

1 Reinforcement Learning and Reactive Search

Reactive Search (RS) [1,2,3] advocates the integration of sub-symbolic machine learning techniques into search heuristics for solving complex optimization problems. The word *reactive* hints at a ready response to events during the search through an internal online feedback loop for the self-tuning of critical parameters. When Reactive Search is applied to local search (Reactive Local Search or RLS), its objective is to maximize a given function $f(x)$ by analyzing the past local search history (the trajectory of the tentative solution in the search space) and by learning the appropriate balance of intensification and diversification. In this manner the knowledge about the task and about the local properties of the *fitness surface* surrounding the current tentative solution can influence the future search steps to render them more effective.

Reinforcement Learning (RL) arises in the different context of machine learning, where there is no guiding teacher, but *feedback signals from the environment* which are used by the learner to modify its future actions. Think about bicycle riding: after some initial trials with positive or negative rewards, in the form of admiring friends or injuries to biological tissues, the goal is accomplished. The

* Work supported by the project CASCADAS (IST-027807) funded by the FET Program of the European Commission.

V. Maniezzo, R. Battiti, and J.-P. Watson (Eds.): LION 2007 II, LNCS 5313, pp. 25–40, 2008.

reinforcement learning context is more difficult than the one of supervised learning, where a teacher gives examples of correct outputs: in RL one has to make a *sequence of decisions* (e.g., about steering wheel rotation). The outcome of each decision is not fully predictable. In addition to an immediate *reward*, each action causes a change in the system state and therefore a different context for the next decisions. To complicate matters the reward is often delayed and one aims at maximizing not the immediate reward, but some form of *cumulative reward* over a sequence of decisions. This means that greedy policies do not always work. In fact, it can be better to go for a smaller immediate reward if this action leads to a state of the system where bigger rewards can be obtained in the future. Goal-directed learning from interaction with an (unknown) environment with trial-and-error search and delayed reward is the main feature of RL.

As it was suggested for example in [4], the issue of learning from an initially unknown environment is therefore shared by RS and RL. A basic difference is that RS optimizes a function and the environment is provided by a fitness surface to be explored, while RL optimizes the long-term reward obtained by selecting actions at the different states. The sequential decision problem and therefore the non-greedy nature of choices is also common. For example, in Reactive Tabu Search (the application of RS in the context of Tabu Search), steps leading to worse configurations need in some cases to be performed to escape from a basin of attraction around a local optimizer. It is therefore of interest to investigate the relationship in more detail, to see whether specific techniques of Reinforcement Learning can be profitably used in Reactive Search.

This paper is organized as follows. First the basics of RL learning and neuro-dynamic programming are summarized. Then the relationship between RL and RS are investigated, also with reference to existing work bridging the border between optimization and RL. Finally, the novel proposal is presented, together with the first obtained experimental results.

2 Reinforcement Learning and Neuro-dynamic Programming Basics

In this section, *Markov Decision Processes* are formally defined and the standard Dynamic Programming technique to determine the optimal policy is introduced in Sec. 2.2. In many practical cases exact solutions must be abandoned in favor of approximation strategies, which are the focus of Sec. 2.4.

2.1 Markov Decision Processes

A standard Markov process is given by a set of states $\mathcal{S}$ with transitions between them described by probabilities $p(i,j)$ (let us note the fundamental property of Markov models: earlier states do not influence the transition probabilities to the next state). Its evolution cannot be controlled, because it lacks the notion of decisions, *actions* taken depending on the current state and leading to a different state and to an immediate *reward*.

A Markov Decision Process (MDP) is an extension of the classical Markov process designed to capture the problem of *sequential decision making under uncertainty*, with states, decisions, unexpected results, and "long-term" goals to be reached. A MDP can be defined as a quintuple $(\mathcal{S}, \mathcal{A}, P, R, \gamma)$, where $\mathcal{S}$ is a set of states, $\mathcal{A}$ a finite set of actions, $P(s, a, s')$ is the probability of transition from state $s \in \mathcal{S}$ to state $s' \in \mathcal{S}$ if action $a \in \mathcal{A}$ is taken, $R(s, a, s')$ is the corresponding reward, and γ is the discount factor, in order to exponentially decrease future rewards. This last parameter is fundamental in order to evaluate the overall value of a choice when considering its consequences on an infinitely long chain. In particular, given the following evolution of a MDP

$$s(0) \stackrel{a(0)}{\rightarrow} s(1) \stackrel{a(1)}{\rightarrow} s(2) \stackrel{a(2)}{\rightarrow} s(3) \stackrel{a(3)}{\rightarrow} \dots \tag{1}$$

the cumulative reward obtained by the system is given by

$$\sum_{t=0}^{\infty} \gamma^t R(s(t), a(t), s(t+1)).$$

Note that state transitions are not deterministic, nevertheless their distribution can be controlled by the action a. The goal is to control the system in order to maximize the expected cumulative reward.

Given a MDP $(\mathcal{S}, \mathcal{A}, P, R, \gamma)$, we define a *policy* as a probability distribution $\pi(\cdot|s) : \mathcal{A} \to [0, 1]$, where $\pi(a|s)$ is the probability of choosing action a when the system is in state s. In other words, π maps states onto probability distributions over $\mathcal{A}$. Note that we are only considering stationary policies. If a policy is deterministic, then we resort to the more compact notation $a = \pi(s)$.

2.2 The Dynamic Programming Approach

The intelligent component goal is to select a policy that maximizes a measure of the total reward accumulated during an infinite chain of decisions (infinite-horizon). To achieve this goal, let us define the *state-action value function* $Q^\pi(s, a)$ of the policy π as the expected overall future reward for applying a specified action a when the system is in state s, in the hypothesis that the ensuing actions are taken according to policy π. A straightforward implementation of the Bellman principle leads to the following definition:

$$Q^\pi(s, a) = \sum_{s' \in \mathcal{S}} P(s, a, s') \Big(R(s, a, s') + \gamma \sum_{a' \in \mathcal{A}} \pi(a'|s') Q^\pi(s', a') \Big). \tag{2}$$

where the sum over $\mathcal{S}$ can be interpreted as an integral in case of a continuous state set. The interpretation is that the value of selecting action a in state s is given by the expected value of the immediate reward plus the value the future rewards which one expects by following policy π from the new state. These have to be discounted by γ (they are a step in the future w.r.t. starting immediately from the new state) and properly weighted by transition probabilities and action-selection probabilities given the stochasticity in the process.

The expected reward of a state/action pair $(s, a) \in \mathcal{S} \times \mathcal{A}$ is

$$R(s, a) = \sum_{s' \in \mathcal{S}} P(s, a, s') R(s, a, s'),$$

so that (2) can be rewritten as

$$Q^{\pi}(s, a) = R(s, a) + \gamma \sum_{s' \in \mathcal{S}} \left(P(s, a, s') \sum_{a' \in \mathcal{A}} \pi(a'|s') Q^{\pi}(s', a') \right)$$

or, in a more compact linear form,

$$Q^{\pi} = \boldsymbol{R} + \gamma \boldsymbol{P} \boldsymbol{\Pi}_{\pi} Q^{\pi} \tag{3}$$

where $\boldsymbol{R}$ is the $|\mathcal{S}||\mathcal{A}|$-entry column vector corresponding to $R(s, a)$, $\boldsymbol{P}$ is the $|\mathcal{S}||\mathcal{A}| \times |\mathcal{S}|$ matrix of $P(s, a, s')$ values having (s, a) as row index and s' as column, while $\boldsymbol{\Pi}_{\pi}$ is a $|\mathcal{S}| \times |\mathcal{S}||\mathcal{A}|$ matrix whose entry $(s, (s, a))$ is $\pi(a|s)$.

Equation (3) can be seen as a non-homogeneous linear problem with unknown Q^{π}

$$(\boldsymbol{I} - \gamma \boldsymbol{P} \boldsymbol{\Pi}_{\pi}) Q^{\pi} = R \tag{4}$$

or, alternatively, as a fixed-point problem

$$Q^{\pi} = \boldsymbol{T}_{\pi} Q^{\pi}, \tag{5}$$

where $\boldsymbol{T}_{\pi} : x \mapsto R + \gamma \boldsymbol{P} \boldsymbol{\Pi}_{\pi} x$ is an affine functional.

If the state set $\mathcal{S}$ is finite, then (3-5) are matrix equations and the unknown Q^{π} is a vector of size $|\mathcal{S}||\mathcal{A}|$.

In order to solve these equations explicitly, a model of the system is required, i.e., full knowledge of functions $P(s, a, s')$ and $R(s, a)$. When the system is too large, or the model is not completely available, approximations in the form of *reinforcement learning* come to the rescue. As an example, if a *generative model* is available, i.e., a black box that takes state and action in input and produces the reward and next state as output, one can estimate $Q^{\pi}(s, a)$ through *rollouts*. In each rollout, the generator is used to simulate action a followed by a sufficiently long chain of actions dictated by policy π. The process is repeated several times because of the inherent stochasticity, and averages are calculated.

The above described state-action value function Q, or its approximation, is instrumental in the basic methods of dynamic programming and reinforcement learning.

2.3 Policy Iteration

A method to obtain the optimal policy π^* is to generate an improving sequence (π_i) of policies by building a policy π_{i+1} upon the value function associated to policy π_i:

$$\pi_{i+1}(s) = \arg\max_{a \in \mathcal{A}} Q^{\pi_i}(s, a). \tag{6}$$

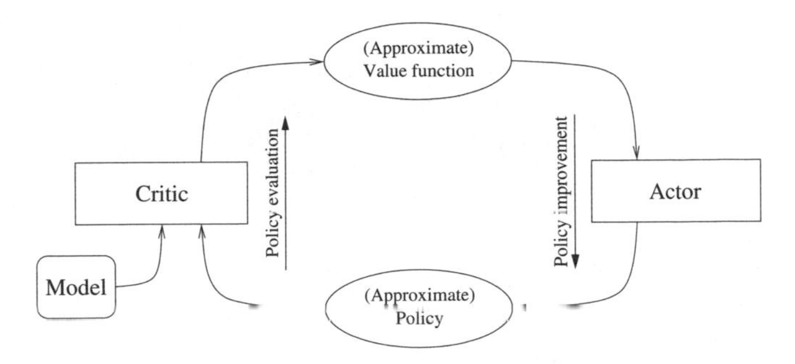

Fig. 1. The Policy Iteration (PI) mechanism

Policy π_{i+1} is never worse than π_i, in the sense that $Q^{\pi_{i+1}} \geq Q^{\pi_i}$ over all state/action pairs.

In the following, we assume that the optimal policy π^* exists in the sense that for all states it attains the minimum of the right-hand side of Bellman's equation, see [5] for more details.

The Policy Iteration (PI) method consists on the alternate computation shown in Fig. 1: given a policy π_i, the *policy evaluation* procedure (also known as the "Critic") generates its state-action value function Q^{π_i}, or a suitable approximation. The second step is the *policy improvement* procedure (the "Actor"), which computes a new policy by applying (6).

The two steps are repeated until the value function does not change after iterating, or when the change between consecutive iterations is less than a given threshold.

2.4 Approximations: Reinforcement Learning and LSPI

To carry out the above discussion by means of exact methods, in particular using (4) as the Critic component, the system model has to be known in terms of its transition probability $P(s, a, s')$ and reward $R(s, a)$ functions. In many cases this detailed information is not available but we have access to the system itself or to a *simulator*. In both cases, we have a black box which given the current state and the performed action determines the next state and reward. In both cases, more conveniently with a simulator, several sample trajectories can be generated, so that more and more information about the system behavior can be extracted aiming at optimal control.

A brute force approach can be that of estimating the system model functions $R(\cdot, \cdot, \cdot)$ and $R(\cdot, \cdot)$ by executing a very large series of simulations. The model-free *Reinforcement Learning* methodology bypasses the system model and directly learns the value function.

Assume that the system simulator (the "Model" box in Fig. 1) generates quadruples in the form

$$(s, a, r, s')$$

where s is the state of the system at a given step, a is the action taken by the simulator, s' is the state in which the system falls after the application of a, and

Variable	Scope	Description
$\mathcal{D}$	In	Set of sample vectors $\{(s, a, r, s')\}$
k	In	Number of basis functions
$\boldsymbol{\Phi}$	In	Vector of k basis functions
γ	In	Discount factor
π	In	Policy
$\boldsymbol{A}$	Local	$k \times k$ matrix
$\boldsymbol{b}$	Local	k-entry column vector
$\boldsymbol{w}^\pi$	Out	k-entry weight vector

1. **function** LSTDQ ($\mathcal{D}$, k, $\boldsymbol{\Phi}$, γ, π)
2. $\quad \boldsymbol{A} \leftarrow 0$;
3. $\quad \boldsymbol{b} \leftarrow 0$;
4. $\quad$ **for each** $(s, a, r, s') \in D$
5. $\quad\quad \boldsymbol{A} \leftarrow \boldsymbol{A} + \boldsymbol{\Phi}(s, a)\big(\boldsymbol{\Phi}(s, a) - \gamma\boldsymbol{\Phi}(s', \pi(s'))\big)^T$
6. $\quad\quad \boldsymbol{b} \leftarrow \boldsymbol{b} + r\boldsymbol{\Phi}(s, a)$
7. $\quad \boldsymbol{w}^\pi \leftarrow \boldsymbol{A}^{-1}\boldsymbol{b}$

Fig. 2. The LSTDQ algorithm [6]

r is the reward received. In the setting described by this paper, the (s, a) pair is generated by the simulator.

A viable method to obtain an approximation of the state-action value function is to approximate it with respect to a functional linear subspace having basis $\boldsymbol{\Phi} = (\phi_1, \ldots, \phi_k)$. The approximation $\hat{Q}^\pi \approx Q^\pi$ is in the form

$$\hat{Q}^\pi = \boldsymbol{\Phi}^T \boldsymbol{w}^\pi.$$

The weights vector $\boldsymbol{w}^\pi$ is the solution of the linear system $\boldsymbol{A}\boldsymbol{w}^\pi = \boldsymbol{b}$, where

$$\boldsymbol{A} = \boldsymbol{\Phi}^T(\boldsymbol{\Phi} - \gamma \boldsymbol{P}\boldsymbol{\Pi}_\pi\boldsymbol{\Phi}) \qquad \boldsymbol{b} = \boldsymbol{\Phi}^T R. \tag{7}$$

An approximate version of (7) can be obtained if we assume that a finite set of samples is provided by the "Model" box of Fig. 1:

$$\mathcal{D} = \{(s_1, a_1, r_1, s'_1), \ldots, (s_l, a_l, r_l, s'_l)\}.$$

In this case, matrix $\mathcal{A}$ and vector $\boldsymbol{b}$ are "learned" as sums of rank-one elements, each obtained by a sample tuple:

$$\boldsymbol{A} = \sum_{(s,a,r,s') \in \mathcal{D}} \boldsymbol{\Phi}(s, a)\Big(\boldsymbol{\Phi}(s, a) - \gamma\boldsymbol{\Phi}(s', \pi(s'))\Big)^T, \qquad \boldsymbol{b} = \sum_{(s,a,r,s') \in \mathcal{D}} r\boldsymbol{\Phi}(s, a).$$

These approximations lead to the Least Squares Temporal Difference for Q (LSTDQ) algorithm proposed in [6], and shown in Figure 2, where the functions $R(s, a)$ and $P(s, a, s')$ are supposed to be unknown and are replaced by a finite sample set $\mathcal{D}$.

Note that the LSTDQ algorithm returns the weight vector that best approximates in the least-squares fixed-point sense (within the spanned subspace and

Variable	Scope	Description
$\mathcal{D}$	In	Set of sample vectors $\{(s, a, r, s')\}$
k	In	Number of basis functions
$\boldsymbol{\Phi}$	In	Vector of k basis functions
γ	In	Discount factor
ϵ	In	Weight vector tolerance
$\boldsymbol{w}_0$	In	Initial value function weight vector
$\boldsymbol{w}'$	Local	Weight vectors in subsequent iterations
$\boldsymbol{w}$	Out	Optimal weight vector

```
1. function LSPI (D, k, Φ, γ, ε, w0)
2.    w' ← w0;
3.    do
4.       w ← w';
5.       w' ← LSTDQ (D, k, Φ, γ, w);
6.    while ||w − w'|| > ε
```

Fig. 3. The LSPI algorithm [6]

according to the sample data) the value function of a given policy π. It therefore acts as the "Critic" component of the Policy Iteration algorithm. The "Actor" component is straightforward, because it is an application of (6). The policy does not need to be explicitly represented: if the system is in state s and the current value function is defined by weight vector $\boldsymbol{w}$, the best action to take is

$$a = \arg\max_{a \in \mathcal{A}} \boldsymbol{\Phi}^T \boldsymbol{w}.$$

The complete LSPI algorithm is given in Fig. 3. Note that, because of the identification between the weight vector $\boldsymbol{w}$ and the ensuing policy π, the code assumes that the previously declared function `LSTDQ()` accepts its last parameter, i.e., the policy π, in form of a weight vector $\boldsymbol{w}$.

3 Reinforcement Learning for Optimization

Many are the intersections between optimization, Dynamic Programming and Reinforcement Learning. Approximated versions of DP/RL contain challenging optimization tasks, let's mention the maximization operations in determining the best action when an action value function is available, the optimal choice of approximation architectures and parameters in neuro-dynamic programming, or the optimal choice of algorithm details and parameters for a specific RL instance.

This paper, however, goes in the opposite direction: which techniques of RL can be used to improve heuristic algorithms for a standard optimization task such as minimizing a function? Interesting summaries of statistical machine learning for large-scale optimization are present in [7].

An application of RL in the area of local search for solving $\max_x f(x)$ is presented in [8]: the rewards from a local search method π starting from an initial configuration x are given by the size of improvements of the best-so-far value f_{best}. In detail, the value function $V^{\pi}(x)$ of configuration x is given by the expected best value of f seen on a trajectory starting from state x and following the local search method π. The curse of dimensionality discourages using directly x for state description: informative *features* extracted from x are used to compress the state description to a shorter vector $\boldsymbol{s}(x)$, so that the value function becomes $V^{\pi}(\boldsymbol{s}(x))$.

A second application of RL to local search is to supplement f with a "scoring function" to help in determining the appropriate search option at every step. For example, different basic moves or entire different neighborhoods can be applied. RL can in principle make more systematic some of the heuristic approaches involved in designing appropriate "objective functions" to guide the search process. An example is the RL approach to job-shop scheduling in [9,10], where a neural-network based $TD(\lambda)$ scheduler is demonstrated to outperform a standard iterative repair (local search) algorithm.

Also, tree-search techniques can profit from ML. It is well known that variable and value ordering heuristics (choosing the right order of variables or values) can noticeably improve the efficiency of complete search techniques, e.g. for constraint satisfaction problems. For example, RLSAT [11] is a DPLL solver for the Satisfiability (SAT) problem which uses experience from previous executions to learn how to select appropriate branching heuristics from a library of predefined possibilities, with the goal of minimizing the total size of the search tree, and therefore the CPU time. Lagoudakis and Littman [12] extend algorithm selection for recursive computation, which is formulated as a sequential decision problem. According to the authors, their work demonstrates that "some degree of reasoning, learning, and decision making on top of traditional search algorithms can improve performance beyond that possible with a fixed set of hand-built branching rules."

A different application is suggested in [5] in the context of constructive algorithms, which build a complete solution by selecting value for a component at a time. Let's assume that K fixed construction algorithms are available for the problem. The application consists of combining in the most appropriate manner the information obtained by the set of construction algorithms in order to fix the next index and value.

In the context of continuous function optimization, [13] uses RL for replacing a priori defined adaptation rules for the step size in Evolution Strategies with a reactive scheme which adapt step sizes automatically during the optimization process. The states are characterized only by the success rate after a fixed number of mutations, the three possible actions consists of increasing (by a fixed multiplicative amount), decreasing or keeping the current step size. SARSA learning with various reward functions is considered, including combinations of the difference between the current function value and the one evaluated at the last reward computation and the movement in parameter space (the distance traveled in the last phase). On-the-fly parameter tuning, or on-line calibration of parameters for evolutionary algorithms by reinforcement learning (crossover, mutation, selection operators, population size) is suggested in [14].

4 Reinforcement Learning for Reactive Tabu Search

This paper investigates a novel application of Reinforcement Learning in the framework of Reactive Tabu Search. An optimization algorithm operates a sequence of elementary actions (*local moves*, e.g., bit flips). The choice of the local move is driven by many different factors, in particular, most algorithms are *parametric*: their behavior (and their efficiency) depends on the values attributed to some free parameters, so that different instances of the same problem, and different configurations within the same instance, may require different parameter values.

This Section describes the proposed application of the LSPI algorithm to MAX-SAT: the Markov Decision Process (MDP) is described in Sec. 4.1, while the design of the basis function is described in Sec.4.2.

4.1 The Markov Decision Process Definition

The effect of a parameter change on the algorithm's behavior can only be evaluated after a significant number of local moves. As a consequence, also for performance reasons, algorithm parameters are not changed too often. We therefore divide the algorithm's trace into *epochs*, each composed of a suitable number of local moves, and to allow parameter changes only between epochs.

If the "state" of the system at the end of an epoch describes the algorithm's behavior during the last epoch, and an "action" is the modification of the algorithm's parameters before it enters the next epoch, then a local search algorithm can be modeled as a Markov Decision Process (MDP) and a Reinforcement Learning method such as LSPI can be used to control the evolution of its parameters.

The "state" should capture all criteria that we consider useful in order to decide how to change parameters in a proper way. Given the subdivision of the Local Search algorithm's trace into a sequence of epochs $(E_1, E_2, \dots)$, we define the state at the end of epoch E_i as a collection of features extracted from the algorithm's execution up to that moment in form of a tuple: $s(E_1, \dots, E_i) \in \mathbb{R}^d$, where d is the number of features that form the state. The features can be adequately normalized for better stability of the system. The cardinality of the action set $\mathcal{A}$ and the semantics of its elements changes according to the parameters required by the LS technique. Variable Neighborhood Search algorithms can define one action for each implemented neighborhood, or define just two actions (to be interpreted, e.g., as "widen" and "reduce") if the neighborhood set is ordered. Simulated Annealing, which basically depends on a continuous parameter T, can define two actions ("increase T" and "decrease T"). Likewise, a Tabu Search algorithm will increase or decrease the prohibition period T.

In this paper we consider a prohibition-based (Tabu) algorithm for the MAX-SAT problem [15]. It takes in input a CNF SAT instance (i.e., a Boolean formula being the conjunction of disjunctive clauses) and each algorithm step simply flips a variable. In particular, every variable is considered for flipping (i.e., non-prohibited) only if it hasn't been changed in the previous T iterations, T being the prohibition parameter to be controlled. At each iteration, the non-prohibited variable causing the largest increase in the number of satisfied clauses (or the lowest decrease, if no increase is possible) is selected for flipping. Ties are broken randomly. In this paper, T is assumed to take values over the interval $[T_{min,T_{max}}]$.

The Reinforcement Learning approach is exploited to adjust the prohibition parameter during the algorithm execution. Assume n and m the number of variables and clauses of the input SAT instance, respectively. Let $f(\boldsymbol{x})$ the score function counting the number of unsatisfied clauses in the truth assignment $\boldsymbol{x}$.

Each state of the MDP is created by observing the behavior of the Tabu search algorithm over an epoch of $2 * T_{max}$ consecutive variable flips. As in a prohibition mechanism with prohibition parameter T, during the first T steps, the Hamming distance keeps increasing and only in the subsequent steps it may decrease, an epoch is long enough to monitor the behavior of the algorithm also in the case of the largest allowed T value.

In particular, let us define the following:

- $\boldsymbol{x}_{\mathrm{bsf}}$ is the "best-so-far" configuration *before* the current epoch;
- T_{f} is the current fractional prohibition value (the actual prohibition period is

$$T = \lfloor nT_{\mathrm{f}} \rfloor \tag{8}$$

  );
- $\overline{f}_{\mathrm{epoch}}$ is the average value of f during the epoch;
- $\overline{H}_{\mathrm{epoch}}$ is the average Hamming distance during the current epoch from the configuration at the beginning of the current epoch itself.

These variables have been chosen because of the Reactive Search paradigm's concern on the trade-off between diversification (the ability to explore new configurations in the search space by moving away from local minima) and bias (the preference for configurations with low objective function values), so that changes in f and the Hamming distance are good representatives of the current state. Many possible choices based on these considerations have been tested. Furthermore, for the purpose of addressing uniformly SAT instances with different number of variables, the fractional prohibition value T_{f} is used rather than the prohibition value T. The compact state representation chosen to describe an epoch is the following triplet:

$$\boldsymbol{s} \equiv \left(\Delta f, \frac{\overline{H}_{\mathrm{epoch}}}{n}, T_{\mathrm{f}} \right), \qquad \text{where} \qquad \Delta f = \frac{\overline{f}_{\mathrm{epoch}} - f(\boldsymbol{x}_{\mathrm{bsf}})}{m}.$$

The first component is the mean change of f in the current epoch with respect to the best value; all components of the state have been normalized.

The actions set is composed by two choices: $\mathcal{A} = \{\text{increase}, \text{decrease}\}$, with the following effects:

- if $a =$ increase: $T_\text{f} \leftarrow \max\{T_\text{f} \cdot 1.1, T_\text{f} + 1/n\}$;
- if $a =$ decrease: $T_\text{f} \leftarrow \min\{T_\text{f}/1.1, T_\text{f} - 1/n\}$.

Changes in T_f are designed in order to ensure variation of at least 1 in the actual prohibition period T. In addition, T_f is bounded between a minimum and a maximum value (0 and .2 in our experiments).

The reward signal is given by the normalized change of the best value achieved in the observed epoch with respect to the "*best so far*" value *before* the epoch: $(f(\boldsymbol{x}_\text{bsf}) - f(\boldsymbol{x}_\text{localBest}))/m$.

4.2 Basis Function Definition

Among the various tests that have been executed, in this paper we concentrate on the following 13-function basis function set:

$$\boldsymbol{\Phi}(s,a) = \begin{pmatrix} I_\text{increase}(a) & I_\text{decrease}(a) \\ I_\text{increase}(a) \cdot \Delta f & I_\text{decrease}(a) \cdot \Delta f \\ I_\text{increase}(a) \cdot \overline{H}_\text{epoch} & I_\text{decrease}(a) \cdot \overline{H}_\text{epoch} \\ I_\text{increase}(a) \cdot \overline{H}_\text{epoch} \cdot \Delta f & I_\text{decrease}(a) \cdot \overline{H}_\text{epoch} \cdot \Delta f \\ I_\text{increase}(a) \cdot (\Delta f)^2 & I_\text{decrease}(a) \cdot (\Delta f)^2 \\ I_\text{increase}(a) \cdot \overline{H}^2_\text{epoch} & I_\text{decrease}(a) \cdot \overline{H}^2_\text{epoch} \\ & T_\text{f} + \frac{I_\text{increase}(a) - I_\text{decrease}(a)}{n} \end{pmatrix}, \tag{9}$$

where I_increase and I_decrease are the indicator functions for the two actions (1 if the action is the indicated one, 0 otherwise), discerning the "state-action" features for the two different actions considered.

5 Experimental Results

In order to test the performance of Reinforcement Learning for on-line parameter tuning in Reactive Tabu Search (RTS), we have implemented C++ functions for the Tabu Search method described in Sec. 4.1 and interfaced them to the Matlab LSPI implementation found in [16].

The experimental work includes the generation of a training set of samples discussed in Sec. 5.1, the generation of an optimal policy and in the preliminary comparison with other relevant SLS heuristics for MAX-SAT in Sec. 5.2

5.1 Training Examples Generation

The training examples are created by running the Tabu search algorithm over selected MAX-3-SAT random instances defined in [17]. In detail, we selected two ($n = 500, m = 5000$) instances and 6 different initial prohibition periods ($T_\text{f} = .01, .02, .05, .1, .15, .2$), and performed 2 runs of the algorithm for each

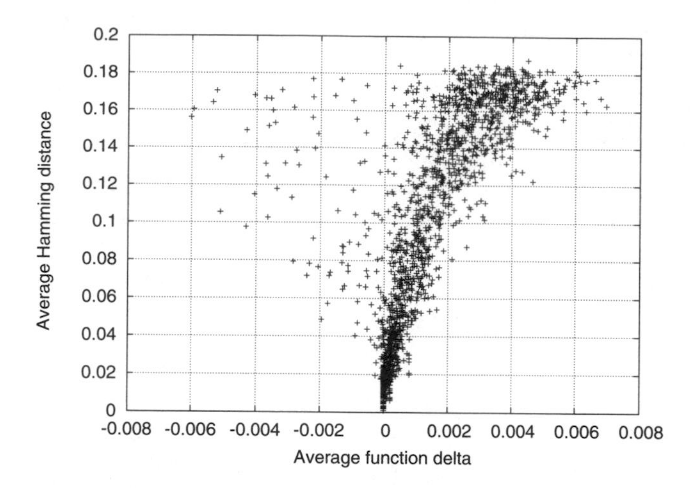

Fig. 4. Distribution of training sample states

combination with different randomly chosen starting truth assignments. Every run has been executed for 50 epochs to generate 50 training examples. The T_{f} parameter has been bounded in $[0, .2]$.

Each epoch is composed of 200 consecutive flips, as $T_{max} = 500 \cdot 0.2$ by Eq. 8.

Fig. 4 shows the distribution of examples states, projected onto the Δf and the $\overline{H}_{\mathrm{epoch}}$ state features.

5.2 Optimal Policy and Comparison

The LSPI algorithm has been applied to the training sample set, and with (9) as approximate space basis. The resulting approximate value function $\hat{Q}(s, a)$ is shown in Fig. 5 for the two actions, thus defining an approximation to the optimal policy. Note that the action "increase" is suggested in cases where the average Hamming distance between the configurations explored in the last epoch and the last local minimum does not exceed a certain value, provided that the current portion of landscape is not much worse than the previously explored regions. This policy is consistent with intuition: a higher value of T causes a larger differentiation of visited configurations (more different variables need to be flipped), and this is desired when the algorithm needs to escape the neighborhood of a local minimum; in this case, in fact, movement is limited because the configuration is trapped at the "bottom of a valley". On the other hand, when the trajectory is not within the attraction basin of a minimum, a lower value of T enables a better exploitation of the neighborhood.

To evaluate our novel MAX-SAT solver based on Reinforcement learning we report here a comparison with some of the best and famous SLS algorithms for MAX-SAT. In particular, the following SLS techniques are considered:

- GSAT/Tabu [18], which enriches the GSAT algorithm [19] via a prohibition-based search criterion;

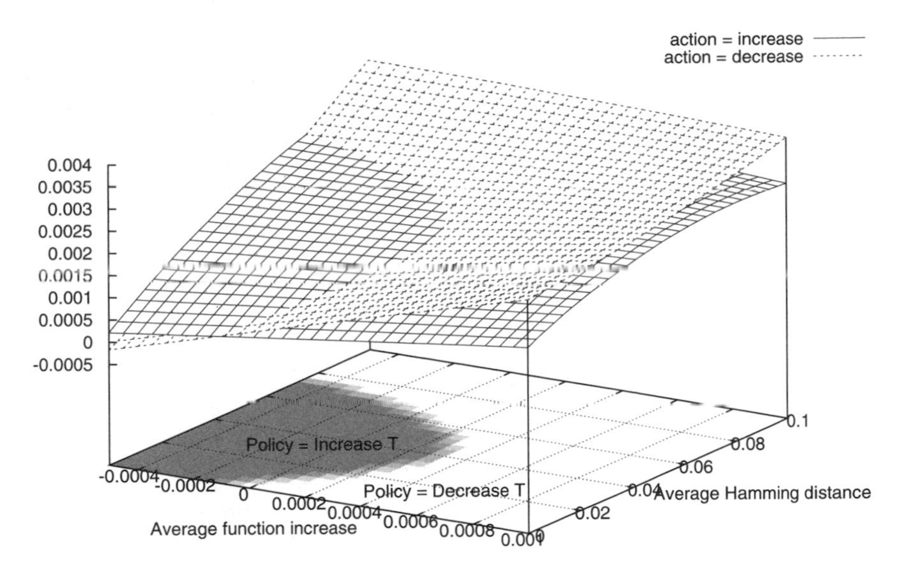

Fig. 5. Value function $\hat{Q}(s, a)$ for the two actions in the significant portion of the state space

- WalkSAT/Tabu [20], that adopts the same score function and the same variables selection mechanism of the WalkSAT/SKC algorithm [21], complemented by Tabu search;
- AdaptNovelty$^+$ [22], that exploits the concept of variable "age" and uses the same scoring function of GSAT.
- RSAPS, a reactive version of the Scaling and Probabilistic Smoothing (SAPS) [23] algorithm, on its turn, an accelerated version of the Exponentiated Subgradient algorithm [24] based on dynamic penalties;
- H_RTS ([25]), a prohibition-based algorithm that dynamically adjusts the prohibition parameter by monitoring the Hamming distance along the search trajectory.

While in this paper we base our comparisons on the solution quality after a given number of iterations, we note that the CPU time required by the proposed algorithm is analogous to that of the basic Tabu Search algorithm, with the overhead of two floating-point 13-element vector (Eq. 9) products in order to compute $\hat{Q}(s, a)$ for the two actions.

For each algorithm, 10 runs with different random seeds are performed for each of the 50 instances taken from the benchmark set described in [17], for a total of 500 tests. Fig. 6 shows the average results as a function of the number of iterations (flips), in the case of $(n = 500, m = 5000)$ instances. Among all the possible values for the prohibition parameter of the WalkSAT/Tabu algorithm, we plot the case $T_{\mathrm{f}} = .01$, as with this setting we obtain the best performance over the considered benchmark. The same for the GSAT/Tabu algorithm, whose curve is drawn for the optimal T_{f} value 0.05 over our benchmark set. Fig. 6

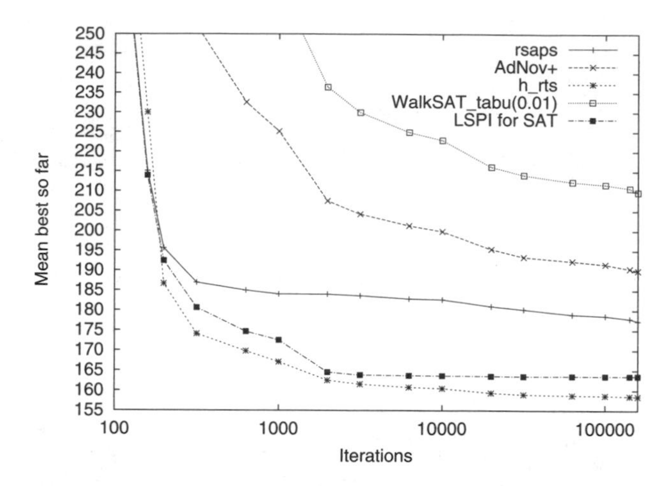

Fig. 6. Comparison among different algorithms

indicates that our RL-based approach is competitive with the other existing SLS MAX-SAT solvers.

6 Conclusions

This paper described preliminary results on the application of Dynamic Programming and Reinforcement Learning techniques to Reactive Search algorithms. In particular, the dependence of the algorithm on the prohibition parameter has been modeled as a Markov Decision Process and solved by means of the LSPI technique, achieving results that are comparable to the best algorithms in the literature.

Possible future improvements include the definition of alternative features for state description and of different reward functions. The optimal policy is currently learnt by means of the off-line generation of sample traces on a small number of instances, and the robustness of the learnt policy with respect to different problem instances has been tested. Another direction of research will cover on-line training where the optimal policy is determined by learning *while* the target optimization task is performed.

References

1. Battiti, R., Tecchiolli, G.: The reactive tabu search. ORSA Journal on Computing 6(2), 126–140 (1994)
2. Battiti, R., Brunato, M.: Reactive search: machine learning for memory-based heuristics. In: Gonzalez, T.F. (ed.) Approximation Algorithms and Metaheuristics, pp. 21–1 – 21–17. Taylor and Francis Books, CRC Press, Washington (2007)

3. Battiti, R., Brunato, M., Mascia, F.: Reactive Search and Intelligent Optimization. In: Operations research/Computer Science Interfaces. Springer, Heidelberg (in press, 2008)
4. Battiti, R.: Machine learning methods for parameter tuning in heuristics. In: 5th DIMACS Challenge Workshop: Experimental Methodology Day, Rutgers University (October 1996)
5. Bertsekas, D., Tsitsiklis, J.: Neuro-Dynamic Programming. Athena Scientific (1996)
6. Lagoudakis, M., Parr, R.: Least-Squares Policy Iteration. Journal of Machine Learning Research 4(6), 1107–1149 (2004)
7. Baluja, S., Barto, A., Boese, K., Boyan, J., Buntine, W., Carson, T., Caruana, R., Cook, D., Davies, S., Dean, T., et al.: Statistical Machine Learning for Large-Scale Optimization. Neural Computing Surveys 3, 1–58 (2000)
8. Boyan, J.A., Moore, A.W.: Learning evaluation functions for global optimization and boolean satisfability. In: Press, A. (ed.) Proc. of 15th National Conf. on Artificial Intelligence (AAAI), pp. 3–10 (1998)
9. Zhang, W., Dietterich, T.: A reinforcement learning approach to job-shop scheduling. In: Proceedings of the Fourteenth International Joint Conference on Artificial Intelligence, vol. 1114 (1995)
10. Zhang, W., Dietterich, T.: High-performance job-shop scheduling with a time-delay TD (λ) network. Advances in Neural Information Processing Systems 8, 1024–1030 (1996)
11. Lagoudakis, M., Littman, M.: Learning to select branching rules in the DPLL procedure for satisfiability. In: LICS 2001 Workshop on Theory and Applications of Satisfiability Testing, SAT 2001 (2001)
12. Lagoudakis, M., Littman, M.: Algorithm selection using reinforcement learning. In: Proceedings of the Seventeenth International Conference on Machine Learning, pp. 511–518 (2000)
13. Muller, S., Schraudolph, N., Koumoutsakos, P.: Step size adaptation in evolution strategies using reinforcementlearning. In: Proceedings of the 2002 Congress on Evolutionary Computation, 2002. CEC 2002, vol. 1, pp. 151–156 (2002)
14. Eiben, A., Horvath, M., Kowalczyk, W., Schut, M.: Reinforcement learning for online control of evolutionary algorithms. In: Brueckner, S.A., Hassas, S., Jelasity, M., Yamins, D. (eds.) ESOA 2006. LNCS (LNAI), vol. 4335. Springer, Heidelberg (2007)
15. Battiti, R., Protasi, M.: Approximate algorithms and heuristics for MAX-SAT. In: Du, D., Pardalos, P. (eds.) Handbook of Combinatorial Optimization, vol. 1, pp. 77–148. Kluwer Academic Publishers, Dordrecht (1998)
16. Lagoudakis, M., Parr, R.: LSPI: Least-squares policy iteration (as of September 1, 2007),`http://www.cs.duke.edu/research/AI/LSPI/`
17. Mitchell, D., Selman, B., Levesque, H.: Hard and easy distributions of SAT problems. In: Proceedings of the Tenth National Conference on Artificial Intelligence (AAAI 1992), San Jose, Ca, pp. 459–465 (July 1992)
18. Steinmann, O., Strohmaier, A., Stutzle, T.: Tabu search vs. random walk. In: KI - Kunstliche Intelligenz, pp. 337–348 (1997)
19. Selman, B., Levesque, H., Mitchell, D.: A new method for solving hard satisfiability problems. In: Proceedings of the Tenth National Conference on Artificial Intelligence (AAAI 1992), San Jose, Ca, pp. 440–446 (July 1992)
20. McAllester, D., Selman, B., Kautz, H.: Evidence for invariants in local search. In: Proceedings of the national conference on artificial intelligence (14), pp. 321–326. John Wiley & sons LTD., USA (1997)

21. Selman, B., Kautz, H., Cohen, B.: Noise strategies for improving local search. In: Proceedings of the national conference on artificial intelligence, vol. 12. John Wiley & sons LTD., USA (1994)
22. Tompkins, D.A.D., Hoos, H.H.: Novelty$^+$ and adaptive novelty$^+$. SAT 2004 Competition Booklet (solver description) (2004)
23. Tompkins, F.H.D., Hoos, H.: Scaling and probabilistic smoothing: Efficient dynamic local search for sat. In: Van Hentenryck, P. (ed.) CP 2002. LNCS, vol. 2470, p. 233. Springer, Heidelberg (2002)
24. Schuurmans, D., Southey, F., Holte, R.: The exponentiated subgradient algorithm for heuristic boolean programming. In: Proceedings of the international joint conference on artificial intelligence, vol. 17, pp. 334–341. Lawrence Erlbaum associates LTD., USA (2001)
25. Battiti, R., Protasi, M.: Reactive search, a history-sensitive heuristic for MAX-SAT. ACM Journal of Experimental Algorithmics 2 (ARTICLE 2) (1997), `http://www.jea.acm.org/`

On Effectively Finding Maximal Quasi-cliques in Graphs*

Mauro Brunato[1], Holger H. Hoos[2], and Roberto Battiti[1]

[1] Dipartimento di Ingegneria e Scienza dell'Informazione,
Università di Trento, Trento, Italy
brunato,battiti@disi.unitn.it
[2] Department of Computer Science,
University of British Columbia, Vancouver, BC, Canada
hoos@bs.ubc.ca

Abstract. The problem of finding a *maximum clique* in a graph is prototypical for many clustering and similarity problems; however, in many real-world scenarios, the classical problem of finding a *complete* subgraph needs to be relaxed to finding an *almost complete* subgraph, a so-called *quasi-clique*. In this work, we demonstrate how two previously existing definitions of *quasi-cliques* can be unified and how the resulting, more general quasi-clique finding problem can be solved by extending two state-of-the-art stochastic local search algorithms for the classical maximum clique problem. Preliminary results for these algorithms applied to both, artificial and real-world problem instances demonstrate the usefulness of the new quasi-clique definition and the effectiveness of our algorithms.

1 Introduction

Finding maximum cliques, *i.e.*, largest complete subgraphs, within a given graph is a well-known NP-hard combinatorial problem that is particularly intractable because of its non-approximability. However, state-of-the-art heuristic search methods, in particular stochastic local search algorithms, are typically able to find maximum or near-maximum cliques surprisingly effectively.

In real-world scenarios, relationships commonly represented by graphs are often subject to noise, resulting in erroneously missing or added edges. This motivates generalisations of the maximum clique problem in which the objective is to find maximum size subgraphs that are almost fully connected — so-called *quasi-cliques*.

Definition 1. *Given an undirected graph* (V, E), *and two parameters* λ *and* γ *with* $0 \leq \lambda \leq \gamma \leq 1$, *the subgraph induced by a subset of the node set* $V' \subseteq V$ *is a* (λ, γ)-quasi-clique *if, and only if, the following two conditions hold:*

* Holger Hoos acknowledges support provided by the Natural Sciences and Engineering Research Council of Canada (nSERC) under Discovery Grant 238788-05; Mauro Brunato and Roberto Battiti acknowledge support by the project CASCADAS (IST-027807) funded by the FET Program of the European Commission.

V. Maniezzo, R. Battiti, and J.-P. Watson (Eds.): LION 2007 II, LNCS 5313, pp. 41–55, 2008.

$$\forall v \in V' : \quad \deg_{V'}(v) \geq \lambda \cdot (|V'| - 1) \tag{1}$$

$$|E'| \geq \gamma \cdot \binom{|V'|}{2}, \tag{2}$$

where $E' = E \cap (V' \times V')$ *and* $\deg_{V'}(v)$ *is the number of elements of* V' *connected to* v.

Note that for $\lambda = \gamma = 1$, classical cliques are obtained; consequently, the problem of finding a maximum (λ, γ)-quasi-clique for arbitrary λ, γ is at least as hard as the maximum clique problem.

Condition (1) enforces a lower bound on the degree of each node within the quasi-clique, while condition (2) poses a lower bound on the overall number of edges. While both constraints have been previously proposed in the literature, to the best of our knowledge they have not previously been combined. In this work we propose this combination and show that by using it, cluster detection for noisy data (as motivated above) can be improved.

The remainder of this work is organised as follows: Section 2 discusses the motivation of our research and places it in the context of related work; Section 3 outlines the two maximum clique finding techniques we chose to adapt to our new problem formulation; Section 4 discusses the modifications that these algorithms require to operate in the more difficult space of quasi cliques; and Section 5 presents an initial experimental study on graphs that have been constructed to capture important characteristics of several real-world applications.

2 Context and Related Work

Relevant examples of quasi-clique applications and related clustering approaches include classifying molecular sequences in genome projects by using a linkage graph of their pairwise similarities [1], analysis of massive telecommunication data sets [2], as well as various data mining and graph mining applications, such as cross-market customer segmentation [3].

Different clustering techniques capable of identifying significant interconnected sub-structures in the presence of random noise have been used recently to analyze complex interconnected networks ranging from autonomous systems in the Internet, protein-protein interaction networks, e-mail and web-of-trust networks, co-authorship networks, and trade relationships among countries [4,5,6,7]. In the area of data mining, Du et al. have recently studied techniques to enumerate all maximal cliques in a complex network [8].

This contribution aims at extending the work done by the authors in efficient clique algorithms, in particular Reactive Local Search (RLS) and Dynamic Local Search for Maximum Clique (DLS-MC). Both algorithms are based on stochastic local search methods. The DLS-MC algorithm for the maximum clique problem is based on the idea of assigning penalties to nodes that are selected to be part of a clique [9]. The RLS algorithm uses a reactive mechanism to control the amount of diversification during the search process by means of prohibitions [10,11]. Both algorithms are outlined in Sec. 3. The objectives of our work are as follows:

- To develop efficient and effective heuristic algorithms for the problem of finding maximum (λ, γ)-quasi-cliques.
- To understand which heuristic components are effective as a function of the relevant graph parameters.
- To understand the effect of the problem parameters (average and minimum connectivity requirements) on the empirical run-time of the algorithms and on the quality of the results they produce.
- To assess the viability of the proposed techniques to discover a wide range of maximal quasi-cliques. Discovering many comparable solutions is critical when a quasi-clique detection module is only a first step towards a compact description of a graph, or when additional requirements or constraints are given by the user in a later phase to select from a large set of quasi-cliques.

3 Clique Finding Heuristics

This section outlines the two state-of-the-art stochastic local search algorithms for the maximum clique problem whose extension to quasi-clique finding is described later in this paper. A description of the data structures used by both algorithms precedes their outlines.

3.1 Data Structures for Classical Clique Search

Both, DLS-MC and RLS use two data structures to efficiently compute local search steps from a current clique, $V' \subseteq V$: the set of nodes Add(V') that can be added to the current clique V' (i.e., of nodes in $V \setminus V'$ that are connected to all nodes in V') and the set Miss(V') of nodes in V that are connected to all nodes but one in the current clique V', as shown in Fig. 1. Note that Add(V') is called PossibleAdd in the original RLS description, and *improving neighbour set* in the DLS-MC paper. The set Miss(V') is called OneMissing in RLS and *level neighbour set* in DLS-MC. Based on these sets, the following types of search steps can be efficiently implemented:

- **Add one node:** Once the node $v \in$ Add(V') to be added has been chosen, all elements of Miss(V') not connected to v are dropped; elements of Add(V') not connected to v are moved to Miss(V').
- **Remove one node:** All nodes $v \in V'$ are eligible for removal. Once a node v to be removed has been chosen, it is added to Add(V'); furthermore, all nodes in Miss(V') that are not connected to v are promoted to Add(V'). The set of nodes from $V \setminus (V' \cup \text{Miss}(V') \cup \text{Add}(V'))$ that are *not* connected to v is then scanned to identify nodes to be added to Miss(V').
- **Plateau move:** A node $v \in$ Miss(V') is added to the current clique, followed by the removal of the node $v' \in V'$ that is not connected to v; the sets Add(V') and Miss(V') are incrementally updated similarly as in the case of add and remove moves.

All three types of moves result in a clique.

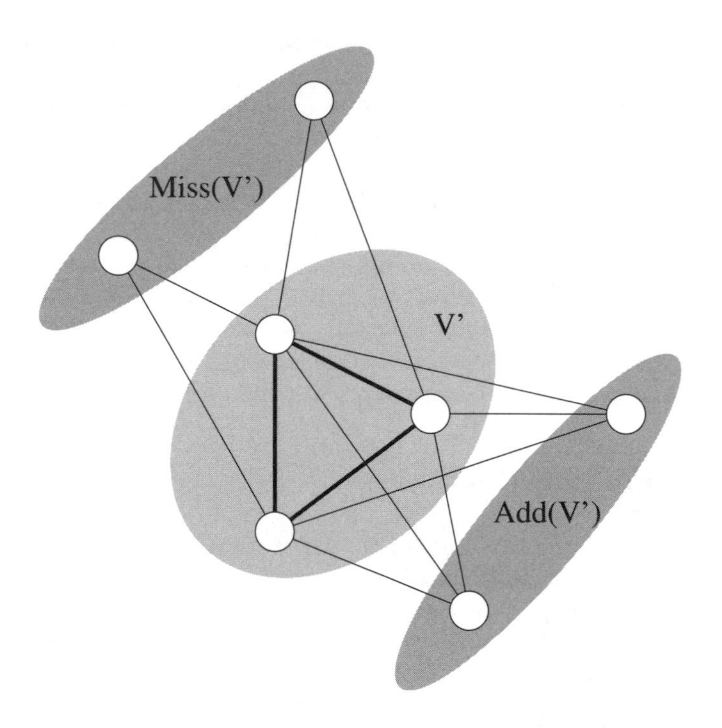

Fig. 1. Node sets involved in a classical clique search

3.2 Reactive Local Search (RLS)

Reactive Local Search (RLS) [10,11] operates by maintaining the current clique V' and modifying it with two basic moves: node addition and node removal. The basic diversification mechanism of RLS is based on Tabu Search [12,13]: every time a node is added to or removed from the current clique, it cannot be considered for removal or addition (it is *prohibited*) for the next T moves. The basic Tabu Search heuristic is complemented by a memory-based reactive scheme that automatically modifies the parameter T in order to adapt it to the problem instance. Important details of RLS are as follows:

Choice of the Best Move. At every step, the algorithm looks for a node to be added to the current clique V' among all non-prohibited nodes in $\text{Add}(V')$. A node $v \in \text{Add}(V')$ with a maximal number of connections within $\text{Add}(V')$ (*i.e.*, one whose addition to the clique leaves as many candidates as possible in the next step) is chosen, with ties broken randomly. If no nodes can be added, a non-prohibited node $v \in V'$ whose removal results in the largest set $\text{Add}(V' \setminus \{v\})$ (containing the candidates for addition to be considered in the following step) is chosen for removal, with ties broken randomly. If all candidates for addition and removal are prohibited, then a node is removed uniformly at random. In all cases, the selected node is prohibited for the following T steps.

Memory Reaction. Different graphs require different degrees of diversification, so a reactive mechanism is used in RLS to adjust the value of T. All visited cliques are mapped to the step of their last appearance as current cliques (using a hash table for fast retrieval). If the current clique has been visited too recently, the prohibition period T is increased. If a predefined number of steps occur without any increase of T, meaning that no clique is repeated too early, then T is decreased.

Restart Mechanism. If the algorithm performs a given number of steps without improving the size of the maximum clique, the search is restarted by resetting the current clique to an unused node of the graph that is chosen uniformly at random. If all nodes have already been used at least once, then the random choice is made from the set of all nodes. All other parameters and data structures are reset to their initial values ($T \leftarrow 1$, the hash table is emptied).

3.3 Dynamic Local Search for Maximum Clique (DLS-MC)

Dynamic Local Search for Maximum Clique (DLS-MC) [9] uses two basic types of search steps to modify the current clique V': node addition and plateau moves.

The basic diversification mechanism of DLS-MC is based on *penalty values* associated with each node. At the beginning of the search process, the current clique is set to a single node that is uniformly chosen at random and all node penalties are set to zero. The algorithm then alternates between two search phases:

- In the *expansion phase*, which continues as long as Add(V') is not empty, a node from Add(V') with minimum penalty is selected (with ties broken randomly) and added to the current clique;
- In the *plateau phase*, which continues as long as Add(V') is empty and Miss(V') contains at least one node, a node in Miss(V') with minimum penalty is selected for addition, and the node of V' not connected to it is removed from the current clique. Moreover, at the beginning of the plateau phase, the current clique is recorded (as V''), and the phase is terminated when $V' \cap V'' = \emptyset$.

In addition, a prohibition mechanism is used to prevent the plateau phase from cycling through a small set of cliques: once a node is chosen for addition, it becomes prohibited until the end of the current plateau phase. At the end of the plateau phase, two actions are taken:

Penalty Update. The penalty values of all nodes in the current clique V' are increased by 1. Additionally, every *pd* update cycles, all nonzero penalties are decremented by 1, so that penalties are 'forgotten' over time.

Clique Perturbation. If *pd* $= 1$ (meaning that every increase in penalties is immediately cancelled, so that penalties always remain equal to zero), a new current clique is generated by adding a new node $v \in V$ chosen uniformly at random and removing from the current clique V' all nodes that are not connected to v. If, on the other hand, $pd > 1$ (penalties are used), then the current clique is reduced to the last node that was added to it.

4 Supporting Data Structures for Quasi-clique Search

When adapting DLS-MC and RLS to the quasi-clique setting, more complex sets of nodes must be maintained in order to support the basic search steps.

The number of edges is always an integer, therefore we can conveniently rewrite constraints (1) and (2) in order to use integer variables to store clique bounds:

$$\forall v \in V' : \quad \deg_{V'}(v) \geq \lceil \lambda \cdot (|V'| - 1) \rceil \tag{3}$$

$$|E'| \geq \left\lceil \gamma \cdot \binom{|V'|}{2} \right\rceil \tag{4}$$

4.1 Adding One Node

Let us define the set of *critical* nodes in a (λ, γ)-clique as those nodes whose degree in V' is high enough to justify their presence, but would fail to satisfy condition (3) if the clique size increased without adding an edge to them:

$$\mathrm{Crit}(V') := \{v \in V' : \deg_{V'}(v) < \lceil \lambda \cdot |V'| \rceil\}.$$

Consider, for example, the case shown in Fig. 2, where $V' = \{1, 5, 6, 7\}$. It is easy to verify that V' is a (λ, γ)-quasi-clique for $\lambda = \gamma = 2/3$. If, however, a new node were added to V', the degree of nodes 1 and 7 would no longer satisfy condition (3), unless edges are added to both of them; therefore, $\mathrm{Crit}(V') = \{1, 7\}$.

The addition of a node should also satisfy the global density constraint (4), so any new node must contribute at least $d_{V'}$ edges to the clique, where

$$d_{V'} := \left\lceil \gamma \cdot \binom{|V'| + 1}{2} \right\rceil - |E'|$$

is the minimum number of edges that must be added in order to maintain constraint (4). In the example from Fig. 2, a new node must contribute at least $d'_V = 2$ edges. Consequently, a node $v \in V \setminus V'$ is eligible for addition to V' if the three following conditions hold:

- v has an adequate degree in V' according to (3);
- v has enough edges to nodes in V' that as a result of adding it, the edge density in $V' \cup \{v\}$ does not fall below threshold (4);
- all critical nodes in V' receive at least one more edge when adding v (i.e., v is connected to all critical nodes).

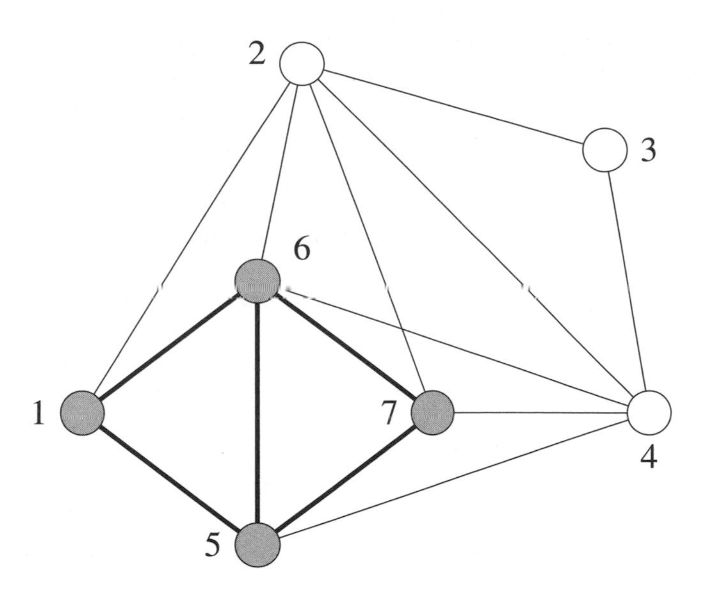

Fig. 2. A graph and a $(2/3, 2/3)$-quasi-clique (shaded nodes)

Based on these conditions, the set of nodes eligible for addition to V' is defined as

$$\text{Add}(V') := \left\{ v \in V \backslash V' : \deg_{V'}(v) \geq \max\{\lceil \lambda \cdot |V'| \rceil, d_{V'}\} \wedge \{v\} \times \text{Crit}(V') \subseteq E \right\}.$$

In the example from Fig. 2, the only eligible node (connected to all critical nodes, and contributing at least 2 edges) is node 2.

4.2 Removing One Node

To be eligible for removal from a (λ, γ)-clique, a node must not be connected by an edge to any *removal-critical* node, where the set of removal-critical nodes is defined as follows:

$$\text{RCrit}(V') := \{v \in V' : \deg_{V'}(v) - 1 < \lceil \lambda \cdot (|V'| - 2) \rceil\}.$$

By losing an edge, such nodes would no longer satisfy constraint (3) for the resulting, smaller quasi-clique. In the example from Fig. 2, nodes 1 and 7 are removal-critical, and therefore nodes connected to them, i.e., 5 and 6, cannot be removed.

Secondly, if a node has sufficiently high degree in V', its removal would cause the global edge density to fall below threshold (4). The maximum number of edges that can be removed from quasi-clique V' without violating the global density constraint is

$$e_{V'} := |E'| - \left\lceil \gamma \cdot \binom{|V'| - 1}{2} \right\rceil.$$

In the example from Fig. 2, up to 3 edges can be removed.

Generally, the set of edges that are eligible for removal is therefore

$$\mathrm{Rem}(V') := \left\{ v \in V' : \left(\{v\} \times \mathrm{RCrit}(V')\right) \cap E = \emptyset \wedge \deg_{V'}(v) \leq e_{V'} \right\}.$$

In the example, only nodes 1 and 7 can be removed, while removing node 5 or 6 would leave nodes 1 and 7 with too small a degree to remain within the clique.

4.3 Plateau Moves

In the classical clique case, a plateau move is a node removal followed by a node addition, and the clique property is maintained throughout the process. Note that removals do not maintain the (λ,γ)-clique property in the general case, but the property could be restored after a node addition; therefore, a plateau move in the quasi-clique domain must be regarded as atomic with respect to the quasi-clique constraints. Despite this atomicity from the quasi-clique's point of view, the removal and addition operations must be performed sequentially in one of the two possible orders. In this work, we only consider the "add, then remove" option; however, the opposite is possible.

We allow a node $v \in V \setminus V'$ to be added to the quasi-clique in the context of a plateau move if, and only if, it is subsequently possible to identify at least one node in V' whose removal would maintain or restore the quasi-clique property. Considering that a node different from v should be removed, the nodes eligible for addition only need to be sufficiently well-connected in V':

$$\mathrm{PAdd}(V') := \{v \in V \setminus V' : \deg_{V'}(v) \geq \lambda \cdot (|V'| - 1)\}.$$

The set of nodes eligible for removal depends on the added node. Note that, in general, $\mathrm{PAdd}_{V'} \supseteq \mathrm{Add}_{V'}$, and once $w \in \mathrm{PAdd}_{V'}$ is chosen, $V' \cup \{w\}$ is not necessarily a (λ,γ)-clique. In order to select the node to be removed, we define a *plateau-critical* set of nodes, depending on V' and on w:

$$\mathrm{PCrit}(V', w) := \{v \in V' \cup \{w\} : \mathit{deg}_{V'}(v) - 1 < \lambda \cdot (|V'| - 1) \wedge (v, w) \notin E\}.$$

This set contains the nodes that would not satisfy the quasi-clique property when removing an edge, unless they receive an additional edge when node w is added. Note that w itself may not belong to this set if its degree is not high enough.

When choosing a node to be removed, we must make sure that it is not connected to a plateau-critical node, and that we don't remove too many edges from $V' \cup \{w\}$. The maximum number of edges we can afford to lose from $V' \cup \{w\}$ in order to guarantee its quasi-clique property is

$$r_{V',w} := |E'| + \deg_{V'}(w) - \gamma \cdot \binom{|V'|}{2}.$$

Considering this, the set of candidates for removal is defined as

$$\mathrm{PRem}(V', w) := \left\{ v \in V' : \deg_{V' \cup \{w\}}(v) \leq r_{V',w} \wedge (\{v\} \times \mathrm{PCrit}(V', w)) \cap E = \emptyset \right\},$$

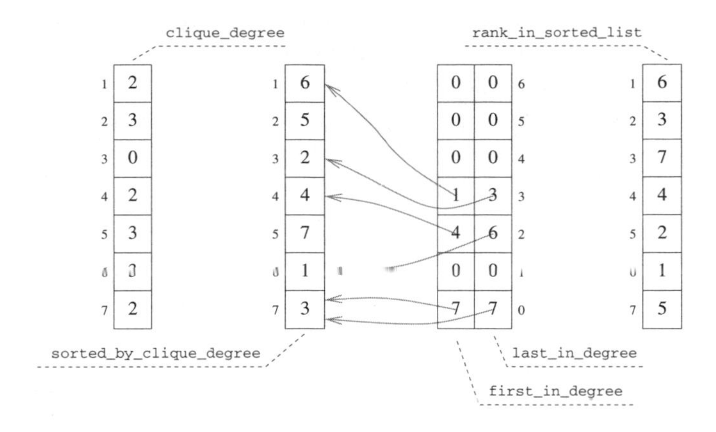

Fig. 3. Data structures used to support quasi-clique search, instantiated for the quasi-clique from Fig. 2

and the only nodes that can actually be added to the quasi-clique in the plateau step are those whose corresponding set of candidates for removal is not empty.

4.4 Support Set Implementation

The previously defined support sets can be maintained efficiently using the array structures illustrated in Fig. 3. In particular, we make use of an array `sorted_by_clique_degree`, where nodes $v \in V$ are sorted by increasing $\deg_{V'}(v)$. Everytime the clique is updated by adding or removing a node, the V''-degree of some nodes will be increased or decreased, and the array must be modified to reflect the new ordering. In order to perform these changes in time $O(\deg_{V'}(v))$, where V' is the clique before updating and v is the node to be added, we maintain three additional arrays:

`rank_in_sorted_list`
: the inverse of `sorted_by_clique_degree`;

`first_in_degree`
: whose i-th element contains the first index in `sorted_by_clique_degree` containing a node with the given degree in the current quasi-clique;

`last_in_degree`
: whose i-th element contains the last index in `sorted_by_clique_degree` containing a node with the given degree in the current quasi-clique.

Note that we assume that all arrays are indexed starting from 1, with the exception of `first_in_degree` and `last_in_degree`, which are indexed by node degrees and therefore start from zero.

In general, it is not possible to incrementally maintain $\mathrm{Add}(V')$ and $\mathrm{Rem}(V')$ in the same way as for the classical maximum clique problem. For instance, $\mathrm{Add}_{V'}$ is monotonic for greedy clique constructions, while for (λ, γ)-cliques, this set can acquire or lose elements at every step. Using the previously mentioned arrays, however, it is possible to restrict the search of eligible nodes to smaller sets of candidate nodes. This is effective, since all conditions for node removal or addition are defined on clique degree ranges; in particular, critical and removal-critical nodes typically fall into a small range of degrees within the current quasi-clique, usually just one or two.

5 Experimental Analysis

In the following we demonstrate how by using our new quasi-clique definition, based on the combined constraint set (1)+(2) with proper non-zero settings of the local and global density parameter, and our generalised versions of RLS and DLS-MC, we can effectively find densely connected subgraphs that are not accessible to classical clique finding approaches.

5.1 Identification of Overlapping Communities

Following the motivating considerations on quasi-clique usage for community identification, a set of benchmark graphs has been designed and generated in order to capture some fundamental features of communities. The generated graphs depend on five parameters: the number of nodes N, number of communities G, intra-community link probability P_{in}, inter-community link probability P_{out} and fraction f of overlap between communities. The last parameter, f, controls the overlap between communities as follows: If N is the number of nodes in our graph and G divides N, we define G groups $g_0, \dots, g_{G-1}$ so that groups g_i and g_{i+1} share $f \cdot N/G$ nodes. To generate such a graph, a random permutation $\pi : \{0, N-1\} \mapsto \{0, N-1\}$ is generated. For every node $n \in \{0, \dots, N-1\}$ and every group index $i \in \{0, \dots, G-1\}$, node n belongs to group g_i if and only if

$$\frac{N}{G} \cdot \left(i - \frac{f}{2}\right) \leq \pi(n) < \frac{N}{G} \cdot \left(i + 1 + \frac{f}{2}\right).$$

Note that if $f = 0$ the G groups form a partition of the node set, while if $f > 1$ overlapping regions extend to non-adjacent groups. Once assignments of nodes to groups are computed, with the permutation function acting as a "scrambling factor," edges are assigned to each pair of nodes according to the number of groups that they have in common. Let h_{ij} represent the number of groups in common between nodes i and j, then:

$$\Pr((i,j) \in E) = \begin{cases} P_{\mathrm{out}} & \text{if } h_{ij} = 0 \\ 1 - (1 - P_{\mathrm{in}})^{h_{ij}} & \text{otherwise.} \end{cases}$$

Figure 4 shows an example with $N = 128$ nodes, numbered from 1 to 128, that are distributed in $G = 4$ communities with overlap factor $f = 1/8$ (groups are formed by nodes 1–33, 30–65, 62–97, 94–128). The probability of edge (i,j) to appear in the graph is $P_{\mathrm{in}} = 0.7$ if i and j belong to the same community,

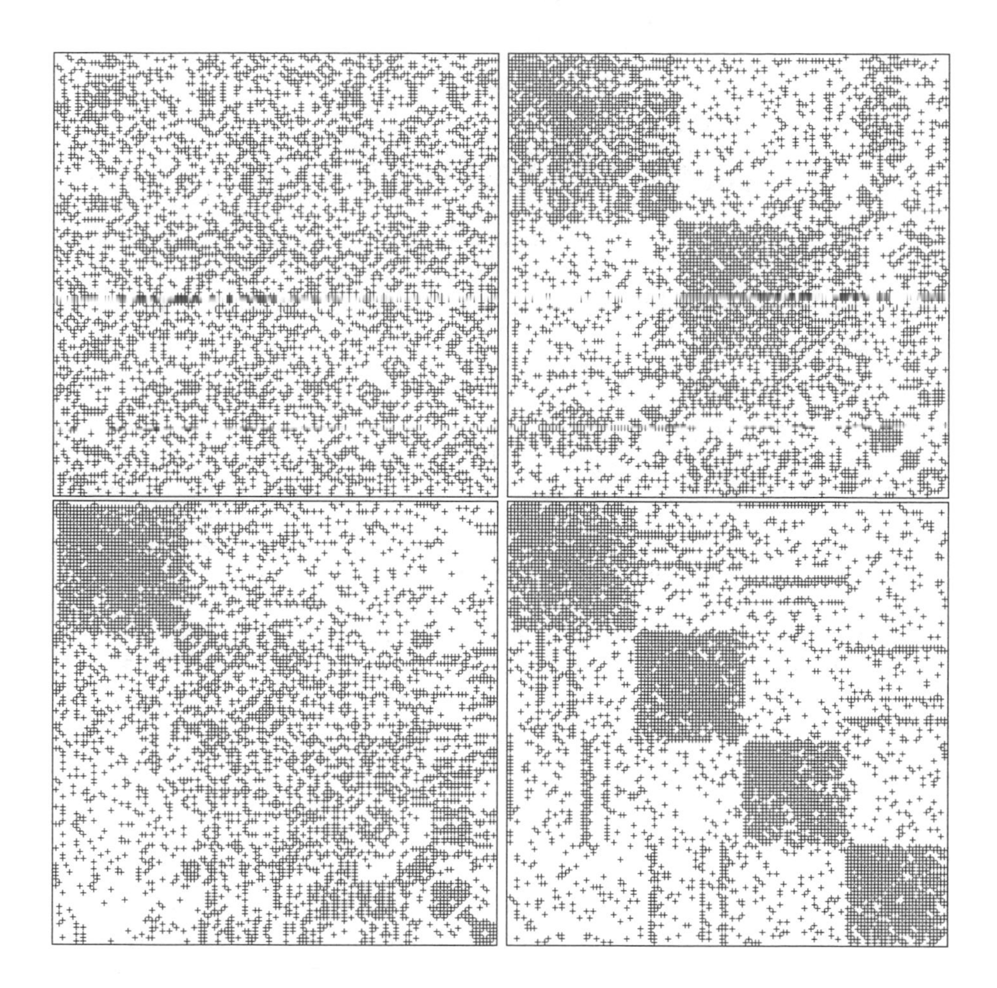

Fig. 4. Adjacency plot of overlapping community graph with nodes sorted by order of appearance in the identified quasi-cliques according to different values of (λ, γ); from top left to bottom right: original ordering, $(0, 0.4)$, $(0.65, 0)$ and $(0.65, 0.4)$. Black crosses represent edges.

$P_{\text{out}} = 0.1$ otherwise. Node labels are then randomly permuted, so that there is no longer a correlation between node labels and community membership.

The top left pane of Figure 4 shows the original adjacency matrix: because of the random permutation, no structure is apparent. Next, we generated a list of the largest maximal $(0.4, 0.65)$-quasi-cliques found by a short run of the quasi-clique extension of RLS. When nodes are sorted in the order in which they appear in this list (so that nodes contained in the largest quasi-clique are grouped together, and so on), the resulting adjacency matrix (bottom right pane of Fig. 4) shows a significant structure: nodes belonging to the same community are grouped together. The adoption of constraint (2) alone (top right pane of Fig.4) and of constraint (1) alone (bottom left pane) are not sufficient to correctly

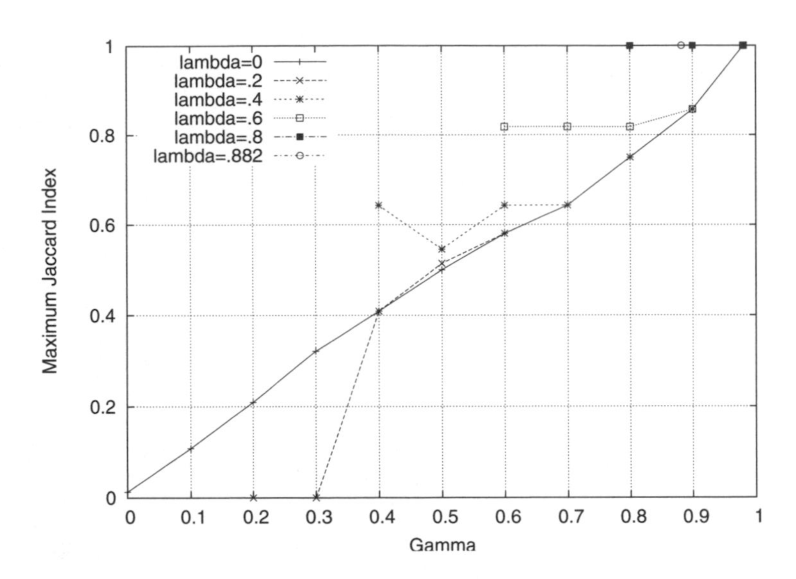

Fig. 5. Differences between maximum (λ, γ)-quasi-cliques in a noisy version of a co-authorship graph and the maximum clique in the original graph (for details, see text)

identify the four clusters. We also note that communities cannot be directly represented by classical cliques.

5.2 Cliques in Noisy Graphs

A second experiment considers noisy versions of graphs, in which edges are removed uniformly at random with some small probability. For appropriate values of λ and γ, the maximum clique of the original graph should appear as a (λ, γ)-quasi-clique in the noisy graph. In Figure 5, a co-authorship graph from a scientific database [4] has been processed by removing each edge with 5% probability; subsequently, the DLS-MC algorithm for maximum quasi-clique finding has been executed 10 times for each of many different values of λ and γ. As a measure of similarity between the maximum quasi-clique found by the algorithm and the maximum (classical) clique in the original graph, the maximum value of the Jaccard index over all maximum quasi-cliques found in the 10 runs for each (λ, γ) has been used. (The Jaccard index of two sets U and V is defined as $J(U, V) := |U \cap V|/|U \cup V|$.)

The results of this experiment, shown in Figure 5, indicate that when γ decreases, the maximum quasi-clique in the noisy graph becomes increasingly different from the maximum clique in the original graph. On the other hand, a higher value of λ tends to keep the similarity value higher, while the global constraint alone (curve with $\lambda = 0$) seems to provide worse results. Note, however, that the curve for $\lambda = 0.2$ is an exception, because it performs worse for lower values of γ. This behavior is not fully understood yet.

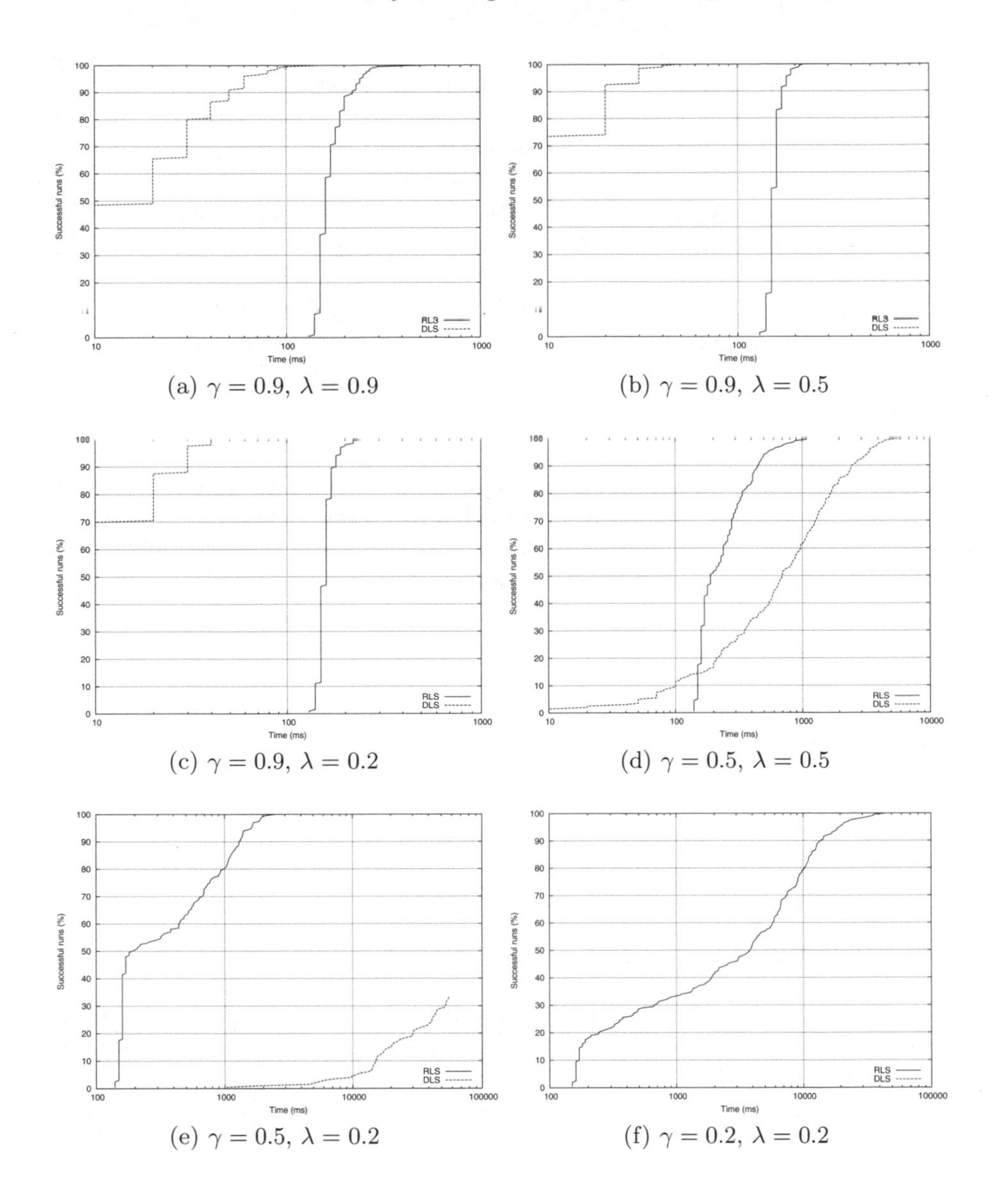

(a) $\gamma = 0.9,\ \lambda = 0.9$

(b) $\gamma = 0.9,\ \lambda = 0.5$

(c) $\gamma = 0.9,\ \lambda = 0.2$

(d) $\gamma = 0.5,\ \lambda = 0.5$

(e) $\gamma = 0.5,\ \lambda = 0.2$

(f) $\gamma = 0.2,\ \lambda = 0.2$

Fig. 6. Performance comparison between RLS and DLS-MC: percentage of successful runs *vs* run-time for finding putative maximum (λ, γ)-quasi-cliques. Notice that the diagrams have different time scales to accommodate for large variations in performance.

5.3 RLS and DLS-MC Run-Time Comparison

To compare the performance of the previously discussed modifications of RLS and DLS-MC, we ran both algorithms on a graph representing protein-protein interaction data for *S. cerevisiae* (yeast). More specifically, we first determined putative maximum (λ, γ)-quasi-clique sizes for various values of λ and γ by performing long (10-minute) runs of each algorithm. Next, we measured empirical run-time distributions for reaching these target clique sizes based on 200 independent runs per algorithm and quasi-clique parameter setting.

Figure 6 shows the resulting cumulative run-time distribution functions for the two algorithms. As can be seen from these results, for high values of the λ and γ, *i.e.*, for dense quasi-cliques, DLS-MC appears to outperform RLS, while RLS appears to perform better on sparse quasi-cliques. The weak performance of DLS-MC for sparser quasi-cliques seems to be caused by a large number of plateau moves executed by the algorithm, and is currently being further investigated.

6 Conclusions

In this work, a new definition for quasi-cliques has been introduced which combines and generalises previously proposed quasi-clique definitions. In order to generalise high-performance stochastic local search algorithms for the maximum clique problem to the more general problem of finding maximum quasi-cliques, we have introduced new data structures, using which basic local search steps (node addition, removal and plateau moves) can be performed efficiently. Preliminary experiments on artificial and real-world graphs have been presented in order to motivate the adoption of this definition and to demonstrate the effectiveness of our generalised algorithms and their supporting data structures.

In future work, we plan to investigate the effectiveness of our new quasi-clique algorithms in more depth and for a wider range of graphs. It would also be interesting to extend further high-performance stochastic local search algorithms for the maximum clique problem, such as PLS [14], to (λ, γ)-quasi-cliques.

Acknowledgements

We gratefully acknowledge contributions by Srinivas Pasupuleti, who helped during the initial stages of this project, Wayne Pullan, who made his implementation of DLS-MC available to us, and Franco Mascia, whose RLS code for maximum clique finding we used as a basis for implementating our RLS extension for quasi-cliques.

References

1. Matsuda, H., Ishihara, T., Hashimoto, A.: Classifying molecular sequences using a linkage graph with their pairwise similarities. Theoretical Computer Science 210(2), 305–325 (1999)
2. Abello, J., Resende, M., Sudarsky, S.: Massive quasi-clique detection. In: Proceedings of the 5th Latin American Symposium on Theoretical Informatics (LATIN), pp. 598–612 (2002)
3. Pei, J., Jiang, D., Zhang, A.: On mining cross-graph quasi-cliques. In: Conference on Knowledge Discovery in Data, pp. 228–238 (2005)

4. Serrano, M., Boguñá, M.: Clustering in complex networks. I. General formalism. Arxiv preprint cond-mat/0608336 (2006)
5. Hopcroft, J., Khan, O., Kulis, B., Selman, B.: Tracking evolving communities in large linked networks (2004)
6. Palla, G., Derényi, I., Farkas, I., Vicsek, T.: Uncovering the overlapping community structure of complex networks in nature and society. Nature 435(7043), 814–818 (2005)
7. Everett, M., Borgatti, S.: Analyzing clique overlap. Connections 21(1), 49–61 (1998)
8. Du, N., Wu, B., Xu, L., Wang, B., Pei, X.: A Parallel Algorithm for Enumerating All Maximal Cliques in Complex Network. In: Sixth IEEE International Conference on Data Mining Workshops, ICDM Workshops 2006, pp. 320–324 (2006)
9. Pullan, W., Hoos, H.H.: Dynamic local search for the maximum clique problem. Journal of Artificial Intelligence Research 25, 159–185 (2006)
10. Battiti, R., Protasi, M.: Reactive local search for the maximum clique problem. Algorithmica 29(4), 610–637 (2001)
11. Battiti, R., Mascia, F.: Reactive local search for maximum clique: a new implementation. Technical Report DIT-07-018, University of Trento (2007)
12. Glover, F.: Tabu search - part i. ORSA Journal on Computing 1(3), 190–260 (1989)
13. Glover, F.: Tabu search - part ii. ORSA Journal on Computing 2(1), 4–32 (1990)
14. Pullan, W.: Phased local search for the maximum clique problem. Journal of Combinatorial Optimization 12(3), 303–323 (2006)

Improving the Exploration Strategy in Bandit Algorithms

Olivier Caelen and Gianluca Bontempi

Machine Learning Group, Département d'Informatique,
Faculté des Sciences, Université Libre de Bruxelles,
Bruxelles, Belgium

Abstract. The K-armed bandit problem is a formalization of the *exploration versus exploitation* dilemma, a well-known issue in stochastic optimization tasks. In a K-armed bandit problem, a player is confronted with a gambling machine with K arms where each arm is associated to an unknown gain distribution and the goal is to maximize the sum of the rewards (or minimize the sum of losses). Several approaches have been proposed in literature to deal with the K-armed bandit problem. Most of them combine a greedy exploitation strategy with a random exploratory phase. This paper focuses on the improvement of the exploration step by having recourse to the notion of *probability of correct selection* (PCS), a well-known notion in the simulation literature yet overlooked in the optimization domain. The rationale of our approach is to perform at each exploration step the arm sampling which maximizes the probability of selecting the optimal arm (i.e. the PCS) at the following step. This strategy is implemented by a bandit algorithm, called ϵ-PCSgreedy, which integrates the PCS exploration approach with the classical ϵ-greedy schema. A set of numerical experiments on artificial and real datasets shows that a more effective exploration may improve the performance of the entire bandit strategy.

1 Introduction

In many real world problems, a decision maker must take decisions in order to maximize some cost function (e.g. the sum of a sequence of rewards). This task is not trivial if the knowledge about the state of the environment is either partial or uncertain. In this context it might be convenient to perform decisions or actions whose goal is not to maximize the reward but rather to reduce the degree of uncertainty.

An example of such a situation is the design of clinical trials [6] for comparing a set of new medical treatments. Here, the goal is to determine the best treatment while minimizing the patients' nuisance. Another example is the economic problem of selecting the best supplier on the basis of incomplete information [2]. The main issue in these two examples is how to trade immediate reward based on the acquired knowledge (exploitation) with the probability of gaining additional insight by performing a suboptimal action (exploration).

The K-armed bandit problem is an instance of this general problem in which a casino player has to decide which arm of a K-slot machine to pull to maximize

V. Maniezzo, R. Battiti, and J.-P. Watson (Eds.): LION 2007 II, LNCS 5313, pp. 56–68, 2008.

the total reward in a series of trials. The player has to define a sequential selection strategy to choose the arms that must be selected. If the player keeps an updated estimate of the arms gains and at each step greedily choices the arm which on average performed the best so far, then we say that the player adopts a pure exploitation strategy. This means that she uses her current knowledge for selecting the apparently best arms without reserving any time to explore apparently inferior arms. A possible alternative to a pure exploitation strategy consists in preserving a fraction of the time (e.g. quantified by a parameter ϵ) for performing random uniform selection. This strategy is called ϵ-greedy [11] and belongs to the family of *semi-uniform* strategies [13]. A *semi-uniform* strategy is characterized by the alternation of two working modes, an exploration mode and an exploitation mode. As far as the exploration is concerned, the semi-uniform ϵ-greedy strategy relies on random choice. In this paper, we advocate that randomly selecting an arm is suboptimal for the exploration. As an alternative we propose to adopt the Probability of Correct Selection (PCS) measure to assess the gain, in term of subsequent exploitation, that could derive from an exploration action. The idea is that an effective exploration step should lead to the largest increase of the probability PCS of correctly selecting the best arm. This means that the exploration action should be the one that maximizes the probability of the consequent selection step. The notion of PCS has been proposed in the Monte Carlo simulation literature in relation to the problem of comparing alternatives by simulation [9,8]. In that case the issue is to decide how many simulation trials should be conducted if we want to have a certain guarantee that the correct selection will be accomplished. Here, we use the PCS notion as a founding principle of a sequential strategy and as a measure of the effectiveness of an exploration step. This paper proposes a formalization of the notion of optimal exploration on the basis of PCS and an algorithm which uses it to implement a semi-uniform strategy.

Note however that the computation of the PCS quantity requires the entire knowledge of the probability distribution of the K arms and the computation of a multivariate integral. Unlike the simulation literature where the PCS term is analytically upper bounded, we propose here a data-driven strategy to estimate the PCS term on the basis of the collected data. In particular our approach has three main characteristics: (i) it makes an hypothesis of normal distributions of the arm rewards, (ii) it adopts a plug-in approach where the mean vector and the covariance matrix are replaced by their sampled counterparts, (iii) it uses a numerical integration algorithm for deriving the PCS from the associated multivariate normal distribution.

This paper is structured as follows: Section 2 formalizes the multi-armed bandit problem. Section 3 presents some state-of-the-art algorithms which will be used for benchmarking our technique. Section 4 introduces the PCS notion while Section 5 presents the ϵ-PCS greedy algorithm which uses the notion of PCS to enhance the exploration step. The experimental results on artificial and real datasets are shown and discussed in Section 6. Possible extensions of the approach are suggested in Section 7.

2 The K-Armed Bandit Problem

The K-armed bandit problem is defined by a set of K random variables $\mathbf{z}_\# = \{\mathbf{z}_1, \mathbf{z}_2, \ldots, \mathbf{z}_K\}$ whose distribution is unknown and where $\mathbf{z}_k$ represents the observed reward of the kth arm. Let $\mu_\# = \{\mu_1, \mu_2, \ldots, \mu_K\}$ be the (unknown) mean vector where $\mu_1 = \max \mu_\#$ and $\sigma_\#$ be the associated variance vector. Let us consider a sequential setting where at the lth step (or round) we first decide which variable to sample and then we collect the associated realization $\mathbf{z}_k^l$ (or reward). The objective is to maximize the sum of the collected rewards after H rounds via a *sequential strategy algorithm* that chooses the next random variable on the basis of the past rounds and the associated rewards.

An effective strategy is expected to solve the well-known *exploration versus exploitation trade-off* [7]. On one hand, since the parameters of the distributions in $\mathbf{z}_\#$ are unknown the strategy should perform a large number of tests on each random variable (*exploration*) in order to improve the estimation of the rewards. On the other hand, since the goal is to maximize the sum of the rewards, the strategy should privilege the tests on the best observed variables (*exploitation*).

The *bandit regret* ρ_B^H of the strategy after H rounds is defined as the difference between the sum of the rewards associated to an optimal strategy and the sum of the collected rewards [13,1],

$$\rho_B^H = H \cdot \mu_1 - \sum_{k=1}^{K} n_k^H \cdot \mu_k \tag{1}$$

where $\mu_1 = \max_k \{\mu_k\}$ is the mean associated to the best r.v. and n_k^H is the number of tests made on $\mathbf{z}_k$ after H rounds. Note that according to this formulation the problem of maximizing the reward is transformed into the dual problem of minimizing the regret.

A sequential strategy where the regret per round tends to zero for any bandit problem and when the horizon tends to infinity $\left(\lim_{H\to\infty} \frac{\rho_B^H}{H} = 0\right)$ is a *zero-regret strategy* [13]. Intuitively, zero-regret strategies are guaranteed to converge to an optimal (not necessarily unique) bandit strategy if enough rounds are played.

3 State-of-the-Art Approaches

3.1 The ϵ-Greedy Algorithm

The ϵ-greedy algorithm [11] is probably the simplest yet the most used strategy in bandit problems. At the lth step the algorithm either chooses a random arm with probability $\epsilon \in [0,1]$ or chooses the arm with the highest sample average $\widehat{\boldsymbol{\mu}}_k^l$. The parameter ϵ is set by the user and represents the exploration/exploitation ratio. If ϵ is equal to 0, the algorithm will always play the estimated best arm. On the other hand, if ϵ equals 1, the algorithm keeps on exploring the alternative arms without taking the collected rewards into account. Note that this approach,

Algorithm 1. The ϵ-greedy algorithm

1: $\mathbf{z}_k^{1,2} \leftarrow$ play arm k twice; $\forall k \in [1, K]$
2: $n_k \leftarrow 2$; $\forall k \in [1, K]$
3: **for** $l = (2K + 1)$ to H **do**
4: $\quad \mathbf{e} \leftarrow U[0, 1]$ *uniform sampling*
5: $\quad$ **if** $\mathbf{e} < \epsilon$ **then**
6: $\quad\quad k \leftarrow$ randomly select an arm
7: $\quad$ **else**
8: $\quad\quad k \leftarrow \arg\max_k \widehat{\boldsymbol{\mu}}_k^l$
9: $\quad$ **end if**
10: $\quad n_k^{l+1} \leftarrow n_k^l + 1$
11: $\quad n_i^{l+1} \leftarrow n_i^l \;\; \forall i \in [1, K]/k$
12: $\quad \mathbf{z}_k^{n_k^{l+1}} \leftarrow$ play arm k
13: **end for**

though naive, is known to be hard to beat [13]. Also, because of the non zero ϵ term, it is not a zero-regret strategy. A pseudo-code of the algorithm follows.

As far as the initialization is concerned (lines 1-2), we will assume that an initialization matrix $\mathbf{z}$ containing two rewards for each arm is available. The same initialization step will be applied to all the following algorithms. For the remaining $H - 2K$ steps, the algorithm samples in a uniform way $\mathbf{e}$ (line 4) and decides the action accordingly. If $\mathbf{e}$ is smaller than ϵ, it tests randomly an arm (line 6), otherwise it tests the current best arm (line 8).

3.2 The Interval Estimation Algorithm

The *Interval Estimation algorithm* was first proposed by Kaelbling et al. in [7]. This algorithm consists in (i) computing the upper-bound $UB[\widehat{\boldsymbol{\mu}}_k^l]$ of the $100(1-\Theta)\%$ confidence interval of the estimate $\widehat{\boldsymbol{\mu}}_k^l$, where $\Theta \in [0, 1]$ and (ii) selecting the arm with the highest $UB[\widehat{\boldsymbol{\mu}}_k^l]$. If we assume that $\mathbf{z}_k$ follows a normal probability distribution (mean μ_k and standard deviation σ_k unknown) then

$$UB[\widehat{\boldsymbol{\mu}}_k^l] = \widehat{\boldsymbol{\mu}}_k^l + t_{1-\Theta/2} \frac{\widehat{\boldsymbol{\sigma}}_k^l}{\sqrt{n_k^l}} \tag{2}$$

where $t_{1-\Theta/2}$ is the *Student's t-function* at confidence level $1-\Theta/2$ with $(n_k^l - 1)$ degrees of freedom,

$$\widehat{\boldsymbol{\mu}}_k^l = \frac{1}{n_k^l} \sum_{i=1}^{n_k^l} \mathbf{z}_k^i, \qquad \widehat{\boldsymbol{\sigma}}_k^l = \sqrt{\frac{1}{n_k^l - 1} \sum_{i=1}^{n_k^l} \left(\mathbf{z}_k^i - \widehat{\boldsymbol{\mu}}_k^l\right)^2} \tag{3}$$

and the set $\left\{\mathbf{z}_k^1, \ldots, \mathbf{z}_k^{n_k^l}\right\}$ contains all the realizations of $\mathbf{z}_k$ from step 1 to step l.

Algorithm 2. The Interval Estimation algorithm

1: $\mathbf{z}_k^{1,2} \leftarrow$ play arm k twice; $\forall k \in [1, K]$
2: $n_k \leftarrow 2$; $\forall k \in [1, K]$
3: **for** $l = (2K+1)$ to H **do**
4: **for** $k = 1$ to K **do**
5: $\widehat{\boldsymbol{\mu}}_k^l \leftarrow \mathit{average}(\mathbf{z}_k)$
6: $\widehat{\boldsymbol{\sigma}}_k^l \leftarrow \mathit{standard\ deviation}(\mathbf{z}_k)$
7: $UB[\widehat{\boldsymbol{\mu}}_k^l] \leftarrow \widehat{\boldsymbol{\mu}}_k^l + \widehat{\boldsymbol{\sigma}}_k^l \cdot \frac{t_{1-\Theta/2}}{\sqrt{n_k^l}}$
8: **end for**
9: $k \leftarrow \arg\max_k UB[\widehat{\boldsymbol{\mu}}_k^l]$
10: $n_k^{l+1} \leftarrow n_k^l + 1$
11: $n_i^{l+1} \leftarrow n_i^l \;\; \forall i \in [1, K]/k$
12: $\mathbf{z}_k^{n_k^{l+1}} \leftarrow$ play arm k
13: **end for**

The rationale of the Interval Estimation algorithm is that infrequently observed arms will have an higher upper-bound and consequently an higher probability of being explored. At the same time, the more an arm is tested, the closer its upper-bound will be to the true mean.

After the initialization phase, at each step the IE algorithm (i) uses (2) to associate an upper-bound to each alternative (line 7) (ii) chooses the arm maximizing $UB[\widehat{\boldsymbol{\mu}}_k^l]$ (line 9) and (iii) updates the set of observed statistics (line 10 to 12).

3.3 The Gittins Index Algorithm

Gittins [5] proposed to solve the bandit problem as a dynamic programming problem [3]. If the rewards of the arms follow a normal probability distribution (unknown mean and standard deviation) the solution of the dynamic programming problem at the l step is the arm which maximizes the *Gittins index*

$$\mathbf{v}_k = \widehat{\boldsymbol{\mu}}_k^l + \widehat{\boldsymbol{\sigma}}_k^l \cdot v_g\left(n_k^l, \mathfrak{D}\right) \tag{4}$$

where $v_g\left(n_k^l, \mathfrak{D}\right)$ is the Gittins index for a standard $\mathbf{z}_k$ (zero mean and unity variance) and $\mathfrak{D}$ $(0 < \mathfrak{D} < 1)$ is a discount factor. The values of $v_g\left(n_k^l, \mathfrak{D}\right)$ have been tabulated by Gittins in his book [5] and can be derived numerically.

The Gittins algorithm returns the optimal solution in the case of an infinite *temporal discount factor* problem. However, if we want to consider time finite problems, a modification of the Gittins algorithm is required. In [10] the authors proposed the following heuristic

$$\mathfrak{D} = \frac{H-1}{H} \tag{5}$$

to make the infinite time solution compliant with a finite time task. The rationale is that, if a system with a discount factor $\mathfrak{D}$ receives, at every step, a reward z, then the total reward will be $z/(1-\mathfrak{D})$. Thus its total reward will be the same as if there is no infinite temporal and it stops after $1/(1-\mathfrak{D})$ loops.

The Gittins algorithm pseudocode follows:

Algorithm 3. The Gittins index algorithm

1: $\mathbf{z}_k^{1,2} \leftarrow$ play arm k twice; $\forall k \in [1, K]$
2: $n_k \leftarrow 2$; $\forall k \in [1, K]$
3: **for** $l = (2K+1)$ to H **do**
4: **for** $k = 1$ to K **do**
5: $\widehat{\boldsymbol{\mu}}_k^l \leftarrow average(\mathbf{z}_k)$
6: $\widehat{\boldsymbol{\sigma}}_k^l \leftarrow standard\ deviation(\mathbf{z}_k)$
7: $\mathbf{v}_k \leftarrow \widehat{\boldsymbol{\mu}}_k^l + \widehat{\boldsymbol{\sigma}}_k^l \cdot v_y\left(n_k^l, \mathfrak{D}\right)$
8: **end for**
9: $k \leftarrow \arg\max_k \mathbf{v}_k$
10: $n_k^{l+1} \leftarrow n_k^l + 1$
11: $n_i^{l+1} \leftarrow n_i^l \ \forall i \in [1, K]/k$
12: $\mathbf{z}_k^{n_k^{l+1}} \leftarrow$ play arm k
13: **end for**

After the initialization (lines 1 and 2), the algorithm computes the Gittins index for each alternative arm (line 7) and then selects the one with the highest value (line 9).

4 The Probability of Correct Selection

The *probability of correct selection* (PCS) [9,8] at the lth step is the probability that a $(\epsilon = 0)$-greedy algorithm will select the best arm (i.e. the arm one in our notation)

$$PCS^l = P\left(\arg\max_{k\in[1\dots K]} \{\mu_k\} = \arg\max_{k\in[1\dots K]} \left\{\widehat{\boldsymbol{\mu}}_k^l\right\}\right) \tag{6}$$

$$= P\left(\ \widehat{\boldsymbol{\mu}}_1^l > \widehat{\boldsymbol{\mu}}_2^l\ ,\ \ \widehat{\boldsymbol{\mu}}_1^l > \widehat{\boldsymbol{\mu}}_3^l\ ,\ \dots\ ,\ \widehat{\boldsymbol{\mu}}_1^l > \widehat{\boldsymbol{\mu}}_K^l\ \right) \tag{7}$$

$$= P\left(\ \widehat{\boldsymbol{\mu}}_1^l - \widehat{\boldsymbol{\mu}}_2^l > 0\ ,\ \ \widehat{\boldsymbol{\mu}}_1^l - \widehat{\boldsymbol{\mu}}_3^l > 0\ ,\ \dots\ ,\ \widehat{\boldsymbol{\mu}}_1^l - \widehat{\boldsymbol{\mu}}_K^l > 0\ \right) \tag{8}$$

$$= P(\ \widehat{\mathbf{r}}_2 > 0\ ,\ \ \widehat{\mathbf{r}}_3 > 0\ ,\ \dots\ ,\ \widehat{\mathbf{r}}_K > 0\ ) \tag{9}$$

where $\widehat{\mathbf{r}}_k = \widehat{\boldsymbol{\mu}}_1^l - \widehat{\boldsymbol{\mu}}_k^l$.

Under the assumption of Gaussianity the vector $(\widehat{\mathbf{r}}_2, \dots, \widehat{\mathbf{r}}_K)^T$ is a multivariate normal random variable with mean vector Γ and covariance matrix Σ

$$\begin{pmatrix} \widehat{\mathbf{r}}_2 \\ \widehat{\mathbf{r}}_3 \\ \vdots \\ \widehat{\mathbf{r}}_K \end{pmatrix} = \begin{pmatrix} \widehat{\boldsymbol{\mu}}_1^l - \widehat{\boldsymbol{\mu}}_2^l \\ \widehat{\boldsymbol{\mu}}_1^l - \widehat{\boldsymbol{\mu}}_3^l \\ \vdots \\ \widehat{\boldsymbol{\mu}}_1^l - \widehat{\boldsymbol{\mu}}_K^l \end{pmatrix} \sim \mathcal{N}[\Gamma, \Sigma] \tag{10}$$

where

$$\Gamma = \begin{pmatrix} \mu_1 - \mu_2 \\ \mu_1 - \mu_3 \\ \vdots \\ \mu_1 - \mu_K \end{pmatrix} \quad \text{and} \quad \Sigma = \begin{pmatrix} \sigma_{\widehat{\mathbf{r}}_2,\widehat{\mathbf{r}}_2} & \sigma_{\widehat{\mathbf{r}}_2,\widehat{\mathbf{r}}_3} & \cdots & \sigma_{\widehat{\mathbf{r}}_2,\widehat{\mathbf{r}}_K} \\ \sigma_{\widehat{\mathbf{r}}_3,\widehat{\mathbf{r}}_2} & \sigma_{\widehat{\mathbf{r}}_3,\widehat{\mathbf{r}}_3} & \cdots & \sigma_{\widehat{\mathbf{r}}_3,\widehat{\mathbf{r}}_K} \\ \vdots & \vdots & \ddots & \vdots \\ \sigma_{\widehat{\mathbf{r}}_K,\widehat{\mathbf{r}}_2} & \sigma_{\widehat{\mathbf{r}}_K,\widehat{\mathbf{r}}_3} & \cdots & \sigma_{\widehat{\mathbf{r}}_K,\widehat{\mathbf{r}}_K} \end{pmatrix} \tag{11}$$

and where since $\text{cov}[\widehat{\boldsymbol{\mu}}_i^l, \widehat{\boldsymbol{\mu}}_j^l] = 0$ for $i \neq j$

$$\sigma_{\widehat{\mathbf{r}}_j,\widehat{\mathbf{r}}_j} = \text{Var}\left(\widehat{\boldsymbol{\mu}}_1^l - \widehat{\boldsymbol{\mu}}_j^l\right) = \frac{\sigma_1^2}{n_1^l} + \frac{\sigma_j^2}{n_j^l} \tag{12}$$

$$\sigma_{\widehat{\mathbf{r}}_i,\widehat{\mathbf{r}}_j} = \sigma_{\widehat{\mathbf{r}}_j,\widehat{\mathbf{r}}_i} = \text{cov}\left[\widehat{\boldsymbol{\mu}}_1^l - \widehat{\boldsymbol{\mu}}_i^l, \widehat{\boldsymbol{\mu}}_1^l - \widehat{\boldsymbol{\mu}}_j^l\right] = \text{E}\left[(\widehat{\boldsymbol{\mu}}_1^l)^2\right] - \mu_1^2 = \frac{\sigma_1^2}{n_1^l} \tag{13}$$

4.1 The PCS Exploration Strategy

Consider a set of K random variables where the means μ_k and the standard deviations σ_k are known and where the vector $n_{\#}^l = \{n_1^l, n_2^l, \ldots, n_K^l\}$ contains the number of tests done up to the lth step. We define the exploration *PCSstrategy* as follows

---

Algorithm 4. *PCSstrategy*$(\mu_{\#}, \sigma_{\#}, n_{\#}, K)$

1: **for** $k = 1$ to K **do**
2: $\quad n_{\#}^{l+1} = \{n_1^l, \ldots, n_k^l + 1, \ldots, n_K^l\}$
3: $\quad PCS_k^{l+1} \leftarrow computePCS\ (\mu_{\#}, \sigma_{\#}, n_{\#}^{l+1}, K)$
4: **end for**
5: $bestK = \arg\max PCS_k^{l+1}$

---

where the function *computePCS* computes the quantity (6). Several techniques have been proposed in literature to compute a multivariate normal integral [12]. In our algorithm we adopt the numerical method proposed by Genz [4].

Note that, so far, we have assumed that the means in $\mu_{\#}$ and the standard deviations in $\sigma_{\#}$ are known. This is of course an unrealistic assumption that will be relaxed in the following section.

The PCS exploration strategy simulates a sampling for each alternative $\mathbf{z}_k$ by increasing the number of tests n_k^l made on this alternative (line 2) and then computes the corresponding PCS_k^{l+1} (line 3). This quantity returns a guess of the probability of correct selection of a greedy strategy at the $l+1$step if the exploration step samples the kth arm. The best candidate for exploration is then the arm which has the highest PCS_k^{l+1}.

In the following we present a simplified version of the PCS strategy when $K = 2$, i.e. the set $\mathbf{z}_{\#} = \{\mathbf{z}_1, \mathbf{z}_2\}$ contains only two random variables. According to the

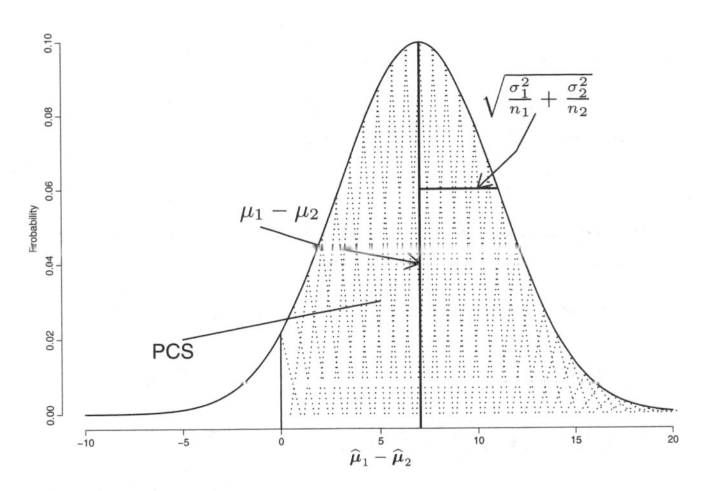

Fig. 1. If K=2, the probability of correct selection (PCS) is the surface under the Gaussian when the abscissa takes his value in the set $[0, +\infty]$

PCSstrategy, the best candidate for exploration is the arm which maximizes the quantity

$$PCS^{l+1} = P\left(\ \widehat{\boldsymbol{\mu}}_1^{l+1} > \widehat{\boldsymbol{\mu}}_2^{l+1} \ \right) \tag{14}$$

$$= P\left(\ \widehat{\boldsymbol{\mu}}_1^{l+1} - \widehat{\boldsymbol{\mu}}_2^{l+1} > 0 \ \right) \tag{15}$$

where, because of the Gaussian assumption

$$\widehat{\boldsymbol{\mu}}_1^{l+1} - \widehat{\boldsymbol{\mu}}_2^{l+1} \sim \mathcal{N}\left(\mu_1 - \mu_2, \frac{\sigma_1^2}{n_1^{l+1}} + \frac{\sigma_2^2}{n_2^{l+1}}\right) \tag{16}$$

The distribution of $\widehat{\boldsymbol{\mu}}_1^{l+1} - \widehat{\boldsymbol{\mu}}_2^{l+1}$ is shown in Figure 1. This figure shows that in order to maximize the probability of correct selection at the $l+1$ step, we have to minimize the variance of $\widehat{\boldsymbol{\mu}}_1^{l+1} - \widehat{\boldsymbol{\mu}}_2^{l+1}$.

$$\mathrm{Var}\left(\widehat{\boldsymbol{\mu}}_1^{l+1} - \widehat{\boldsymbol{\mu}}_2^{l+1}\right) = \frac{\sigma_1^2}{n_1^{l+1}} + \frac{\sigma_2^2}{n_2^{l+1}} \tag{17}$$

Let $\mathrm{Var}^{T_1}\left(\widehat{\boldsymbol{\mu}}_1^{l+1} - \widehat{\boldsymbol{\mu}}_2^{l+1}\right)$ be the variance at step $(l+1)$ if $\mathbf{z}_1$ is tested and $\mathrm{Var}^{T_2}\left(\widehat{\boldsymbol{\mu}}_1^{l+1} - \widehat{\boldsymbol{\mu}}_2^{l+1}\right)$ the variance at step $(l+1)$ if $\mathbf{z}_2$ is tested. Let us define $\Delta\mathrm{Var}^T$ as the difference between the two variances at step $(l+1)$.

$$\Delta\mathrm{Var}^T = \mathrm{Var}^{T_1}\left(\widehat{\boldsymbol{\mu}}_1^{l+1} - \widehat{\boldsymbol{\mu}}_2^{l+1}\right) - \mathrm{Var}^{T_2}\left(\widehat{\boldsymbol{\mu}}_1^{l+1} - \widehat{\boldsymbol{\mu}}_2^{l+1}\right) \tag{18}$$

$$= \frac{\sigma_1^2}{n_1^l + 1} + \frac{\sigma_2^2}{n_2^l} - \frac{\sigma_1^2}{n_1^l} - \frac{\sigma_2^2}{n_2^l + 1} \tag{19}$$

$$= \frac{n_\Delta}{d_\Delta} \tag{20}$$

where

$$n_\Delta = n_2^l n_1^l \left(n_2^l + 1\right) \sigma_1^2 + \left(n_1^l + 1\right) n_1^l \left(n_2^l + 1\right) \sigma_2^2 \quad (21)$$

$$- \left(n_1^l + 1\right) n_2^l \left(n_2^l + 1\right) \sigma_1^2 - \left(n_1^l + 1\right) n_2^l n_1^l \sigma_2^2 \quad (22)$$

$$= n_1^l \left(\sigma_2^2 - \sigma_1^2\right) \left(n_1^l + 1\right) + \sigma_1^2 \left(n_1^l - n_2^l\right) \left(n_1^l + n_2^l + 1\right) \quad (23)$$

$$d_\Delta = \left(n_1^l + 1\right) n_2^l n_1^l \left(n_2^l + 1\right) \quad (24)$$

Since d_Δ is positive and the goal of the *PCSstrategy* is to minimize the variance at $l+1$ the resulting strategy for the exploration step is

$$\begin{cases} \text{if } n_\Delta < 0 \text{ then explore } \mathbf{z}_1 \\ \text{if } n_\Delta > 0 \text{ then explore } \mathbf{z}_2 \\ \text{if } n_\Delta = 0 \text{ then explore either } \mathbf{z}_1 \text{ or } \mathbf{z}_2. \end{cases}$$

5 The ϵ-PCSgreedy Algorithm

This section introduces an ϵ-greedy bandit algorithm which integrates the *PCSstrategy* (Algorithm 4) in order to perform the exploration step. An ϵ-greedy algorithm randomly alternates an exploitation phase (selecting the current best arm for the test) with an exploration phase (randomly selecting an arm for the test). In the previous section we have shown that a better than random exploration can be performed by adopting a *PCSstrategy* if the distribution of each arm reward is known. Since, by definition, these distributions are unknown we propose a plug-in strategy to perform a *PCSstrategy* in a real setting. This means that the vectors $\mu_\#$ and $\sigma_\#$ are replaced by their plug-in estimators $\widehat{\boldsymbol{\mu}}_\#$ and $\widehat{\boldsymbol{\sigma}}_\#$. The resulting algorithm, called ϵ-PCSgreedy is described by the following pseudo-code Note how this algorithm differentiates from the conventional

Algorithm 5. The ϵ-PCSgreedy algorithm

1: $\mathbf{z}_k^{1,2} \leftarrow$ play arm k twice; $\forall k \in [1, K]$
2: $n_k \leftarrow 2$; $\forall k \in [1, K]$
3: **for** $l = (2K+1)$ to H **do**
4: $\quad \mathbf{e} \leftarrow U[0,1]$
5: $\quad$ **if** $\mathbf{e} < \epsilon$ **then**
6: $\quad\quad$ **for** $k = 1$ to K **do**
7: $\quad\quad\quad \widehat{\boldsymbol{\mu}}_k^l \leftarrow average(\mathbf{z}_k)$
8: $\quad\quad\quad \widehat{\boldsymbol{\sigma}}_k^l \leftarrow standard\ deviation(\mathbf{z}_k)$
9: $\quad\quad$ **end for**
10: $\quad\quad k \leftarrow PCSstrategy(\widehat{\boldsymbol{\mu}}_\#^l, \widehat{\boldsymbol{\sigma}}_\#^l, n_\#^l, K)$
11: $\quad$ **else**
12: $\quad\quad k \leftarrow \arg\max_k(\widehat{\boldsymbol{\mu}}_k^l)$
13: $\quad$ **end if**
14: $\quad n_k^{l+1} \leftarrow n_k^l + 1$
15: $\quad n_i^{l+1} \leftarrow n_i^l \ \forall i \in [1, K]/k$
16: $\quad \mathbf{z}_k^{n_k^{l+1}} \leftarrow$ play arm k
17: **end for**

ϵ-greedy only for what concerns the exploration step, while the exploitation step is the same.

5.1 ϵ-PCSgreedy Is Not a *Zero-Regret Strategy*

A zero-regret strategy is a strategy where the regret per round tends to zero when the horizon tends to infinity [13] :

$$\frac{\rho_B^H}{H} \xrightarrow[H\to\infty]{} 0 \quad \Leftrightarrow \tag{25}$$

$$\frac{H \cdot \mu_1 - \sum_{k=1}^{K} n_k^H \cdot \mu_k}{H} \xrightarrow[H\to\infty]{} 0 \quad \Leftrightarrow \tag{26}$$

$$\mu_1 - \frac{n_1^H \cdot \mu_1}{H} - \frac{\sum_{k=2}^{K} n_k^H \cdot \mu_k}{H} \xrightarrow[H\to\infty]{} 0 \tag{27}$$

$$\Rightarrow \begin{cases} \frac{n_1^H}{H} \xrightarrow[H\to\infty]{} 1 \\ \frac{n_k^H}{H} \xrightarrow[H\to\infty]{} 0 \ \ \forall k \in [2, \ldots, K] \end{cases} \tag{28}$$

A zero-regret strategy is thus a strategy which bounds the number of tests on suboptimal arms asymptotically. A simple reasoning by contradiction shows that ϵ-PCSgreedy is not a zero-regret strategy for a fixed ϵ. According to eq. (11), for a given $\mu_{\#}$ and $\sigma_{\#}$, the PCS covariance matrix Σ is

$$\Sigma = \begin{pmatrix} \frac{\sigma_1^2}{n_1} + \frac{\sigma_2^2}{n_2} & \frac{\sigma_1^2}{n_1} & \cdots & \frac{\sigma_1^2}{n_1} \\ \frac{\sigma_1^2}{n_1} & \frac{\sigma_1^2}{n_1} + \frac{\sigma_3^2}{n_3} & \cdots & \frac{\sigma_1^2}{n_1} \\ \vdots & \vdots & \ddots & \vdots \\ \frac{\sigma_1^2}{n_1} & \frac{\sigma_1^2}{n_1} & \cdots & \frac{\sigma_1^2}{n_1} + \frac{\sigma_K^2}{n_K} \end{pmatrix} \tag{29}$$

If the number of tests on the best arm tends to infinity, the matrix Σ tends to the diagonal matrix :

$$\Sigma \xrightarrow[n_1\to\infty]{} \Sigma^{\infty} = \begin{pmatrix} \frac{\sigma_2^2}{n_2} & 0 & \cdots & 0 \\ 0 & \frac{\sigma_3^2}{n_3} & \cdots & 0 \\ \vdots & \vdots & \ddots & \vdots \\ 0 & 0 & \cdots & \frac{\sigma_K^2}{n_K} \end{pmatrix} \tag{30}$$

Thus there exists an N such that $\forall k \in [2, \ldots, K]$, $PCS_1^N < PCS_k^N$. In this situation, *PCSstrategy* will not select the best arm during the exploration step.

Notwithstanding, the strategy like the one proposed in [1] (ϵ goes to zero with a certain rate $\epsilon_l = \min\left\{1, \frac{cK}{d^2 l}\right\}$, where $c > 0$ and $0 < d \leq \min_{j\in[2,\ldots,K]} \{\mu_1 - \mu_j\}$) could be employed to make *PCSstrategy* zero-regret.

6 Bandit Experiments

This section assesses the performance of the ϵ-PCSgreedy algorithm by benchmarking it against the state-of-the-art approaches discussed in Section 3. Six bandit methods are tested : two ϵ-greedy instances ($\epsilon = 0.05$ and $\epsilon = 0.10$), two ϵ-PCSgreedy instances ($\epsilon = 0.05$ and $\epsilon = 0.10$), the Gittins method and the Interval Estimation (IE) ($\Theta = 0.05$) method. We consider four test problems: the first two are based on synthetically generated datasets, the others on a real networking dataset.

The two synthetic benchmarks have an horizon $H = 1000$ and are called *10-armed tasks* and *15-armed tasks*, respectively. Each synthetic benchmark is made of 1000 randomly generated K-armed bandit tasks obtained by uniformly sampling the means ($\mu_k \sim U[0,1]$) and the standard deviations ($\sigma_k \sim U[0,2]$). For each task the rewards are normally distributed ($\mathbf{z}_k \sim N(\mu_k, \sigma_k)$). The performance of a bandit algorithm on these tasks is obtained by averaging over the ensemble of the tasks.

The two real benchmarks are based on a sequential task described in [13]. In this task an agent wants to recover data through different network sources. At each step, the agent selects one source and waits until the data is received. The goal of the agent is to minimize the sum of the waiting time for the successive tests. In order to simulate the delay, a dataset was built by accessing the home pages of 700 universities (every 10 minutes for about 10 days) and storing the time delay (in milliseconds)[1]. If we interpret this task as a bandit problem, each university home page plays the role of an arm and each delay the role of (negative) reward. In our experiments in order to generate a sufficient number of problem instances, we randomly selected 1000 times $K = 10$ or $K = 20$ universities and computed the performance of the methods over an horizon of $H = 500$ tests. The resulting bandit benchmarks are denoted *10-Univer tasks* and *20-Univer tasks*.

Table 1 and 2 report the regrets (to be minimized) obtained for the four datasets by the six methods with different values of H. Table 1 refers to the results in the synthetic tasks while Table 2 refers to the real networking task.

In both tables the first column indicates the horizon H while the other columns indicate the average regret of the six considered methods. The last row, called "mean", reports the regret of the methods averaged over all the horizons. A *paired t-test* is used to compare the average regret of the ϵ-greedy instances with the respective ϵ PCS-greedy alternative having the same ϵ value. A boldface font is used to denote the reward when the p-value of the t-test is smaller than 0.05.

The results show that the ϵ-PCSgreedy strategies significantly outperforms their ϵ-greedy counterparts 9 times out of 20 in the synthetic tasks and 12 times out of 20 in the real tasks, while the contrary never happens. As expected, the improvement is more evident in the case of the largest ϵ since in this case the enhanced exploration strategy has a bigger probability to show its added value.

As far as the other techniques are considered they appear to be largely less effective than the ϵ-greedy approaches.

---

[1] The dataset can be downloaded from `http://sourceforge.net/projects/bandit`

Table 1.

10-armed tasks						
H	ϵ-greedy.05	PCSgreedy.05	ϵ-greedy.1	PCSgreedy.1	Gittins	IE.05
200	21.50	21.28	24.01	23.71	38.07	36.19
400	39.49	38.61	44.1	**42.75**	52.42	53.37
600	56.06	54.59	62.21	**59.47**	60.70	64.44
800	71.22	69.37	78.93	**74.71**	66.78	72.77
1000	85.20	83.16	94.59	**89.14**	71.69	79.44
mean	54.69	**53.40**	60.77	**57.96**	57.93	61.24
15-armed tasks						
H	ϵ-greedy.05	PCSgreedy.05	ϵ-greedy.1	PCSgreedy.1	Gittins	IE.05
200	21.81	21.78	24.95	24.90	48.14	44.82
400	38.97	38.75	45.61	**44.44**	72.16	70.70
600	54.53	54.01	64.46	**62.15**	85.14	87.55
800	69.31	68.25	82.16	**78.38**	94.13	100.08
1000	83.47	**81.64**	98.9	**93.53**	100.79	109.71
mean	53.62	**52.89**	63.22	**60.68**	80.07	82.57

Table 2.

10-Univer tasks						
H	ϵ-greedy.05	PCSgreedy.05	ϵ-greedy.1	PCSgreedy.1	Gittins	IE.05
100	7080	6781	9147	**8298**	20618	17892
200	15310	14424	20198	**17852**	27749	26700
300	22978	23124	31426	**28289**	33219	33660
400	30914	30642	42963	**38300**	38166	39164
500	38583	37220	53658	**46551**	43219	43747
mean	22973	22438	31479	**27858**	32594	32232
20-Univer tasks						
H	ϵ-greedy.05	PCSgreedy.05	ϵ-greedy.1	PCSgreedy.1	Gittins	IE.05
100	5516	**4763**	7032	**5977**	31294	23675
200	12650	**11107**	17168	**14806**	43566	38457
300	20055	18472	27866	**24485**	51505	48273
400	27394	25832	38798	**34019**	58691	57063
500	34551	33280	49438	**43300**	65551	64581
mean	20033	**18691**	28060	**24518**	50122	46410

7 Conclusion and Future Works

The paper contribution is a new exploration strategy for ϵ-greedy algorithms to solve the bandit problem. The approach is based on the notion of PCS, borrowed from the Monte Carlo simulation literature on the best selection problem in a stochastic environment. We used this notion to propose an improved version of the classic ϵ-greedy algorithm where the uniform random exploration phase is replaced by a PCS informed strategy. Future works will focus on three major extensions of the approach: a non-parametric version of the PCS strategy to remove the Gaussian assumption, the use of a confidence interval on the PCS estimation to take into account the uncertainty related to the plug-in approach and the use of a PCS based criterion to tune the value of ϵ to the problem under examination.

References

1. Auer, P., Cesa-Bianchi, N., Fischer, P.: Finite-time analysis of the multiarmed bandit problem. Machine Learning 47(2/3), 235–256 (2002)
2. Azoulay-Schwartz, R., Kraus, S., Wilkenfeld, J.: Exploitation vs. exploration: choosing a supplier in an environment of incomplete information. Decision support systems 38(1), 1–18 (2004)
3. Bertsekas, D.P.: Dynamic Programming - Deterministic and Stochastic Models. Prentice-Hall, Englewood Cliffs (1987)
4. Genz, A.: Numerical computation of multivariate normal probabilities. Journal of Computational and Graphical Statistics (1), 141–149 (1992)
5. Gittins, J.C.: Multi-armed Bandit Allocation Indices. Wiley, Chichester (1989)
6. Hardwick, J., Stout, Q.: Bandit strategies for ethical sequential allocation. Computing Science and Statistics 23, 421–424 (1991)
7. Kaelbling, L.P., Littman, M.L., Moore, A.P.: Reinforcement learning: A survey. Journal of Artificial Intelligence Research 4, 237–285 (1996)
8. Kim, S., Nelson, B.: Selecting the Best System. In: Handbooks in Operations Research and Management Science. Elsevier Science, Amsterdam (2006)
9. Kim, S.-H., Nelson, B.L.: Selecting the best system: theory and methods. In: WSC 2003: Proceedings of the 35th conference on Winter simulation, pp. 101–112 (2003)
10. Schneider, J., Moore, A.: Active learning in discrete input spaces. In: Proceedings of the 34th Interface Symposium (2002)
11. Sutton, R.S., Barto, A.G.: Reinforcement Learning: An Introduction. MIT Press, Cambridge (1998)
12. Tong, Y.L.: The Multivariate Normal Distribution. Springer, Heidelberg (1990)
13. Vermorel, J., Mohri, M.: Multi-armed bandit algorithms and empirical evaluation. In: Gama, J., Camacho, R., Brazdil, P.B., Jorge, A.M., Torgo, L. (eds.) ECML 2005. LNCS, vol. 3720, pp. 437–448. Springer, Heidelberg (2005)

Learning from the Past to Dynamically Improve Search: A Case Study on the MOSP Problem

Hadrien Cambazard[1] and Narendra Jussien[2]

[1] Cork Constraint Computation Centre
Department of Computer Science
University College Cork, Ireland
h.cambazard@4c.ucc.ie
[2] École des Mines de Nantes – LINA CNRS UMR 6241
4 rue Alfred Kastler – BP 20722
F-44307 Nantes Cedex 3, France
narendra.jussien@emn.fr

Abstract. This paper presents a study conducted on the minimum number of open stacks problem (MOSP) which occurs in various production environments where an efficient simultaneous utilization of resources (stacks) is needed to achieve a set of tasks. We investigate through this problem how classical look-back reasonings based on explanations could be used to prune the search space and design a new solving technique. *Explanations* have often been used to design intelligent backtracking mechanisms in Constraint Programming whereas their use in nogood recording schemes has been less investigated. In this paper, we introduce a generalized nogood (embedding explanation mechanisms) for the MOSP that leads to a new solving technique and can provide explanations.

1 The Minimum Number of Open Stacks Problem

The Minimum number of Open Stacks Problem (MOSP) has been recently used to support the IJCAI 2005 constraint modeling challenge [14]. This scheduling problem involves a set of products and a set of customer's orders. Each order requires a specific subset of the products to be completed and sent to the customer. Once an order is started (*i.e.* its first product is being made) a *stack* is created for that order. At that time, the order is said to be *open*. When all products that an order requires have been produced, the stack/order is closed. Because of limited space in the production area, the maximum number of stacks that are used simultaneously, *i.e.* the number of customer orders that are in simultaneous production, should be minimized.

Therefore, a solution for the MOSP is a total ordering of the products describing the production sequence that minimizes the set of simultaneously opened stacks. This problem is known to be NP-hard and is related to well known graph problems such as the minimum path-width or the vertex separation problems. Moreover, it arises in many real life problems as packing, cutting or VLSI design. Table 1 gives an example instance of this problem.

V. Maniezzo, R. Battiti, and J.-P. Watson (Eds.): LION 2007 II, LNCS 5313, pp. 69–80, 2008.

The explanations provided in this paper for the MOSP are related to the *removal explanations* of Ginsberg which are the basis of dynamic backracking [4]. It does not provide natural language explanations for a user but is intended to improve the search by avoiding redundant computation. Indeed by carefully *explaining* why a failure occur, one can avoid a number of other failures that would occur for the same reason. It provides at the same time a justification of optimality for an expert. In the case of the MOSP we propose to define an explanation as a subset of products that would account for the minimal number of open orders. As long as those products are processed, a planner knows that there is no way to reduce the minimum number of simultaneous stacks needed. In other words, it provides insights about the bottleneck in the optimal planning.

The following notations will be used throughout the paper:

Table 1. A 6×5 instance of MOSP with an optimal solution of value 3 – no more than 3 ones can be seen at the same time on the Stacks representation. The *instance* and *optimal ordering* parts of the table are to be read as for example, product P_3 has been ordered by customers c_1 and c_2. The *stacks* part shows that for example the order of custumer c_3 is open from the production of product P_1 to the production of product P_3 (in this part of the table, all 1s are consecutive representing the open nature of the stack representing the order). Another example is to consider the order of customer c_2 which remains open only during the production of product P_2 as this order is composed only with this product.

	Instance						Optimal ordering						Stacks					
	P_1	P_2	P_3	P_4	P_5	P_6	P_1	P_2	P_6	P_4	P_3	P_5	P_1	P_2	P_6	P_4	P_3	P_5
c_1	0	0	1	0	1	0	0	0	0	0	1	1	-	-	-	-	1	1
c_2	0	1	0	0	0	0	0	1	0	0	0	0	-	1	-	-	-	-
c_3	1	0	1	1	0	0	1	0	0	1	1	0	1	1	1	1	1	-
c_4	1	1	0	0	0	1	1	1	1	0	0	0	1	1	1	-	-	-
c_5	0	0	0	1	1	1	0	0	1	1	0	1	-	-	1	1	1	1

- P : the set of m available products and C, the set of n orders.
- $P(c)$ is the set of products required by order c. $C(p)$ is the set of orders requiring product p. A natural extension of this last notation is used for sets of products ($C(s_P)$ is the set of orders that requires at least one product in s_P).
- $O^K(S)$ denotes the set of open orders implied by a subset $S \subseteq K$ of products: $O^K(S) = |C(S) \cap C(K - S)|$. $O(S)$ is a short notation for $O^P(S)$. The open orders therefore refer to a certain point in time where a set of products have already been processed.
- $f(S)$ is the minimum number of stacks needed to complete a set S and $f^A(S)$ is the number of stacks needed to complete set S assuming a set A of initially active (opened) orders.
- p_j denotes the product assigned to position j in the production sequence and $open_j$ expresses the number of open orders at time j. Those variables take their value in a set called their original domain ($D^{orig}(x)$ for variable x).

This set is reduced to a singleton thanks to the solving process (using propagation). The current domain is denoted $D(x)$.

Using those notations, one can provide a mathematical model of the problem:

$$\begin{array}{ll} \min(\max_{j<m} open_j) & \text{s.t.} \\ \quad \forall j,\ 0 < j \leq m, & p_j \in [1..m] \\ \quad \forall j,\ 0 < j \leq m, & open_j \in [1..n] \\ \quad \text{alldifferent}(\{p_1, \ldots, p_m\}) & \\ \quad open_j = |C(\{p_1, \ldots, p_j\}) \cap C(\{p_j, \ldots, p_m\})| & \end{array}$$

2 First Insights in Solving the MOSP

We give a rapid review of the main results obtained for solving the MOSP during the IJCAI 2005 challenge.

2.1 Search Techniques

A wide variety of approaches were proposed for solving this problem during the IJCAI 2005 constraint modelling challenge [14]. One of the most efficient has been identified by [3] and [2]. It is based on dynamic programming (DP): consider a set S of products that have been placed chronologically up to time t from the beginning of the sequence ($|S| = t-1$ and products from S are set from slot 1 to slot $t-1$). Then, one can notice that $f^{O(S)}(P-S)$ remains the same for any permutation of S. Indeed, problem $P - S$ is only related to problem P by the active set of orders at time t: $O(S)$ which does not depend on any particular order of S (an order c is indeed open if $P(c) \cap S \neq \varnothing$ and $P(c) \cap (P-S) \neq \varnothing$). This fact gives a natural formulation of the problem in DP and the objective function can be recursively[1] written as: $f(P) = min_{j \in P}(max(f(P - \{j\}), |O(P - \{j\})|)$. The strong advantage of this approach is to switch from a search space of size $m!$ to one of size 2^m because one only need to explore the subsets of P. From a constraint programming point of view, if S is a **nogood**, *i.e.* a set of products that has been proven as infeasible (according to the current upper bound), any permutation of products of S will lead to an infeasible subproblem $P-S$. Storing such nogoods during a chronological enumeration of the production sequence leads to the same search space size of 2^m.

2.2 Preprocessing and Lower Bounds

A useful preprocessing step can be applied by removing any product p such that $\exists p', C(p) \subseteq C(p')$. This can even be done during search: if S is the current set of already chronologically assigned products up to time t, then one can assign immediately after S the products p such that $C(p) \subseteq O(S)$.

Lower bounds are often based on the *co-demand graph* G which has been defined in the literature in [1]. The nodes of G are associated to orders and an

[1] We consider that if $|P| = 1$ with $P = \{p\}$, $f(\{p\}) = |C(p)|$.

Algorithm 1. NogoodRecMOSP($\{p_1, \ldots, p_{t-1}\}$)

```
1.If t − 1 ≠ m do
2.  For each i ∈ D(p_t) then
3.    p_t ← i; ∀k > t, remove i from D(p_k);
4.    S ← {p_1, ..., p_t};
5.    Try
6.      filter(S,p_{t+1});
7.      NogoodRecMOSP(S);
8.    Catch (Contradiction c)
9.      add nogood {p_1, ..., p_t};
10.   EndTryCatch;
11. EndFor
12.Else store new best solution, update ub;
13.throw new Contradiction();
```

edge (i, j) is added if and only if orders i and j share at least one product. Several lower bounds can be defined from this graph, and we use the size of a clique obtained as a minor of G [1] by edge contraction as it appeared the most effective in our experimentation. These bounds may also be used during search on the problem restricted to $P - S$ by taking into account the current open orders. We used the ones given in [3].

2.3 Our Solution

None of the proposed approaches during the IJCAI 05 Challenge involved look-back techniques (intelligent backtrackers or explanation-based techniques). We intend to show in this paper that the MOSP is a good candidate for these approaches because it is a structured problem, and the optimal number of stacks is often related to small kernels of products.

We will first introduce the simple nogood recording scheme and the main ideas of look-back reasonings for the MOSP. Second, we will define formally the generalized nogood and the two related backjumping algorithms. Experimental results finally show that the proposed look-back techniques perform well on the challenge instances.

3 Simple Nogood Recording

The nogood recording approach (see [11,7] for a general description of the nogood recording technique) is simply based on the chronological enumeration of p_i from p_1 to p_m. Algorithm 1 takes as input a partial sequence of assigned products and tries to extend it to a complete sequence. If the sequence has not been completed yet (line 1), all remaining products for slot p_t will be tried (line 2). A filtering step is then performed and if no contradiction occurs, the algorithm goes on (recursive call line 7). The filtering applied in line 6 is minimal and only prunes the next

Algorithm 2. filter($S = \{p_1, \ldots, p_t\}$, p_{t+1})

```
1.For each i ∈ dom(p_{t+1}) then
2.  If |O^P(S) ∪ C(i)| ≥ ub or S ∪ {i} is a nogood
3.     remove i from D(p_{t+1});
4.EndFor
```

time-slot according to the current upper bound ub (*i.e* the value of the best solution found so far) and the known nogoods (Algorithm 2). Once a sequence, $p_1, \ldots, p_k$ is proved as infeasible (line 9), it is stored so that all permutations will be forbidden in future search. [12] outlines this fact while finally choosing another enumeration scheme. Line 13 is called to backtrack when the domain of p_t has been emptied by search or when a new solution is found (line 12) to prove its optimality. Computation of lower bounds and the use of dominance rules on including products should be included in line 6 and a heuristic to order the products in line 2. This basic algorithm will be improved step by step in the following.

4 Learning from Failures

Nogoods and explanations have long been used in various paradigms for improving search [4,11,10,6]. In the MOSP, from a given nogood S, we can try to generalize it and design a whole class of equivalent nogoods to speed up search.

4.1 Computing Smaller Nogoods

The idea is to answer the following question: once the fact that the minimum number of stacks needed to complete the set of remaining products $P - S$ to schedule considering the open orders $O(S)$ due to the sequence of products S is greater that the current upper bound (in other words that $f^{O(S)}(P - S) \geq ub$ – line 9 of Algorithm 1), what are the conditions on S under which this proof remains valid? In other words, what are the conditions (subsets) in the already scheduled orders that makes this combination a not optimal one considering the current upper bound?

As the optimal value $f^{O(S)}(P - S)$ depends on $P - S$ and $O(S)$, removing a product from S that does not decrease $O(S)$ provides another valid nogood. Indeed adding the corresponding product to $P-S$ can only increase $f^{O(S)}(P-S)$. We can therefore compute some minimal subsets of S that keep $O(S)$ by applying the Xplain algorithm [13,5]. Table 2 gives an example.

4.2 Computing Equivalent Nogoods

The main question now becomes: once $f^{O(S)}(P - S) \geq ub$ has been proven, what are the conditions on $P - S$ under which this proof remains valid? Can

Table 2. Example of nogood reduction: we consider here a new instance. We only consider here the first 5 products and give the open stacks ($O(S)$ column (1 if the order is still open, 0 otherwise) when considering the sequence $S = \{P_1, P_2, P_3, P_4, P_5\}$. Suppose that S is a nogood. It is quite obvious that we can remove production (they will be produced later) the production of products P_1, P_2 and P_3 without modifying the open stacks at the end of sequence of the remaining products ($\{P_4, P_5\}$). Indeed, for example for customer c_1 the order will remain open as at least one of its products has been produced, for customer c_2 as the whole set of products is removed, the stack will remain closed, etc. Therefore, here, $\{P_4, P_5\}$ is also a nogood.

	P_1	P_2	P_3	P_4	P_5	O(S)	...
c_1	1	0	0	1	0	1	
c_2	0	1	1	0	0	0	
c_3	0	0	1	0	1	1	...
c_4	1	1	0	0	0	0	
c_5	0	0	0	1	0	1	
c_6	0	0	0	0	1	1	

we build from those conditions larger sets of nogoods? In other words, what are the conditions (subsets) on the products remaining to schedule that makes the current situation on not optimal one considering the current upper bound.

This problem relies on explanations. Instead of computing some conditions S_1 on S which can be seen as the decisions made so far, we compute some conditions S_2 on $P - S$ which can be seen as original constraints of the problem. A contradiction on S is therefore logically justified by $S_1 \cup S_2$. Only S_2 needs really to be stored within the explanation because S_1 can be computed from scratch at each failure and is resolved by search.

Definition 1. *Let $S = p_1, \ldots, p_{j-1}$ be a sequence of products and $S' = p_1, \ldots,$ p_j a sequence that extends S with $p_j = i$. An explanation for the removal of a value i from p_j, $expl(p_j \neq i)$ is defined by a set E, $E \subseteq P - S$ such that $|O^{S \cup E}(S) \cup C(i)| \geq ub$ or $f^{O(S')}(E - \{i\}) \geq ub$ (in other words, the remaining problem reduced to E is infeasible).*

All filtering mechanisms must now be explained. In the simple case of Algorithm 2, a value i can be removed from $D(p_j)$ if $open_j$ is incompatible with the current upper bound ub. An explanation is therefore only a subset of the remaining products that keep open the open orders at time j. If $S = \{p_0, .., p_{j-1}\}$, $E = expl(p_j \neq i)$ is in this case defined as :

$$|O^{(S \cup E)}(S) \cup C(i)| \geq ub$$

As $open_{j-1}$ is compatible with ub, once S is proved infeasible (both by search and pruning), $expl(p_{j-1} \neq k) = \bigcup_{v \in D^{orig}(p_j)} expl(p_j \neq v)$.

Example: Consider the first example in table 3. $S = \{P_1, P_2\}$, $P - S = \{P_3, P_4, P_5, P_6, P_7\}$, $O(S) = \{c_2, c_3, c_4\}$. On step 1, the upper bound is currently

Table 3. Example of explanation computation

	Example 1																	Example 2				
	Step 1								Step 2													
	P_1	P_2	$O(S)$	P_3	P_4	P_5	P_6	P_7	P_1	P_3	$O(S)$	P_2	P_4	P_5	P_6	P_7		P_1	P_2	P_3	$O(S)$	P_4...
c_1	0	0	0	1	0	0	1	1	0	1	1	0	0	0	1	1	c_1	1	0	0	1	1
c_2	0	1	1	0	0	0	1	0	0	0	0	1	0	0	1	0	c_2	0	1	1	0	0
c_3	1	0	1	1	1	0	0	0	1	1	1	0	1	0	0	0	c_3	1	0	1	0	0 ...
c_4	1	1	1	0	0	1	0	0	1	0	1	1	0	1	0	0	c_4	0	1	0	1	0
c_5	0	0	0	0	1	1	0	1	0	0	0	0	1	1	0	1	c_5	0	0	1	1	1

4 and $p_2 \neq P_2$ because $f^{O(S)}(\{P_3, P_4, P_5, P_6, P_7\}) \geq 4$. This is however also true as long as $O(S)$ is unchanged. It is the case with $\{P_3, P_5, P_6\}$ or $\{P_4, P_5, P_6\}$. All values v of p_3 are removed by filtering and $\{P_4, P_5, P_6\}$ is recorded for each $expl(p_3 \neq v)$ so that $expl(p_2 \neq P_2) = \{P_4, P_5, P_6\}$. Going a step further, the search tries $p_2 = P_3$ and an explanation such as $\{P_4, P_5, P_6\}$ or $\{P_2, P_5, P_7\}$ is computed. The first set leads to $expl(p_2 \neq \{P_2, P_3\}) = \{P_4, P_5, P_6\}$ and the process goes on.

For a filtering due to a nogood N, an explanation $expl(N)$ has already been recorded. A contradiction raised by the lower bound needs also to be *explained.* Xplain can be again applied on the products of $P - S$ to find a subset that respects the needed property.

For each infeasible sequence S, by explaining the proof made on $P - S$, one may first incriminate only a subset $O^P(S)$ that could be used to derive a more accurate subset of S leading to the same contradiction (and a more relevant point to backtrack). Second, it can be useful to generalize the nogood based on the products that are *not involved* in the explanation. In the second example of Table 3, P_4 can be exchanged with $\{P_1, P_3\}$ if P_4 is not needed to prove that $\{P_1, P_2, P_3\}$ is a nogood. Therefore, $\{P_4, P_2\}$ is also a nogood. Equivalent sets to S provided by explanations as well as subsets could therefore allow the pruning of future paths of the search.

Explanations rely on the idea that independency and redundancy among P can lead to small subsets of P having the same optimal value. Explanations provide a way to take advantage of these structures.

5 Generalized Nogoods for the MOSP

A *classical* nogood, defined as a partial assignment that can not be extended to a solution, becomes useless as soon as one of its subset becomes a nogood. This is however not true for the nogoods presented above for the MOSP. The nogood $\{P_1, P_3, P_4\}$ is a subset of $\{P_1, P_2, P_3, P_4\}$ but does not forbid the sequence $\{P_1, P_3, P_2, P_4\}$. A MOSP nogood is indeed a sequence of products that forbids to *start* the production sequence by any of its *permutations.* With the subsets of a set S denoted by $\mathcal{P}(S)$, the nogoods considered for our problem are defined as follows:

Definition 2. *A generalized nogood N is defined by a pair of sets (R, T) (root and tail) which forbids to start the production sequence by any permutation of a set belonging to $\{R \cup T_i, T_i \in \mathcal{P}(T)\}$.*

This definition provides a way of factorizing information regarding a set of nogoods (into the *tail* part). It is meant to capture a large number of identified nogoods. The following proposition is used to characterize generalized nogoods when confronted to infeasible sequences of products.

Proposition 1. *If S is an infeasible sequence of products and $expl(S) \subseteq P - S$ an explanation of this fact then a pair (R, T) such that,*

- $(R \cup T) \cap expl(S) = \emptyset$,
- $O^{S \cup expl(S)}(S) \subseteq O(R)$,

is a valid generalized nogood.

Proof: As S is a nogood, $f^{O(S)}(P - S) \geq ub$. Moreover, $expl(S)$ is a subset of $P - S$ such that, after assigning chronologically S, the remaining problem restricted to $expl(S)$ is infeasible. This leads to $f^{O^{S \cup expl(S)}(S)}(expl(S)) \geq ub$. Due to $O^{S \cup expl(S)}(S) \subseteq O(R)$ and $R \cap expl(S) = \emptyset$ the previous inequality becomes $f^{O(R)}(expl(S)) \geq ub$. This inequality shows that $(R, S - R)$ is a valid generalized nogood even if $P - S$ is restricted to $expl(S)$. Moreover, adding products not included in $expl(S)$ to the tail of $(R, S - R)$ cannot decrease $O(R)$. Each order of $O(R)$ is indeed active because of at least one of the product of $expl(S)$. So (R, T) remains a valid nogood as long as $T \cap expl(S) = \emptyset$. □

In practice, such nogoods are obtained by applying to S the reasonings presented in the previous section. This leads to the algorithms detailed in the following.

5.1 Generalized Nogood Recording

To implement the above idea, lines 8-10 of Algorithm 1 are modified to introduce the computation of the generalized nogood and the backjumping feature. The following pseudo-code of algorithm 3 assumes that a contradiction c is labeled by the *level* where it occurs (c.level in line 9a), in other words, infeasibility of $p_1, \dots p_k$ proved by the empty domain of p_{k+1} would raise a contradiction labelled by $k + 1$ (throw new Contradiction(k+1)).

The function $minimize(S, O_p)$ computes S', a subset of S, such that $O_p \subseteq O(S')$ based on the Xplain technique. Moreover the order of S is used to guide the generation of the subset of S. If $S = p_1, \dots p_i$, it ensures that $argmax_k(p_k \in S')$ is minimal[2]. Two nogoods, based on the roots R^1 and R^2, are recorded at each contradiction. The purpose of R^1 is to provide the best backjumping point (as the latest product within R_1 will be as early as possible) whereas R^2 is the one with the best chance of being minimal (as the contradiction may only involve recent products, S is reversed to focus the subset on the last added products). Backjumping is ensured in line 9h by raising immediately a contradiction if the guilty level is not reached.

[2] There is no subsequence of S with a product p_j s.t $j < k$.

Algorithm 3. Extends lines 8-10 of algorithm 1.

```
8.  Catch (Contradiction c)
9a.   If t < c.level
9b.     R^1 ← minimize(S, O^P(S));
9c.     R^2 ← minimize({p_t, p_{t-1}, ..., p_1}, O^P(S));
9d.     ∀j ∈ {1,2} ad nogood (R^j, S − R^j);
9e.     newLevel ← argmax_k(p_k ∈ R_1)
9f.     If newLevel < t
9g.       throw new Contradiction(newLevel);
9h.   Else if (t > c.level)
9i.     throw new Contradiction(c.level);
10. EndTryCatch;
```

Algorithm 4. Filter($S = \{p_1, \ldots, p_t\}$, p_{t+1})

```
1. For each i ∈ D(p_{t+1}) do
2.   If |O^P(S) ∪ C(i)| ≥ ub
4.     remove i from p_{t+1};
5.     expl(p_{t+1} ≠ i) ← E s.t |O^{S∪E}(S) ∪ C(i)| ≥ ub;
6.   Else If S ∪ {i} is a nogood N;
7.     remove i from p_{t+1};
8.     expl(p_{t+1} ≠ i) ← expl(N);
9. EndFor
```

5.2 Explanation-Based Generalized Nogood Recording

Let us go a step further to develop the above ideas. First, Algorithm 4 replaces Algorithm 2 to explain the pruning due to ub and already known nogoods (whose explanations are computed line 9j of Algorithm 5). We also assume that a contradiction c is labeled by its explanation (c.exp).

Algorithm 1 is extended again when getting a contradiction to deal with explanations leading to algorithm 5. A contradiction explanation is computed in line 9b from the empty domain of p_{t+1}. This explanation will be recorded to explain the removal of the value i that has been tried for p_t (refer to Algorithm 1 for i, p_t and S) when the corresponding level is reached (lines 9n, 9o). Then, at most 4 nogoods are recorded. R^1, R^2 are the same as previously except that $S \cup E$ can be more precise than P for $O^{S \cup E}(S)$ (lines 9c, 9d). However, this should be very rare without more advanced filtering[3] and the main improvements of explanations are to further generalize the nogoods. This generalization occurs in lines 9f, 9g and 9i when using $\overline{E}$ to build the roots R^3,R^4 and $\{\overline{E} \cup S\} - R^i$ to build the tail.

The generalized nogood of Definition 2 corresponds to an exponential number of simple nogoods used in DP and it is impossible to store them all individually. We use a simple form of a finite automaton called a TRIE [8] to store them

[3] This may occur due to the lower bound for example.

Algorithm 5. Extends lines 8-10 of algorithm 1.

8. **Catch (Contradiction c)**
9a. **If** t < c.level
9b. $E \leftarrow \bigcup_{j \in D^{orig}(p_{t+1})} expl(p_{t+1} \neq j)$;
9c. $R^1 \leftarrow minimize(S, O^{S \cup E}(S))$;
9d. $R^2 \leftarrow minimize(\{p_t, p_{t-1}, \ldots, p_1\}, O^{S \cup E}(S))$;
9e. $\overline{E} \leftarrow P - S - E$;
9f. $R^3 \leftarrow minimize(R^1 \cup \overline{E}, O^{S \cup E}(S))$;
9g. $R^4 \leftarrow minimize(R^2 \cup \overline{E}, O^{S \cup E}(S))$;
9h. **for each** $j \in \{1, 2, 3, 4\}$
9i. add nogood $N_j = (R^j, \{\overline{E} \cup S\} - R^j)$;
9j. $expl(N_j) \leftarrow E$;
9k. $newLevel \leftarrow argmax_k(p_k \in R^1)$
9l. **If** $newLevel < t$
9m. **throw** $new\ Contradiction(newLevel, E)$;
9n. **Else** $expl(p_t \neq i) \leftarrow E$
9o. **Else if** ($t = c.level$) $expl(p_t \neq i) \leftarrow c.exp$;
9p. **Else throw** $new\ Contradiction(c.level, c.exp)$;
10. **EndTryCatch;**

and perform the pruning. Storing and efficiently managing nogoods is always a challenging problem. SAT solvers [9] provides interesting results in that matter that are however difficult to apply in our case.

6 Experimental Results

Our experiments are performed on the challenge instances [4] on a laptop MacBook, 2Ghz processor with 2Gb of RAM. The algorithms have been implemented in Java.

We first analyze the accuracy of explanations by looking at the explanation of optimality. It gives a subset of P such that the problem reduced to this set will have the same optimal value as the original problem. For comparison reasons, the problem was also iteratively solved within an Xplain scheme and both approaches were tried. Xplain was able to derive shorter explanations with 35.1% of products removed on average against 21.1% for our explanation based approach but is unpractical on larger instances. These rates can reach 51.8% and 40% on simonis20_20 benchmark for example demonstrating that some problems can be very structured.

Secondly, results are given for the three approaches: NR (the existing simple nogood recording which is closely related to DP), GNR (Algorithm 3) and EXP (Algorithm 5). All instances except the last $SP2$, $SP3$ and $SP4$ instances are solved optimally, and smaller instances than 15_30 such as 20_20 ones are all

[4] http://www.dcs.st-and.ac.uk/~ipg/challenge/

Table 4. Average and maximum results (time and backtracks) on hard benchmarks from the challenge for NR, GNR and EXP. OptAvg represents the average of the optimum.

Inst	OptAvg	TAvg (s)	Tmax	BkAvg	BkMax
		NR			
simonis_30_30	28.32	1.5	14.5	67617.3	708845
wbo_30_30	22.56	2.1	16.2	99586.9	760602
wbop_30_30	23.84	2.3	18.9	109465.9	829886
wbp_30_30	24.46	2.7	35.5	125704.7	1618700
		GNR			
simonis_30_30	28.32	1.6	14.8	23246.3	196229
wbo_30_30	22.56	2.1	14.2	44181.0	283491
wbop_30_30	23.84	2.8	22.6	59874.1	402049
wbp_30_30	24.46	3.0	35.5	51621.3	504488
		EXP			
simonis_30_30	28.32	8.4	64.3	18126.0	167322
wbo_30_30	22.56	13.0	88.0	35818.9	238380
wbop_30_30	23.84	28.0	177.1	52267.2	361704
wbp_30_30	24.46	25.5	256.9	43836.2	431215

solved in less than one second in the worst case. Average and maximum measures of time (in seconds) and search effort (backtracks) are indicated for the hardest instances of the challenge in Table 4. The search space reduction implied by the backjumping and the generalized nogood recording (GNR) is huge (on average by 55, 5% and at most by 64%) with however similar time results. This clearly shows the accuracy of the technique and time improvements could be obtained by improving the nogoods computation and management which remain currently quite naive. The use of explanations (EXP) is clearly more costly. The search space is nevertheless reduced again confirming the accuracy of explanations analyzed on smaller problems. But this technique remains competitive with the best pure constraint programming approaches while computing an explanation at the same time. It is therefore able to highlight hard subsets of products responsible for the minimum number of open stacks.

7 Conclusion: Beyond the MOSP

The main assumptions for the results given here on the MOSP are related to the chronological enumeration and the nature of the objective function. We believe that side constraints could be naturally added (*e.g.* precedences among orders) and any propagation scheme could be performed as long as it is *explained* on $P - S$.

We investigated on the MOSP, how classical look-back reasonings based on explanations could be used to prune the search space. We focused our attention on deriving a set of conditions which generalize a given failure to a whole class of failures. The experimental results demonstrate the interest of such an approach for the MOSP even if no time improvements have been obtained yet. We believe that the dynamic programming-based approaches could therefore be improved by the ideas presented in the paper. There exist many ways to speed up GNR and EXP which could eventually lead to a time gain. The current data structure storing the nogoods is a critical component which could be vastly improved by allowing incremental propagation. This remains to be done as we first investigate the accuracy of explanations for this problem. Moreover, many nogood generation schemes could be designed.

References

1. Becceneri, J.C., Yannasse, H.H., Soma, N.Y.: A method for solving the minimization of the maximum number of open stacks problem within a cutting process. C&OR 31, 2315–2332 (2004)
2. Benoist, T.: A dynamic programming approach. In: IJCAI 2005 Fifth Workshop on Modelling and Solving Problems with Constraints (2005)
3. Garcia de la Banda, M., Stuckey, P.J.: Using dynamic programming to minimize the maximum number of open stacks. INFORMS Journal of Computing (2007)
4. Ginsberg, M.: Dynamic backtracking. Journal of Artificial Intelligence Research 1, 25–46 (1993)
5. Junker, U.: QuickXplain: Conflict detection for arbitrary constraint propagation algorithms. In: IJCAI 2001 Workshop on Modelling and Solving problems with constraints (2001)
6. Jussien, N., Lhomme, O.: Local search with constraint propagation and conflict-based heuristics. Artificial Intelligence 139(1), 21–45 (2002)
7. Katsirelos, G., Bacchus, F.: Generalized nogoods in csps. In: National Conference on Artificial Intelligence (AAAI 2005), pp. 390–396 (2005)
8. Knuth, D.E.: Sorting and Searching. The Art of Computer Programming, vol. 3, pp. 492–512. Addison-Wesley, Reading (1997)
9. Moskewicz, M.W., Madigan, C.F., Zhao, Y., Zhang, L., Malik, S.: Chaff: Engineering an efficient SAT solver. In: Proceedings of the 38th Design Automation Conference (DAC 2001) (2001)
10. Prosser, P.: MAC-CBJ: maintaining arc consistency with conflict-directed backjumping. Technical Report Research Report/95/177, Dept. of Computer Science, University of Strathclyde (1995)
11. Schiex, T., Verfaillie, G.: Nogood recording for static and dynamic constraint satisfaction problem. IJAIT 3(2), 187–207 (1994)
12. Shaw, P., Laborie, P.: A constraint programming approach to the min-stack problem. In: IJCAI 2005 Fifth Workshop on Modelling and Solving Problems with Constraints (2005)
13. De Siqueira, J.L., Puget, J.F.: Explanation-based generalisation of failures. In: ECAI 1988, pp. 339–344 (1988)
14. Smith, B., Gent, I.: Constraint modelling challenge 2005. In: IJCAI 2005 Fifth Workshop on Modelling and Solving Problems with Constraints (2005)

Image Thresholding Using TRIBES, a Parameter-Free Particle Swarm Optimization Algorithm

Yann Cooren, Amir Nakib, and Patrick Siarry

Laboratoire Images, Signaux et Systèmes Intelligents, LiSSi, E.A 3956
Université de Paris 12, 61 avenue du Général de Gaulle, 94010 Créteil, France
{cooren,nakib,siarry}@univ-paris12.fr

Abstract. Finding the optimal threshold(s) for an image with a multimodal histogram is described in classical literature as a problem of fitting a sum of Gaussians to the histogram. The fitting problem has been shown experimentally to be a nonlinear minimization problem with local minima. In this paper, we propose to reduce the complexity of the method, by using a parameter-free particle swarm optimization algorithm, called TRIBES which avoids the initialization problem. It was proved efficient to solve nonlinear and continuous optimization problems. This algorithm is used as a "black-box" system and does not need any fitting, thus inducing time gain.

1 Introduction

The image segmentation process is defined as the extraction of the important objects from an input image. Image segmentation is considered by many authors to be an essential component of any image analysis system, therefore many methods exist to solve this kind of problem. A survey of most segmentation methods may be found in [1]. Image thresholding is one of the most popular segmentation approaches to extract objects from images since it is straightforward to implement. It is based on the assumption that the objects can be distinguished by their gray levels. The automatic fitting of this threshold is one of the main challenges in image segmentation. As this image segmentation approach can be formulated as an optimization problem, many metaheuristics were used to solve it, for instance the segmentation problem was solved using simulated annealing [2] or using an hybrid PSO [3]. With all these methods the initialization problem is not solved.

Particle Swarm Optimization (PSO) is a population-based optimization technique proposed by Kennedy and Eberhart in 1995 [4]. Like other "metaheuristics", PSO shows the drawback of comprising many parameters which have to be defined. The problem is that it is difficult and time consuming to find the optimal combination of parameter values. One aim of researchers is to propose adaptive PSO algorithms of which parameters values change according to results found by the algorithm. The parameter-free algorithm acts as a "black-box" and the user has just to define his problem and the stopping criterion. Clerc has developed a parameter-free algorithm for PSO, called TRIBES [5]. The method incorporates rules defining how the structure of the swarm must be modified and also how a given particle must behave, according to the information gradually collected during the optimization process.

V. Maniezzo, R. Battiti, and J.-P. Watson (Eds.): LION 2007 II, LNCS 5313, pp. 81–94, 2008.

In Section 2, we present the global behavior of the Particle Swarm Optimization algorithm. TRIBES is described in Section 3. Section 4 is dedicated to the presentation of the image thresholding method. Experimental results are discussed in Section 5. A conclusion is given in Section 6.

2 Particle Swarm Optimization

PSO is a simple algorithm that is easy to be coded and implemented. The simplicity of PSO implies that the algorithm is inexpensive in terms of memory requirement and CPU time [5]. All these characteristics have made the popularity of PSO in the field of metaheuristics.

PSO is a population algorithm. It starts with a random initialization of a swarm of particles in the search space. Each particle is modeled by its position in the search space and its velocity. At each time step, all particles adjust their positions and velocities, thus their trajectories, according to their best locations and the location of the best particle of the swarm, in the global version of the algorithm, or of the neighbors, in the local version. Indeed, each individual is influenced not only by its own experience, but also by the experience of other particles.

In a D-dimensional search space, the position and the velocity of the ith particle can be represented as $\vec{X}_i = \left[x_{i,1}, x_{i,2}, ..., x_{i,D}\right]$ and $\vec{V}_i = \left[v_{i,1}, v_{i,2}, ..., v_{i,D}\right]$ respectively. Each particle has its own best location $\vec{p}_i = \left[p_{i,1}, p_{i,2}, ..., p_{i,D}\right]$, which corresponds to the best location reached by the ith particle at time t. The global best location is named $\vec{g} = \left[g_1, g_2, ..., g_D\right]$, which represents the best location reached by the entire swarm. From time t to time $t+1$, each velocity is updated using the following equation:

$$v_{i,j}(t+1) = w.v_{i,j}(t) + c_1.r_1.(p_{i,j} - x_{i,j}(t)) + c_2.r_2.(g_j - x_{i,j}(t)),\ j \in [1:D] \tag{1}$$

where w is a constant called *inertia factor*, c_1 and c_2 are constants called *acceleration coefficients*, r_1 and r_2 are two independent random numbers uniformly distributed in [0,1] and are sampled for each dimension. w controls the influence of the previous direction of displacement. c_1 controls the attitude of the particle of searching around its best location and c_2 controls the influence of the swarm on the particle's behavior. The combination of the values of w, c_1 and c_2 may either favor intensification or diversification. In the first PSO version, the value of each component in V_i was clamped in a range $[-V_{max}, V_{max}]$ to control excessive moves of the particles outside the search space.

The computation of the position at time $t+1$ is derived from (1) using:

$$x_{i,j}(t+1) = x_{i,j}(t) + v_{i,j}(t+1),\ j \in [1:D] \tag{2}$$

In [6], Clerc and Kennedy show that the convergence of PSO may be insured by the use of a constriction factor. Using the constriction factor emancipates us to define V_{max} but also insures a good balance between intensification and diversification. In this case, equation (1) becomes :

$$v_{i,j}(t+1) = K(v_{i,j}(t) + \phi_1.r_1.(p_{i,j} - x_{i,j}(t)) + \phi_2.r_2.(g_j - x_{i,j}(t))),\ j \in [1:D] \tag{3}$$

with:

$$K = \frac{2}{\phi - 2 + \sqrt{\phi^2 - 4.\phi}} \ with\ \phi = \phi_1 + \phi_2,\ \phi > 4 \tag{4}$$

The convergence characteristic of the system can be controlled by ϕ. Namely, Clerc and *al.* [6] found that the system behavior can be controlled so that it has the following features:

- the system does not diverge in a real value region and finally can converge,
- the system can search different regions efficiently by avoiding premature convergence.

Unlike other evolutionary computation methods, PSO with constriction factor ensures the convergence of the search procedure based on the mathematical analysis. The convergence is ensured but it is not ensured that the algorithm converges to the global optimum.

Standard PSO procedure can be summarized through Algorithm 1.

Initialize a population of particles with random positions and velocities.

For each individual i, $\vec{p}_i$ is **initialized** at $\vec{X}_i$.

Evaluate the objective function for each particle and compute $\vec{g}$.

Do

Update the velocities and the positions of the particles.

Evaluate the objective function for each individual.

Compute the new $\vec{p}_i$ and $\vec{g}$.

While the stopping criterion is not met

Algorithm 1. Original PSO procedure.

Generally, the stopping criterion is either a predefined acceptable error or a maximum "reasonable" number of evaluations of the objective function.

3 TRIBES

As it was said in the Introduction, this study deals with algorithms comprising a reduced number of "free" parameters, i.e. parameters to be fitted by the user. In such a framework, the word "parameter" may have two significations:

- "parameter": every component of the algorithm; generally numerical values, but it can also be probability distributions, a strategy, a topology of information links, etc.
- "user-parameter": every "parameter" of the algorithm the user can be led to modify, according to the treated problem.

In all this paper, the word "parameter" is used in the sense "user-parameter".

This section briefly presents TRIBES. For more details, TRIBES is completely described in Clerc's book [5]. Numerical results of TRIBES on real-life problems can be found in [7], describing an application of TRIBES to the flow shop scheduling problem, in [8], describing an application of TRIBES to UMTS radio network modelling, or in [9], describing an application of TRIBES to image segmentation.

3.1 Swarm's Structure

The swarm is structured in different "tribes" of variable size. The aim is to simultaneously explore several promising areas, generally local optima, and to exchange results between all the tribes in order to find the global optimum. This implies two different types of communication: intra-tribe communication and inter-tribes communication.

Each tribe is composed by a variable number of particles. Relations between particles in a tribe are similar with that defined in basic PSO. It is to say that each particle of the tribe stores the best location it has met and knows the best (and the worst) particle of the tribe, i.e. the particle which has met the best (or the worst) location in the search space. This is intra-tribe communication.

Even if each tribe is able to find a local optimum, a global decision must be taken to decide which of these optima is the best one. Each tribe is related to the others in order to take a global decision through its best particle. This is inter-tribes communication.

The most time consuming part of PSO algorithm is the evaluation of the objective function. In order to have execution times of the algorithm as small as possible, it is interesting to carry out the least number of evaluations of the objective function. Consequently, particles are removed of the swarm as soon as possible, in the hope of not affecting the final result. By the way, if a tribe has a good behavior, it is considered that the worst particle of the tribe is useless and, then, it is removed from the swarm. At the opposite, if some tribes have bad performances, new particles will be generated, forming a new tribe, and the "bad" tribes will try to use the information provided by these new particles to improve their performances. Details about the removing and generating processes are available in [5].

To summarize, each particle is informed by itself (best position p), by all the particles of its tribe (internal informers) and, if the particle is a "shaman" (i.e. the best particle of a tribe), by the "shamans" of the other tribes (external informers). All these positions are called the "informers". Then, the best informer of a particle is the informer for which the value of the objective function is lower (resp. higher) in case of minimization (resp. maximization). So, the swarm is composed of a related network of tribes which are themselves dense networks. Fig. 1 illustrates this idea. Arrows symbolize inter-tribes communications and lines symbolize intra-tribe communication. Black particles symbolize the shamans of the different tribes. This structure must be generated and modified automatically, by means of creation, evolution, and removal of the particles. Adaptation rules are described in [5].

At the beginning, the swarm is composed of only one particle which represents a single tribe. If, at the first iteration, this particle does not improve its location, new ones are created, forming a second tribe. At the second iteration, the same process is applied and so on.

The swarm's size will grow up until promising areas are found. The more the swarm grows, the longer the time between two adaptations will be. By this way, the swarm's exploratory capacity will grow up, but the adaptations will be more and more spaced in time. Then, the swarm has more and more chances to find a good solution between two adaptations. At the opposite, once a promising area is found, each tribe will gradually remove its worst particle, possibly until it disappears. Ideally, when convergence is confirmed, each tribe will be reduced to a single particle.

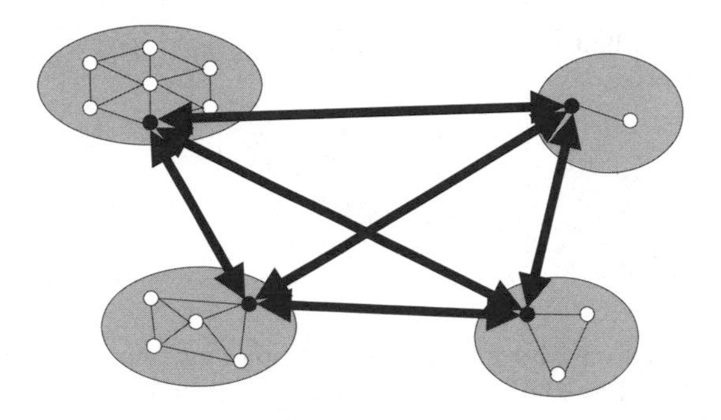

Fig. 1. Intra-tribe and inter-tribes communication

3.2 Swarm's Behavior

In the previous sections, the first way of adaptation of the algorithm was described. The second way in view of adapting the swarm to the results found by the particles is to choose the strategy of displacement of each particle according to its recent past. It will enable a particle with a good behavior to have an exploration of greater scope, with a special strategy for very good particles, which can be compared to a local search. According to this postulate, the algorithm will choose to call the best displacement's strategy in view of moving the particle to a better area of the search space.

There are three possibilities of variation for a particle: deterioration, status quo and improvement, i.e. the current location of the particle is worse, equal or better than its last position. These three statuses are denoted by the following symbols: - for deterioration, = for status quo and + for improvement. The history of a particle includes the

Table 1. Strategies of displacement

History of the particle	Stragegy	Equation
(= +), (+ +)	Independent Gaussians	$X_j = g_j + alea_{normal}(g_j - X_j, \lVert g_j - X_j \rVert),\ j \in [1:D]$ (4)
(- -), (= -), (+ -), (- =), (= =)	Pivot	$\vec{X} = c_1.alea(H_p) + c_2.alea(H_g)$ $c_1 = \frac{f(\vec{p})}{f(\vec{p}) + f(\vec{g})}$ $c_2 = \frac{f(\vec{g})}{f(\vec{p}) + f(\vec{g})}$ (5)
(+ =), (- +)	Disturbed Pivot	$\vec{X} = c_1.alea(H_p) + c_2.alea(H_g)$ $b = N\left(0, \frac{f(\vec{p}) - f(\vec{g})}{f(\vec{p}) + f(\vec{g})}\right)$ $\vec{X} = (1+b).\vec{X}$ (6)

two last variations of its performance. For example, an improvement followed by a deterioration is denoted by (+ -). There are nine possibilities of history. However, we will be satisfied by gathering them in three groups representative of the rule defined above. The three used strategies are defined in Table 1. Let us denote by $\vec{p}$ the best location of the particle, $\vec{g}$ the best position of the informers of the particle and *f* the objective function.

alea(H_p) is a point uniformly chosen in the hyper-sphere of center $\vec{p}$ and radius $\lVert \vec{p} - \vec{g} \rVert$ and alea(H_g) a point uniformly chosen in the hyper-sphere of center $\vec{g}$ and radius $\lVert \vec{p} - \vec{g} \rVert$. $alea_{normal}(g_j - X_j, \lVert g_j - X_j \rVert)$ is a point randomly chosen with a gaussian distribution of center $g_j - X_j$ and radius $\lVert g_j - X_j \rVert$.

3.3 TRIBES Algorithm

Algorithm 2 shows a pseudo-code which summarizes TRIBES process. g_i is the best informer of the ith particle and the *p*'s are the best locations for each particle. *NL* is the number of information links at the last swarm's adaptation and *n* is the number of iterations since the last swarm's adaptation.

```
Initialize a population of particles with random positions and velocities.
For each individual i, p_i is initialized at X_i.
Evaluate the objective function for each particle and compute g_i.
Do
Choice of the displacement strategies
Update the velocities and the positions of the particles.
Evaluate the objective function for each particle.
Compute the new p_i and g_i.
If n<NL
Swarm's adaptations (adding/removing particles, reorganizing the information network)
Computation of NL
End if
While the stopping criterion is not met
```

Algorithm 2. TRIBES algorithm

4 Image Thresholding Method

The image segmentation using the thresholding approach is based on the assumption that the valley between two modes of the image histogram corresponds to a transition between the background and one object. For instance, in the case of bi-level thresholding, the image histogram is usually assumed to have one threshold. In our approach, the thresholding procedure consists in approximating the image histogram h by a sum of Gaussians.

4.1 Gaussian Curve Fitting

For the multimodal histogram $H(x)$ of an image, where x is the gray level, we address the problem of finding the optimal thresholds to be used to separate the modes. We fit the histogram to a sum of d probability density functions (pdf's) [10]. The case where the Gaussian pdf's are used is defined by:

$$PDF(x) = \sum_{i=1}^{d} P_i \exp\left[-\frac{(x-\mu_i)^2}{\sigma_i^2} \right] \tag{7}$$

where P_i is the amplitude of Gaussian pdf on μ_i, μ_i is the mean and σ_i^2 is the variance of mode i.

The histogram is normalized by the following expression:

$$normH(i) = \frac{h(i)}{\sum_{j=0}^{L-1} h(j)} \tag{8}$$

where $h(i)$ is the number of the occurrences of gray-level i over a given image range $[0, L-1]$, and L is the total number of gray-levels.

Our goal is to find a set of parameters, Θ, that minimizes the following objective function J [11]:

$$J = \frac{\sum_i \left| normH(i) - PDF(\Theta, x_i) \right|}{\sum_i normH(i)} \tag{9}$$

The set of parameters defining the Gaussian pdf's and the probabilities is given by:

$$\Theta = \{P_i, \mu_i, \sigma_i; \qquad i = 1, 2, \ldots, d\} \tag{10}$$

J is the objective function to be minimized with respect to Θ. The standard process of setting the partial derivatives to zero results in a set of non-linear coupled equations, the system usually being solved through numerical techniques.

4.2 Overall Probability of Error Criterion

We assume that the histogram is correctly fitted using the Gaussian curve fitting procedure. Then, the optimal threshold is determined by minimizing the overall probability of error. For two successive Gaussian pdf's, it is given by:

$$E(T_i) = P_i \int_{-\infty}^{T_i} p_i(x)\, dx + P_{i+1} \int_{T_i}^{+\infty} p_{i+1}(x)\, dx,\ i \in [1 : d-1] \tag{11}$$

with respect to the threshold T_i, where $p_i(x)$ is the ith pdf [12].

The minimization of this error requires differentiating $E(T_i)$ with respect to T_i (*using the rule of Leibniz*) and equalizing the result to 0. It gives:

$$P_i p(T_i) = P_{i+1} p(T_{i+1}) \tag{12}$$

Applying this result to our case (Gaussian density), the solution of the problem is reduced to solve the second order equation given by:

$$AT_i^2 + BT_i + C = 0 \tag{13}$$

with:

$$\begin{aligned} A &= \sigma_i^2 + \sigma_{i+1}^2 \\ B &= 2(\mu_i \sigma_{i+1}^2 - \mu_{i+1} \sigma_i^2) \\ C &= \mu_{i+1}^2 \sigma_i^2 - \mu_i^2 \sigma_{i+1}^2 + 4\sigma_i^2 \sigma_{i+1}^2 \ln(\sigma_{i+1} P_i / \sigma_i P_{i+1}) \end{aligned} \quad (13)$$

A quadratic equation has two possible solutions, but only one of them is a feasible solution [12].

4.3 Thresholding Procedure

The number of Gaussians (d) is supposed to be known *a priori* and is equal to the number of classes in the image (NC). As a consequence, NC-1 is the number of thresholds. The approximated histogram h' is expressed by Equation (7). Then, the optimal thresholds are localized at the intersections of the different gaussians.

We propose to solve this problem by TRIBES. The number of evaluations of the objective function is used as a stopping criterion. Looking at our experiments, the fitness value does not decrease significantly after 10000xNC evaluations of the objective function. Then, we fixed the maximum number of evaluations of the objective function at 10000xNC.

5 Experimental Results

In this section, we present and discuss the experimental results of the proposed method through two examples of image segmentation. In the first example, the well known image Lena (Fig.2 (a)) is used, and in the second example the Screw (Fig.3 (a)) image is used.

Both images are of size 256x256 and L=256. The procedure of Gaussian curve fitting is performed using TRIBES.

The performances of TRIBES are compared to those provided by the Standard PSO 2006 (SPSO), a constricted PSO algorithm. The fitting of SPSO used in our experiments is as follows: random neighbourhood search, $w = 0.5(\log 2)^{-1}$, the number of particles is calculated by: $S = \left\lfloor 10 + 2\sqrt{D} \right\rfloor$, where D is problem dimension, and $c1 = c2 = 0.5 + \log 2$ [13].

The results obtained using TRIBES in the cases of Lena and Screw images, with NC=3, 4 and 5, are presented in Table 2 and Table 3, respectively. Those obtained using SPSO in the cases of Lena and Screw images, with NC=3, 4 and 5, are presented in Table 4 and Table 5, respectively.

The obtained segmentation results, in the cases of 3 and 4 classes, show that TRIBES and SPSO provide similar results. However, TRIBES does not need any parameter fitting. In the case of 5 classes and more, the tow algorithms do not provide the same results, this is due to the increase of the problem dimension.

In Fig.2 we illustrate the segmentation results for the Lena image through TRIBES in the cases of 3 and 4 classes. Fig. 2 (b) and (c) illustrate the results of the

Table 2. Experimental results for Lena image with TRIBES algorithm

Number of classes	*Parameters of Gaussian curves*	*Thresholds*	*Final value of Fitness*
3	P:(0.2624, 0.7685, 0.8153) μ:(198, 23, 100) σ:(14.0789, 8.3667, 66,4781)	T:(41, 179)	2.0761
4	P:(0.3541, 0.2902, 0.7758, 0.7547) μ:(133,196, 94, 23) σ:(8.9784, 16.8357, 64.3285, 8.6236)	T:(41, 119, 174)	1.2141
5	P:(0.3370, 0.2877, 0.7868, 0.3420,0.7536) μ:(65, 196, 23, 133, 98) σ:(4.7779, 16.3593, 9.3536, 8.5447, 63.0961)	T:(42, 75, 120, 175)	0.7023

Table 3. Experimental results for Screw image with TRIBES algorithm

Number of classes	*Parameters of Gaussian curves*	*Thresholds*	*Final value of Fitness*
3	P:(0.570, 0.036, 0.529) μ:(34, 98, 251) σ:(7.383, 100.0, 0.614)	T:(55, 250)	0.3701
4	P:(0.035, 0.528, 0.845, 0.505) μ:(100, 251, 32,35) σ:(100.0, 0.616, 0.272, 7.866)	T:(34, 57, 250)	0.1311
5	P:(0.038, 0.537, 0.018, 0.639, 0.506) μ:(79, 251, 216, 32, 34, 57) σ:(56.781, 0.552, 100.0, 0.189, 7.891)	T:(34, 56, 170, 250)	0.1199

Table 4. Experimental results for Lena image with SPSO algorithm

Number of classes	*Parameters of Gaussian curves*	*Thresholds*	*Final value of Fitness*
3	P:(0.262, 0.764, 0.815) μ:(198, 23, 100) σ:(14.078, 8.835, 66.480)	T:(41, 179)	2.0761
4	P:(0.775, 0.754, 0.290, 0.354) μ:(94, 23, 196, 133) σ:(64.328, 8.623, 16.835, 8.978)	T:(40, 119, 172)	1.2141
5	P:(0.337, 0.342, 0.786, 0.287, 0.753) μ:(65, 133, 23, 196, 98) σ:(4.777, 8.544, 9.353, 16.359, 63.095)	T:(42, 75, 119, 174)	0.7023

Table 5. Experimental results for Screw image with SPSO algorithm

Number of classes	*Parameters of Gaussian curves*	*Thresholds*	*Final value of Fitness*
3	P:(0.5707, 0.0366, 0.5294) μ:(34, 98, 251) σ:(7.38, 100.0, 0.6141)	T:(54, 250)	0.3701
4	P:(0.5293, 0.0358, 0.5252, 0.5058) μ:(251, 101, 32, 35) σ:(0.6135, 100.0, 0.1752, 7.8581)	T:(34, 57, 250)	0.1311
5	P:(0.5284, 0.0182, 0.5070, 0.0384, 0.7295) μ:(251, 214, 34, 79, 32,) σ:(0.5644, 100.0, 7.8907, 56.5071, 0.2354)	T:(34, 56, 177, 250)	0.1199

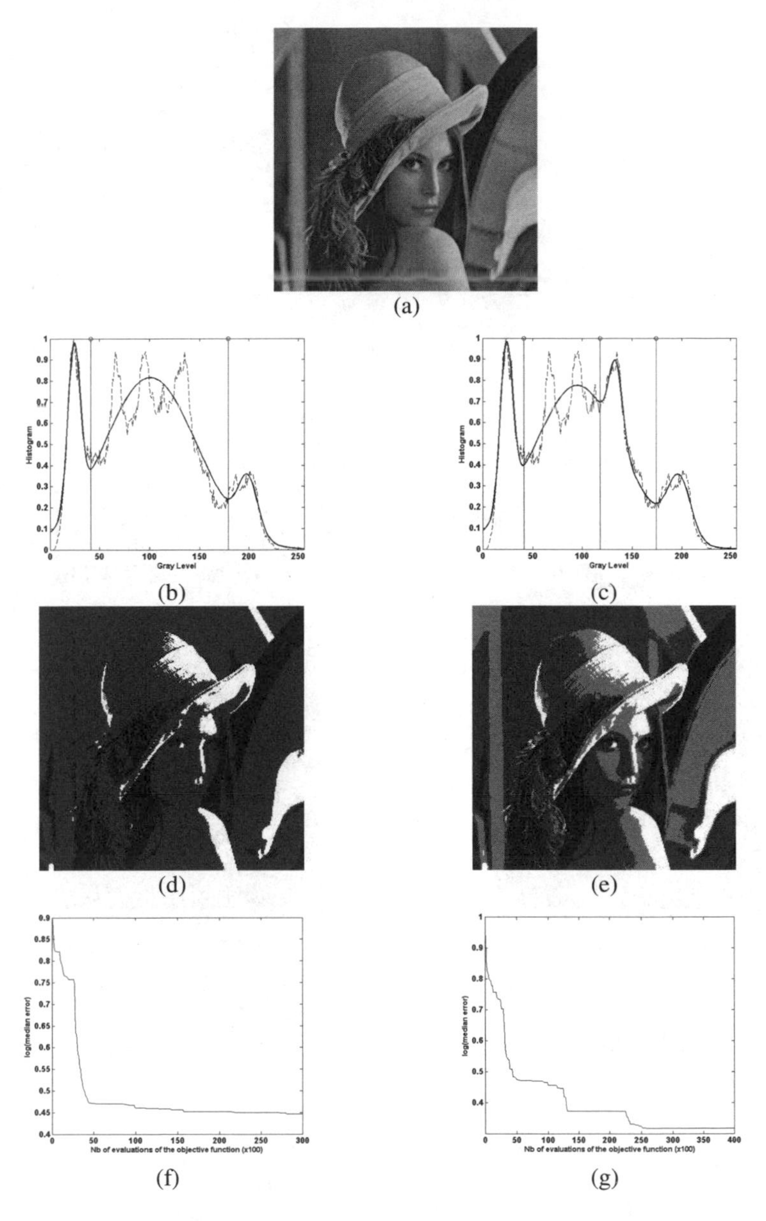

Fig. 2. Image segmentation results for the Lena image, (a) original image, (b) original and fitted histograms with d=3, (c) original and fitted histograms with d=4, (d) segmented image T:(41, 179) , (e) segmented image T:(41, 119, 174), (f) convergence curve for d=3. (g) convergence curve for d=4. Dashed and continuous lines correspond to original and fitted histograms.

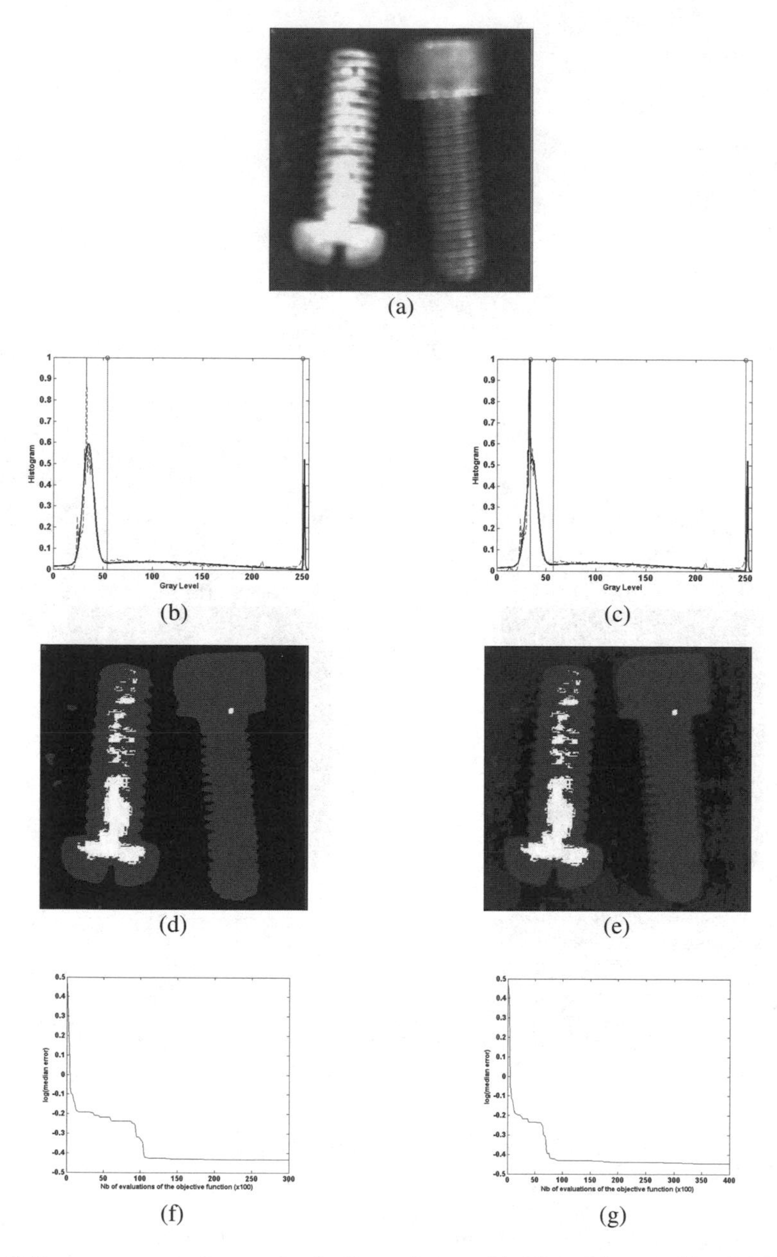

Fig. 3. Image segmentation results for Screw image, (a) original image, (b) original and fitted histograms with d=3, (c) original and fitted histograms with d=4, (d) segmented image T:(55, 250) , (e) segmented image T:(34, 57, 250), (f) convergence curve for d=3. (g) convergence curve for d=4. Dashed and continuous lines correspond to original and fitted histograms.

approximation of the image histogram (Lena) through 3 and 4 Gaussians, respectively. Fig2 (d) and (e) present the segmentation results on 3 and 4 classes. The segmentation result in the case of 3 classes is not good enough, probably since the Lena image has more than 3 classes.

The corresponding TRIBES convergence curves are presented in Fig. 2 (f) and (g) in the two cases, respectively. These figures represent the variations of the logarithm of the median error over the number of evaluations of the objective function. Looking at these figures, the used stopping criterion allows to have the optimal results.

Fig. 3 presents the experimental results in the case of the Screw image. The goal of the segmentation in this case is to extract the Screws from the background. In Fig. 3 (a) and (b) the approximations of the histogram for NC=3 and 4 are presented, respectively.

The corresponding segmented images are presented in Fig. 3 (d) for the segmentation in 3 classes and Fig. 3 (e) in the case of 4 classes. One can observe that the segmentation result in 3 classes allows to extract the two screws. The curves of convergence of TRIBES are presented in Fig. 3 (f) and (g) for NC=3 and 4, respectively. The convergence curves confirm the efficiency of the stopping criterion.

6 Conclusion

In this paper, we proposed a new simple approach to find the optimal thresholds of an image, based on Gaussian curve fitting. The fitting problem was seen as an optimization problem and solved using TRIBES, a parameter-free Particle Swarm Optimization algorithm. Experimental results show that the presented method leads to convincing segmentations with competitive computational times and without any particular initialization. Our study in progress consists in adding other segmentation criteria to further improve the segmentation quality and accelerate the optimization.

References

[1] Sahoo, P.K., Soltani, S., Wong, A.K.C., Chen, Y.C.: A survey of thresholding techniques. Comput. Vis. Graphics Image Process 41, 233–260 (1988)

[2] Nakib, A., Oulhadj, H., Siarry, P.: Image Histogram Thresholding based on multiobjective optimization. Signal Processing 87, 2516–2534 (2007)

[3] Zahara, E., Fan, S.S., Tsai, D.: Optimal multi-thresholding using a hybrid optimization approach. Pattern Recognition Letters 26(8), 1082–1095 (2005)

[4] Kennedy, J., Eberhart, R.C.: Particle Swarm Optimization. In: Proc. IEEE Int. Conf. On Neural Networks, WA, Australia, pp. 1942–1948 (1995)

[5] Clerc, M.: Particle Swarm Optimization. International Scientific and Technical Encyclopaedia (2006)

[6] Clerc, M., Kennedy, J.: The particle swarm: explosion, stability, and convergence in multi-dimensional complex space. IEEE Transactions on Evolutionary Computation 6, 58–73 (2002)

[7] Onwubolu, G.C., Babu, B.V.: TRIBES application to the flow shop scheduling problem. In: New Optimization Techniques in Engineering, ch. 21, pp. 517–536. Springer, Heidelberg (2004)
[8] Nawrocki, M., Dohler, M., Aghvami, A.H.: Understanding UMTS radio network modelling, Theory and Practice. Wiley, Chichester (2006)
[9] Nakib, A., Cooren, Y., Oulhadj, H., Siarry, P.: Magnetic resonance image segmentation based on two-dimensional exponential entropy and a parameter free PSO. In: Proceedings of the 8th International Conference on Artificial Evolution, Tours, France, October 29-31 (2007)
[10] Synder, W., Bilbro, G.: Optimal thresholding: A new approach. Pattern Recognition Letters 11, 803–810 (1990)
[11] Romanenko, S.V., Stromberg, A.G.: Resolution of the overlapping peaks in the case of linear sweep anodic stripping voltametry via curve fitting. Chemo. and Intelligent Lab. Systems 73, 7–13 (2004)
[12] Gonzales, R.C., Woods, R.E.: Digital image processing. Prentice Hall, Upper Sadler River (2002)
[13] Particle Swarm Central (2006), `http://www.particleswarm.info/Standard_PSO_2006`

Explicit and Emergent Cooperation Schemes for Search Algorithms

Teodor Gabriel Crainic[1,2] and Michel Toulouse[1]

[1] CIRRELT, Université de Montréal,
C.P. 6128, Succursale Centre-Ville, Montreal H3C 3J7, Canada
{theo,toulouse}@crt.umontreal.ca
[2] Department of Management and Technology, ESG UQAM,
C.P. 8888, succ. Centre-ville, Montreal QC Canada H3C 3P8

Abstract. Cooperation as problem-solving and algorithm-design strategy is widely used to build methods addressing complex discrete optimization problems. In most cooperative-search algorithms, the explicit cooperation scheme yields a dynamic process not deliberately controlled by the algorithm design but inflecting the global behaviour of the cooperative solution strategy. The paper presents an overview of explicit cooperation mechanisms and describes issues related to the associated dynamic processes and the emergent computation they often generate. It also identifies a number of research directions into cooperation mechanisms, strategies for dynamic learning, automatic guidance, and self-adjustment, and the associated emergent computation processes.

1 Introduction

Cooperation as problem-solving and algorithm-design strategy is widely used in many fields, including but not restricted to ad hoc wireless networks, swarm robotics, multi-agent systems, constraint programming, and exact and meta-heuristic methods addressing complex discrete optimization problems. While the cooperation paradigm may take different forms, they all share two features: a set of highly autonomous programs (*APs*), each implementing a particular solution method, and a cooperation scheme combining these APs into a single problem-solving strategy. This strategy is different from the "cooperation" found in distributed algorithms, which is predetermined by the decomposition of an algorithm into concurrent processes. Cooperative-search algorithms combine independent problem-solving strategies, that is, each element of the cooperation is a stand-alone method able, in most cases, to address, "solve", the problem instance considered. Consequently, cooperation mechanisms must be designed explicitly, which constitutes the core of writing a cooperative algorithm.

In most cooperative-search algorithms, the explicit, deliberately designed, cooperation scheme yields an associated stream of correlated interactions, i.e., reactive actions with a potential for emergent computation and, eventually, cooperation. This dynamic process, active simultaneously with the deliberate, optimization-oriented cooperative-search algorithm, is neither controlled by the

V. Maniezzo, R. Battiti, and J.-P. Watson (Eds.): LION 2007 II, LNCS 5313, pp. 95–109, 2008.

algorithm design, nor spontaneously oriented toward the optimization of the problem at hand, but contributes to determine the global behaviour of the co-operative solution strategy.

Correlated and indirect interactions are similar to ripple (or side) effects in computing systems, defined as coherent processing activities partly independent of algorithmic rules. In distributed computation, for example, a delay to execute some instructions (due to workload variations of shared resources such as CPUs) can generate a ripple effect on the ordering of the execution of several other concurrent operations. Distributed algorithms are, in fact, notoriously difficult to debug because ripple effects make faulty behaviour almost impossible to reproduce. Ripple effects are a form of adaptive response of the processing activities to events not explicitly accounted for in the algorithm design.

In the context of cooperative-search methods, correlated asynchronous interactions occur according to system conditions, as well as the internal state of the cooperating APs and the information exchanged. Though they escape the direct control of cooperation schemes, indirect interactions influence the subsequent logical steps of the APs whether they interact with the computing environment or with other search programs. Now the question is: could indirect interactions, which are inevitable in cooperative search, be harnessed to yield a better performing method? Under which conditions could this behaviour be obtained? Could we understand and eventually "design" an implicit cooperation strategy emerging out of these interactions?

We believe two important research issues stand out in this context: On the one hand, the development of good cooperation mechanisms, including strategies for dynamic learning, automatic guidance, and self-adjustment; On the other hand, the study of the associated dynamic processes focusing on the possible emergence of a second, implicit, layer of cooperation among the APs and on how to harness it to obtain a "better" search method for the problem at hand.

The goal of this paper is to contribute to address these issues. We give an overview of the main explicit cooperation mechanisms encountered in the literature and describe issues related to the associated dynamic processes and possible emergent computation. We also identify a research agenda focused on these issues that we believe important and timely given the current interest not only in the parallelization of exact and meta-heuristic solution methods, but also in novel algorithmic schemes (e.g., the so-called swarm methods).

The paper is organized as follows. Section 2 briefly recalls the main cooperative parallel meta-heuristics mechanisms. Section 3 discusses the reactive component of cooperative-search algorithms and its potential for emergent cooperation. Section 4 revisits cooperative parallel meta-heuristics mechanisms focusing on dynamic learning, automatic guidance, and self-adjustment strategies as a partial response to emergent computation issues. Research avenues are summarized in Section 5 and we conclude in Section 6.

2 Explicit Cooperation Schemes in Meta-heuristics

Providing a full-length literature review of contributions to cooperation and meta-heuristics is beyond the scope of this paper. Our objective is to synthesize the state-of-knowledge of the field as we see it. The interested reader may consult a number of survey papers on parallel meta-heuristics that dedicate (under various forms and names) significant space to cooperative methods, including two recent books [1,30] that collect chapters on many issues in parallel computing for combinatorial optimization and [7,8,14,17].

Parallelism in general, and cooperative strategies in particular, imply that both the individual APs and the resulting global search proceed most of the time with incomplete knowledge regarding the status of the search. The design of the information exchange mechanisms is thus a key element to the good performance of cooperative methods. Important cooperation design issues include its *content* (what information to exchange), *timing* (when to exchange it), *connectivity* (the logical inter-processor structure), *mode* (synchronous or asynchronous communications), *exploitation* (what each AP does with the received information), as well as its *scope*, that is whether new information and knowledge is to be extracted from the exchanged data to guide the search.

The importance of these issues is reflected in most parallel meta-heuristic taxonomies. To illustrate, consider the classification of Crainic and Nourredine [13], which generalizes that of [16,7]; [33,17] present classifications that proceed of the same spirit), where two of the three dimensions relate to cooperation issues. Thus, the *Search Control Cardinality* dimension examines how the global search is controlled: either by a single process or collegially by several processes that may collaborate or not. Cooperative methods belong to the second category denoted *p-control (pC)*. The *Search Control and Communications* dimension then addresses the issue of information exchanges according to four classes to reflect the quantity and quality of the information shared, as well as the additional knowledge derived from these exchanges (if any): *Rigid (RS)* and *Knowledge Synchronization (KS)* and, symmetrically, *Collegial (C)* and *Knowledge Collegial (KC)*. As for the third dimension, *Search Differentiation*, it reflects the diversity of the initial solutions and search strategies: *SPSS, Same initial Point/Population, Same search Strategy*; *SPDS, Same initial Point/Population, Different search Strategies*; *MPSS, Multiple initial Points/Populations, Same search Strategies*; *MPDS, Multiple initial Points/Populations, Different search Strategies* (where "point" is used for neighbourhood-based methods).

The most crude form of cooperation involves (almost) no cooperation at all and is identified as *pC-RS* with any of the previous search differentiation strategies. Such *independent* multi-search methods start several processes, using the same or different solution strategies, from different initial configurations. No attempt is made to take advantage of the multiple APs running in parallel, other than to identify the best overall solution once all processes stop. This parallelization of the classic sequential multi-start heuristic is easy to implement and may offer satisfactory results in terms of search acceleration.

pC-KS strategies obtain cooperation by adopting the same general approach as in the independent search case but taking advantage of the parallel exploration by synchronizing the APs at pre-determined intervals. An information exchange mechanism then determines the best current overall solution and the search is restarted from that point. The mechanism may use a designated process to gather information, extract the best solution, and broadcast it to all search processes. Alternatively, each AP may be empowered to initiate synchronization (e.g., using a broadcast) of all or a pre-specified subset of processes (e.g., processes that run on neighbouring processors). Here, as in the more advanced cooperation mechanisms indicated bellow, *migration* is the term used to identify information exchanges in population-based parallel algorithms.

Synchronization was seen as a means to re-create a state of complete knowledge to share among all participating individual methods, and it was hopped that performances, in terms of computing efficiency and solution quality, would be improved. This did not materialize, however. In fact, compared to independent and most asynchronous strategies, synchronous cooperative methods display larger computational overheads, appear less reactive to the evolution of the global parallel search, and conduct to the premature convergence of the associated dynamic process. It has been shown, for example, that frequent broadcasting of new solutions that stop individual methods from continuing to explore improving sequences leads to either a random search or premature convergence.

Controlled, parsimonious, and timely exchanges of meaningful information are thus characteristic of successful cooperative strategies. Asynchronous methods belong to this group and may be characterized according to the quantity and quality of the information exchanged and, eventually, the "new" knowledge inferred based on these exchanges. *pC-C* asynchronous cooperative methods exchange "good" solutions only or, when a memory mechanism exists, implement simple strategies to extract solutions from memory to pass to APs. More advanced designs, denoted *pC-KC*, add procedures to create new information and solutions based on the solutions exchanged, and implement guiding mechanisms based on this information.

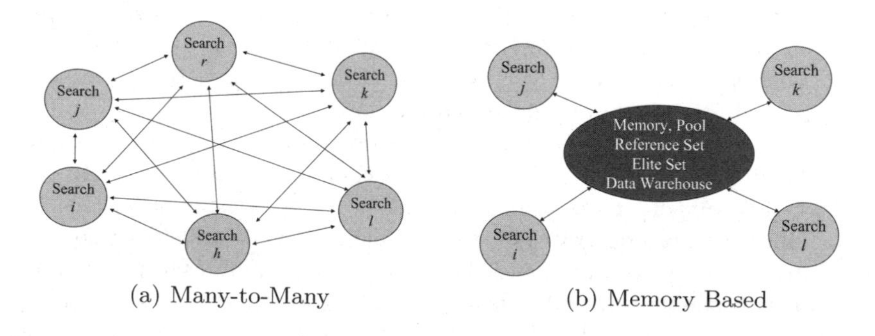

(a) Many-to-Many (b) Memory Based

Fig. 1. Direct Communication Schemes

Communications may be undertaken either directly or indirectly. Strategies based on the evolutionary paradigm generally use direct communications. The population is divided into subsets, each assigned to a processor (alternatively, relatively small populations are generated for each processor), and a genetic algorithm runs on each. An individual population and a genetic algorithm form a so-called *island*. Each island may potentially communicate with any of the other islands, as illustrated in Figure 1a. Then, according to an exchange protocol (e.g., on demand from an island with low population diversity), a migration operator sends a "good" individual to another island. This parallel cooperative strategy is known as *coarse grained*. Islands (processors) may also be allowed to communicate with a limited number of other islands (processors) only, as illustrated in Figure 2a. Such limitations are generally the result of particular topologies of the processor network, e.g., the 2-D torus of Figure 2a. Communications then take place only among adjacent processors according to a so-called *diffusion* mechanism. Notice that, islands tend to have very small populations in this case and the strategy to be denoted *fine grained*. When populations are down to single individuals, the genetic operators are applied to individuals on adjacent islands.

Many cooperative developments outside the evolutionary community are based on indirect communications and, currently, the largest number use some form of *memory* for inter-process communications (the terms *pool* and *solution warehouse* are also used; due to the role assigned to the elements it contains, the terms "reference" and "elite set" are also sometimes used, while the artificial intelligence community uses a similar concept under the name "blackboard"). The individual heuristic or exact methods are generally assigned each to a processor, as illustrated in Figure 1b. In the literature, so-called *adaptive-memory* methods [29] store partial elements of good solutions and combine them to create new complete solutions that are then improved by the cooperating programs, while *central-memory* approaches [15] exchange complete elite solutions that are then used to steer the search and, eventually, create new information.

Cooperation is achieved through asynchronous exchanges of information through the pool (which may share a processor with an AP or be assigned

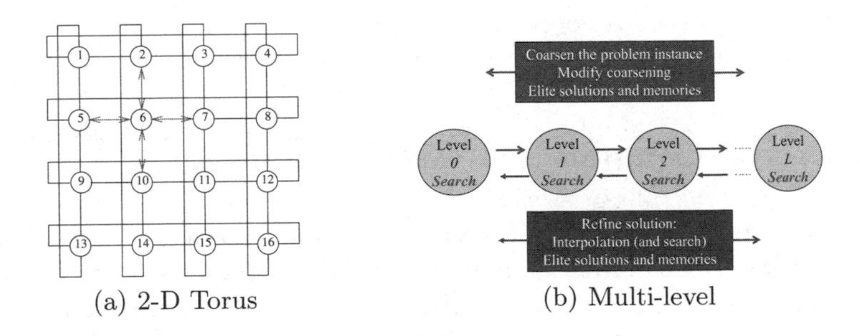

(a) 2-D Torus (b) Multi-level

Fig. 2. Diffusion Communication Schemes

a particular one). Whenever a program desires to send out information (e.g., when a new local optimum is identified), it sends it to the pool. Similarly, when a program needs to access outside information (e.g., to diversify the search), it reaches out and takes it from the pool. Communications are initiated exclusively by the APs, irrespective of their role as senders or receivers of information. No broadcasting is taking place and there is no need for complex mechanisms to select the programs that will receive or send information and to control the cooperation. The pool is thus an efficient implementation device that allows for a strict asynchronous mode of exchange, with no predetermined connection pattern, where no process is interrupted by another for communication purposes, but where any AP may access at all times the data previously sent out by any other AP.

Multi-level cooperative search [32] offers a different *pC-KC* cooperation approach based on controlled diffusion of information principles (Figure 2b). Each AP works at a different level of aggregation of the original problem (one AP works on the original problem) and communicates exclusively with the APs working on the immediate higher and lower aggregation levels. Improved solutions are exchanged asynchronously at moments dynamically determined by each AP according to its own logic, status, and search history. Received solutions are used to modify the search at the receiving level. An incoming solution will not be transmitted further until a number of iterations have been performed, thus avoiding the uncontrolled diffusion of information.

Strict and knowledge-synchronous mechanisms yield a rather strict control of the global search, the trajectory of each AP in the cooperation changing according to the state of all other APs, resulting in no or little emergent behaviour being observed. On the other hand, however, these approaches have been shown experimentally to generally yield inferior results to those of collegial and knowledge-collegial strategies. The behaviour of APs in the latter contexts depends on the information exchanged and, in the case of memory-based cooperation, on the information stored and its management. Moreover, the evolution of *pC-KS* cooperative systems creates new knowledge, new solutions, targets, and statistics, based on the information stored in the memory structure and uses this new knowledge to influence the trajectory of each AP in the cooperation and, thus, the trajectory of the global search. The information propagation inherent to these asynchronous cooperation mechanisms yields significant emergent behaviour as discussed in the next section.

3 Emergent Computation and Cooperation

All cooperation search mechanisms presented in the previous section involve *explicit* exchanges of information among the APs. These exchanges are defined by the design of the cooperation scheme, which details exactly what and when information is to be shared, as well as how this information is to be used. This explicit design exists even when exchanges are performed asynchronously and indirectly through a pool.

The information collected and communicated through an explicit exchange by one AP, the sender, generally modifies information available to, and thus the search trajectory of, at least one other program: the receiver. For example, the sender may communicate its newly improved best solution x and the receiver may re-initialize its search from it. In other words, the control and behaviour of the search performed by the receiver AP is modified by the search and information sharing behaviour of the sender AP. This phenomenon is denoted *direct* (explicit) AP *interaction*.

Explicit information exchanges are designed to improve the performance of the global search performed by the APs involved in the cooperation compared to their individual performances. This goal if often attained but not always or not always at the level hoped for [31]. This is largely due to the fact that, in most cooperative search systems, APs interact not only through explicit information exchanges, but also indirectly through correlated cooperation actions. Information among cooperating APs is thus also shared implicitly through a propagation (or diffusion) process not explicitly defined by their design or that of the cooperation mechanism.

Figure 3 describes a simple propagation process. AP_1 sends information x to AP_2 via a direct interaction $1 \xrightarrow{x} 2$. AP_2 receives this information and uses it to guide its exploration. Later, it sends information $y = f_2(x)$ to AP_3 via a direct interaction $2 \xrightarrow{y} 3$. The notation indicates that the information sent by AP_2 to AP_3, y, results, at least partially, from an interaction between AP_1 and AP_2, which modified the trajectory executed by the search heuristic f_2 of AP_2. There is implicit information propagation because the second interaction is triggered by the modification to the search behaviour of AP_2 following an interaction with AP_1. The second interaction is correlated to the occurrence of the first one.

Correlated interactions propagate control actions of one program onto other programs. In the example of Figure 3, the search activities of AP_1 modify the search behaviour of AP_3, as indicated by the arrow between the two APs. We identify this control as *indirect* (implicit) AP *interaction*.

When direct interactions occur asynchronously according to the internal state of the interacting APs, chains of correlated interactions build up spontaneously among the APs. Sequences of bold arrows in Figure 4 illustrate such occurrences of correlated interactions under a 2-D torus interconnection network,

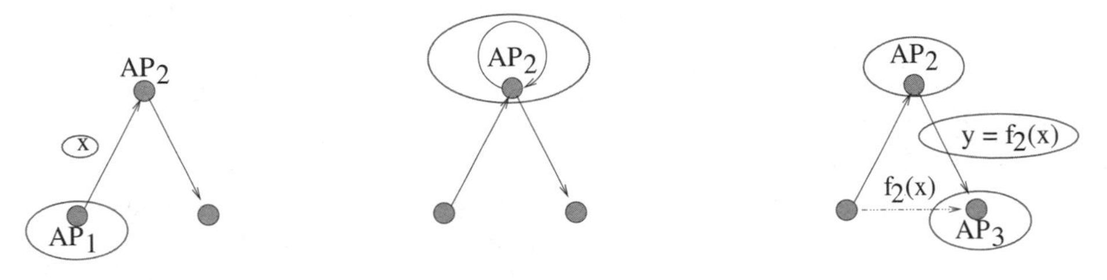

Fig. 3. Information Propagation Process

while dashed arrows represent some of the associated indirect control activities. Figure 4(a) pictures a chain of correlated interactions generated by the sequence of direct interactions 5 → 6, 6 → 10, 10 → 14, 14 → 15, 15 → 3, 3 → 4, and 4 → 8, together with one of several associated indirect interactions: 5 → 8. Figure 4(b) illustrates the case where two chains of correlated interactions develop concurrently, while Figure 4(c) displays indirect interactions forming loops inside a network of correlated interactions. These are only illustrative examples, of course. Yet, they help getting a sense of the complexity of the control activities associated to indirect interactions, the spontaneity of this control, the inter-connectivity of the programs in a cooperative search, and the dependence of the search performed by each AP on these emerging control structures that are the correlated interactions. In a central memory-based systems, where chains of correlated interactions are not restricted by the logical topology of the interconnection network, this self-organization of the search through correlated interactions is even more striking.

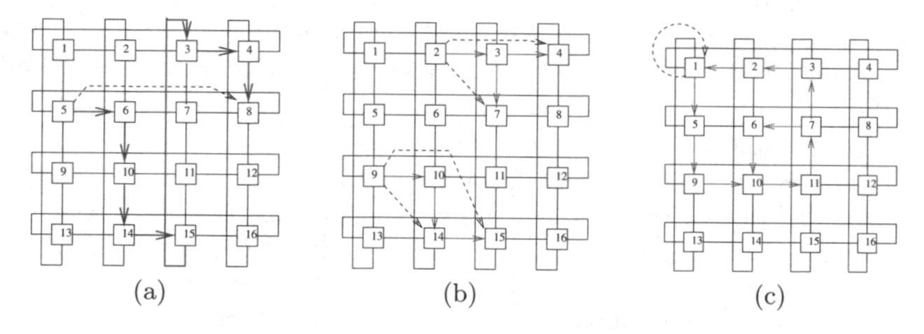

Fig. 4. An Illustration of Indirect Information and Control Propagation

Chains of correlated interactions are the ripple effects of cooperative search and constitute the means by which information propagates among APs, information different from and in addition to that resulting from the interactions specified by the cooperation scheme. It must be emphasized that, unlike explicit information exchanges, the sharing of information through propagation is **not** specified in the cooperation scheme. Nor are specified the search control activities that emerge spontaneously from these information-propagation processes. Moreover, this spontaneous organization of the control activities plays a more important role in the global exploration of the search space performed by cooperation mechanisms implementing asynchronous *pC-C* and *pC-KC* strategies, where the cooperation scheme is executed independently and asynchronously by each cooperating AP.

Interesting questions arise from the realization that emergent control behaviour occurs in systems of cooperating APs and that the global exploration of the search space performed by the cooperating APs is thus the result of the interplay between the two types of AP interactions, the direct ones specified by the design of the cooperation and the complex system of indirect interactions emerging from them. Could indirect interactions act similarly to direct ones and, by modifying the search behaviour of the receiving AP improve its performance? Could indirect interactions emerge as an implicit cooperation scheme, that is, could indirect interactions support a global cooperative exploration of the search space? Does or could this emergent control help improve the exploration of the search space? In what circumstances could the answer to some of these questions by "yes" ? Understanding how such global behaviour may emerge and how (if) it could be harnessed to support the search for good solutions to the problem in hand, could prove important in enhancing the performance of cooperating search methods.

4 Advanced Cooperation Mechanisms and Learning

Studies on emergent cooperation issues have been performed within a number of scientific fields (Section 5) but, with the notable exception of the multi-level paradigm designed to control the indirect diffusion of information among co-operating APs, few efforts were dedicated to these issues within the operations research or parallel meta-heuristic communities. Most efforts were rather directed toward improving the cooperative meta-heuristic mechanisms to enhance their optimization capabilities. Learning was and continues to play a central role in these processes, particularly for memory-based mechanisms.

Consider the adaptive-memory approach where the pool contains solution components (e.g., tours) of good solutions identified by APs (e.g., multi-tours for VRPTW found by tabu searches) that are ranked according to attribute values, including the objective values of their respective solutions.. Each APs then probabilistically selects components in the memory, constructs a new initial solution (e.g., by solving a set-covering heuristic), improves it, and returns the components of its best solution to the memory.. The learning mechanism of adaptive-memory approaches is thus composed of the partial solutions kept in the memory together with the continuously updated rank values, combined to a new-solution creation feature.

Algorithms based on the central-memory approach keep full solutions and attributes sent by the APs involved in cooperation. APs may construct new solutions, execute a neighbourhood-based improving meta-heuristic, implement a population-based meta-heuristic, or perform post-optimization procedures on solutions in the pool. Improving meta-heuristics aggressively explore the search space, while population-based methods (e.g., genetic algorithms [10,21] and path relinking [9]) contribute toward increasing the diversity of solutions exchanged among the co-operating methods. Exact solution methods may participate to the cooperation either to build solutions or to seek out optimal ones (on restricted versions of the problem, eventually). The information exchanged among cooperating APs has to be meaningful, in the sense that it has to be useful for the decision process of the receiving programs, the evolution of the shared data, and thus the evolution of the global search, or both. Information indicative of the current status of the global search or, at least, of some individual search program is, in this sense, meaningful. The basic mechanisms thus only implement exchanges of local good solutions (local optima) together with ranking procedures and probabilistic solution-extraction procedures in the central memory, thus implementing the same learning mechanisms described previously.

More advanced mechanisms may involve exchanges of good solution together with their respective context (e.g., memories recording recent behaviour of solution attributes), or a comprehensive history search. Memories recording the performance of individual solutions or solution components may be added to the pool, as well as procedures to generate new solutions or to compute various statistics on solutions, solution components, individual AP performance, and the trajectory of the global search. This information may then be used to build guidance mechanisms or even to feed external learning programs, e.g., neural networks. Not all these ideas have been thoroughly developed and included in the latest methods found in the literature. They constitute an active field of research, however, as illustrated by the two following examples. First, Le Bouthiller, Crainic, and Kropf proposed a dynamically-adaptive learning and guidance mechanism based on atomic elements (e.g., the arcs present in the routes of VRPTW solutions in the pool [22]). Patterns of arcs present in good or bad solutions in the pool are built and are then sent to the individual APs to intensify or diversify the global search. The particular pattern and guidance directive depends upon the stage of the search as measured by the evolution of the elite population in the pool. The mechanism is general in the sense that, being based on atomic elements, it is independent of any particular problem structure. Second, Crainic *et al.* have shown for the first time the capability of memory-based cooperative search to handle complex, multi-characteristic problems [9]. Their study of the design of third-generation wireless networks aims to optimize the number and configuration (location, power and number of antennas, plus the orientation and tilt of each antenna) of base stations to guarantee level of service and minimize the impact of the electromagnetic emissions of the system on human health. The cooperative system involves several tabu search APs to explore particular parts of the solution space where only a few of the configuration parameters are allowed to vary. A genetic algorithm and a path relinking method are then used to combine partial solutions into complete ones and generate new solutions for the pool.

Many interesting research challenges may be identified in relation to these issues. A first group continues the work in designing intelligent cooperation and learning mechanisms to enhance the optimization capabilities of cooperative meta-heuristics. Promising avenues include, but are not limited to, the integration of adaptive and central memory principles, the enhancement of atomic-based guidance, the integration of memory and multi-level search concepts, and the development of advanced learning mechanisms that 1) build a dynamic image of the performance of each AP to, eventually, modify its search parameters or principles, 2) combine statistics (memories) and artificial intelligence methods (e.g., neural networks), and 3) integrate the distributed, i.e., the APs', memories to the global search knowledge and guidance. A second direction aims to build on the capabilities of such intelligent cooperation and learning mechanisms to build characterizations of the solution space already visited and the dynamic performance of the search. This could then be used to steer the global search accordingly as well as to study the linkages between explicit cooperation and emergent computation. Last but not least, it would be very interesting to develop a theory of learning within the context of multi-AP cooperation. This would provide the means to go beyond the limits of experimental settings in designing effective parallel cooperation search methods for difficult problems. The next section identifies some of these latter possibilities in more detail.

5 Research Directions in Emergent Computation and Cooperative Search

We believe that research directions on the dynamics and emergent computation behaviour of cooperative search should be inspired by research conducted in other fields, as well as build on the learning mechanisms described in the previous section and on experimental and empirical validation processes for particular problems.

The programming of dynamics in computing systems could offer a first direction. As indicated earlier, the performance of the global exploration of the search space by cooperating APs is obtained from the interplay of a complex system of direct and indirect interactions. So far, however, the design of cooperative algorithms has focused on the components of such systems, e.g., the APs, the cooperation scheme, and the interconnection network, taken individually. Little effort, if any, has been dedicated to developing design strategies of cooperative algorithms by considering, "programming", the system as a whole. This is a bit surprising since dynamic interactions among autonomous computing elements and their potential for emergent computation have been investigated for several computing system contexts. Thus, recurrent neural networks with emergent search behaviour are programmed as a whole by directly adjusting their parameters (e.g., the logical interconnection network, the weights on the interconnection links, and the transition functions on the nodes) based on learning algorithms and adjustment procedures built from the problem optimization model. For cooperative search, this translates into working directly on the logical interconnection network, the cooperation mechanism, the APs, and the asynchronous mode of information exchange, that is work directly on the optimization logic of the cooperation. The methods to perform this global design are an important research topic *per se*, which, for us, is strongly related to the learning issues identified previously.

We thus turn to research areas such as multi-robot systems [2,3,6,23], reactive multi-agent systems [5,27], artificial life [18,25,19] and ad hoc implementations of cooperative algorithms in various computer science applications. Emergent cooperation behaviour has been synthesized in the field of behaviour-based robotics [2,3,23] using a methodology denoted behaviour-Based AI, derived from theories on the modular decomposition of intelligence [4,20]. Behaviour-decomposition methods are first used to analyze observed emergent cooperation behaviours in natural social systems made up of individuals displaying relatively simple behaviours (e.g., ants). Similar cooperative behaviours are then synthesized in societies of robots [24]. For example, the foraging behaviour of ants has been synthesized into an object-gathering behaviour in social robotics by combining two basic forms of direct interactions among robots (APs), dispersion (interact to diversify the exploration relative to the others) and homing (aim for the goal by, for example, sharing the good solutions) [23]. These results could provide the theoretical foundations for cooperation mechanisms that include more than one form of information sharing strategy among APs which, hopefully, will display global dynamics that adequately approximate the desired emergent problem-solving strategy.

More ideas for research directions come from the field of computing systems where ad hoc bottom-up and emergent computing strategies are increasingly being proposed with significant success for various applications in telecommunication network routing, reliability and power supply limitations in wireless networks, access security to computer systems, and so on. For example, self-assembly in nanothechonology, a bottom-up nano-fabrication process in which components self-assemble based on shape complementarity, could provide the basic strategies to specify which interactions are allowed to occur at run time and, thus, to specify the logical interconnection networks. Similarly, the research efforts in autonomic computing, which aims to be able to tell a system **what** to do and let it to find **how** to do it, could maybe inspire a new definition of problem solving for cooperative meta-heuristics where it is the emergent behaviour of the system that finds its way to how address a given problem.

Unlike cellular automata and artificial neural networks, logical interconnection network topologies in cooperative algorithms tend to vary widely. So far, empirically, the best cooperative search results have been obtained through interconnection networks that are configured dynamically at run time, e.g., the memory-based approaches. This suggests adaptive approaches to program system dynamics by adjusting dynamically the network topology of cooperative algorithms to meet the desired attributes of global cooperation. Turning to the learning mechanisms of the previous section, we believe that these or similar learning schemes could be applied to system dynamics. Thus, for example, learning could be used to identify system attractors (regions of the search space to which the search returns often) and the path ways leading to them. The mechanism could then be used to adjust dynamically the interconnection topology by blocking these path ways, and thus prevent the occurrence of some chains of correlated interactions that appear to attract the search is the same region of the search space. Learning could also be used to develop interaction policies that favor those that can lead to the emergence of cooperation and block those that are harmful to it. Intelligent control of the system dynamics of cooperative search through learning mechanisms is certainly one of the most promising research avenues in the effort to obtain a cooperative exploration of the search space through spontaneous interactions among APs.

Research on self-organization often analyze natural systems with observed self-organized behaviours to discover strategies for designing distributed systems with particular globally emergent behaviours. However, it is well known that computing systems yield spontaneous activities, e.g., the ripple effects described earlier on. This raises the issue whether we should study the dynamic behaviour of cooperative search systems *per se*, as a research object, similarly to biological systems through comprehensive laboratory-based simulations. An associated question is whether we should seek to understand the complex behaviour of cooperative systems in order to sustain emergent cooperation or, rather, should we seek to discover new search heuristics based on occasionally occurring coherent search behaviour at the global level? A combination of these two methodological approaches has proved to be successful. Thus, the study of locally emergent cooperation has led to ideas to design a new cooperation mechanism, which yielded the highly successful multi-level cooperative search method. The final challenge, obviously, is to bring together the research on explicit and implicit cooperation mechanisms and behaviours and apply the resulting methodology within the context of various solution methods. Even though the research in this area is still at the very beginning, this challenge has been met with some success, producing new methods out of the study of cooperative search [28,12,26,11,22].

6 Conclusion

Cooperative algorithms as computerized problem-solving strategies are becoming ubiquitous in several problem domains, in particular for addressing complex discrete optimization problems. Cooperation has well-known advantages, short development cycle through the re-utilization of existing exact or heuristic methods and high adaptability to different problems and problem characteristics, in particular. Yet, these systems also generate series of indirect interactions that may reduce their performance. We have described cooperation mechanisms, discussed the relations between direct and indirect interactions, and have identified a number of challenges and interesting research directions for the development of the next generation of cooperative search methods.

Acknowledgments

This work was financially supported through the Industrial Research Chair and Discovery grant programs of the Natural Sciences and Engineering Research Council of Canada (NSERC).

References

1. Alba, E. (ed.): Parallel Metaheuristics. A New Class of Algorithms. John Wiley & Sons, Hoboken (2005)
2. Arkin, R.C.: Behavior-Based Robotics. MIT Press, Cambridge (1998)

3. Brooks, R.A.: Intelligence without Representation. Intelligence without Representation 47(1-3), 139–159 (1991)
4. Brooks, R.A.: Cambrian Intelligence: The arly History of the New AI. MIT Press, Cambridge (1999)
5. Brooks, R.S.: A robust layered control system for a mobile robot. In: Readings in Uncertain Reasoning, pp. 204–213. Morgan Kaufmann Publishers Inc., San Francisco (1990)
6. Cao, U.Y., Fukunaga, A.S., Kahng, A.B.: Cooperative Mobile Robotics: Antecedents and Directions. Autonomous Robots 4(1), 7–23 (1997)
7. Crainic, T.G.: Parallel Computation, Co-operation, Tabu Search. In: Rego, C., Alidaee, B. (eds.) Metaheuristic Optimization Via Memory and Evolution: Tabu Search and Scatter Search, pp. 283–302. Kluwer Academic Publishers, Norwell (2005)
8. Crainic, T.G.: Parallel Solution Methods for Vehicle Routing Problems. In: Golden, B., Raghavan, S., Wasil, E. (eds.) The Vehicle Routing Problem: Latest Advances and New Challenges. Springer, Heidelberg (to appear, 2007)
9. Crainic, T.G., Di Chiara, B., Nonato, M., Tarricone, L.: Tackling Electrosmog in Completely Configured 3G Networks by Parallel Cooperative Meta-Heuristics. IEEE Wireless Communications 13(6), 34–41 (2006)
10. Crainic, T.G., Gendreau, M.: Towards an Evolutionary Method - Cooperating Multi-Thread Parallel Tabu Search Hybrid. In: Voß, S., Martello, C., Roucairol, C., Osman, I.H. (eds.) Meta-Heuristics 1998: Theory & Applications, pp. 331–344. Kluwer Academic Publishers, Norwell (1999)
11. Crainic, T.G., Gendreau, M.: Cooperative Parallel Tabu Search for Capacitated Network Design. Journal of Heuristics 8(6), 601–627 (2002)
12. Crainic, T.G., Li, Y., Toulouse, M.: A First Multilevel Cooperative Algorithm for the Capacitated Multicommodity Network Design. Computers & Operations Research 33(9), 2602–2622 (2006)
13. Crainic, T.G., Nourredine, H.: Parallel Meta-Heuristics Applications. In: Alba, E. (ed.) Parallel Metaheuristics, pp. 447–494. John Wiley & Sons, Hoboken (2005)
14. Crainic, T.G., Toulouse, M.: Parallel Strategies for Meta-heuristics. In: Glover, F., Kochenberger, G. (eds.) Handbook in Metaheuristics, pp. 475–513. Kluwer Academic Publishers, Norwell (2003)
15. Crainic, T.G., Toulouse, M., Gendreau, M.: Parallel Asynchronous Tabu Search for Multicommodity Location-Allocation with Balancing Requirements. Annals of Operations Research 63(2), 277–299 (1996)
16. Crainic, T.G., Toulouse, M., Gendreau, M.: Towards a Taxonomy of Parallel Tabu Search Algorithms. INFORMS Journal on Computing 9(1), 61–72 (1997)
17. Cung, V.-D., Martins, S.L., Ribeiro, C.C., Roucairol, C.: Strategies for the Parallel Implementations of Metaheuristics. In: Ribeiro, C.C., Hansen, P. (eds.) Essays and Surveys in Metaheuristics, pp. 263–308. Kluwer Academic Publishers, Norwell (2002)
18. Dagaeff, T., Chantemargue, F.: Performance of Autonomy-based Systems: Tuning Emergent Cooperation. Technical Report 98-20, Computer Science Department, University of Fribourg (1998)
19. Engelbrecht, A.P.: Fundamentals of Computational Swarm Intelligence. John Wiley & Sons, Chichester (2006)
20. Fodor, J.A.: The Modularity of Mind. MIT Press, Cambridge (1983)
21. Le Bouthillier, A., Crainic, T.G.: A Cooperative Parallel Meta-Heuristic for the Vehicle Routing Problem with Time Windows. Computers & Operations Research 32(7), 1685–1708 (2005)

22. Le Bouthillier, A., Crainic, T.G., Kropf, P.: A Guided Cooperative Search for the Vehicle Routing Problem with Time Windows. IEEE Intelligent Systems 20(4), 36–42 (2005)
23. Mataric, M.J.: Designing Emergent Behaviors: From Local Interactions to Collective Intelligence. In: Proceedings of the Second International Conference on From Animals to Animats 2: Simulation of Adaptive Behavior, pp. 432–441. MIT Press, Cambridge (1993)
24. Mataric, M.J.: Issues and Approaches in the Design of Collective Autonomous Agents. Robotics and Autonomous Systems 16(2-4), 321–331 (1995)
25. Nitschke, G.: Emergence of Cooperation: State of the Art. Artificial Life 11(3), 367–396 (2005)
26. Oduntan, I.O., Toulouse, M., Baumgartner, R., Bowman, C., Somorjai, R., Crainic, T.G.: A Multilevel Tabu Search Algorithm for the Feature Selection Problem in Biomedical Data Sets. Computers & Mathematics with Applications 55(5), 1019–1033 (2008)
27. Oliveira, E.C., Fischer, K., Stepánková, O.: Multi-agent systems: which research for which applications. Robotics and Autonomous Systems 27(1-2), 91–106 (1999)
28. Ouyang, M., Toulouse, M., Thulasiraman, K., Glover, F., Deogun, J.S.: Multilevel cooperative search for the circuit/hypergraph partitioning problem. IEEE Transactions on Computer-Aided Design of Integrated Circuits and Systems 21(6), 685–694 (2002)
29. Rochat, Y., Taillard, É.D.: Probabilistic Diversification and Intensification in Local Search for Vehicle Routing. Journal of Heuristics 1(1), 147–167 (1995)
30. Talbi, E.-G. (ed.): Parallel Combinatorial Optimization. Wiley-Interscience, Wiley & Sons, Hoboken (2006)
31. Toulouse, M., Crainic, T.G., Sansó, B., Thulasiraman, K.: Self-Organization in Cooperative Search Algorithms. In: Proceedings of the 1998 IEEE International Conference on Systems, Man, and Cybernetics, Omnipress, Madisson, WI, pp. 2379–2385 (1998)
32. Toulouse, M., Thulasiraman, K., Glover, F.: Multi-Level Cooperative Search: A New Paradigm for Combinatorial Optimization and an Application to Graph Partitioning. In: Amestoy, P., Berger, P., Daydé, M., Duff, I., Frayssé, V., Giraud, L., Ruiz, D. (eds.) Euro-Par 1999. LNCS, vol. 1685, pp. 533–542. Springer, Heidelberg (1999)
33. Verhoeven, M.G.A., Aarts, E.H.L.: Parallel Local Search. Journal of Heuristics 1(1), 43–65 (1995)

Multiobjective Landscape Analysis and the Generalized Assignment Problem

Deon Garrett and Dipankar Dasgupta

University of Memphis, Memphis, TN, 38122, USA
jdgarrtt@memphis.edu, dasgupta@memphis.edu

Abstract. The importance of tuning a search algorithm for the specific features of the target search space has been known for quite some time. However, when dealing with multiobjective problems, there are several twists on the conventional notions of fitness landscapes. Multiobjective optimization problems provide additional difficulties for those seeking to study the properties of the search space. However, the requirement of finding multiple candidate solutions to the problem also introduces new potentially exploitable structure. This paper provides a somewhat high-level overview of multiobjective search space and fitness landscape analysis and examines the impact of these features on the multiobjective generalized assignment problem.

1 Why Landscape Analysis

One of the foremost questions facing designers of metaheuristic algorithms for any sort of problem is how the structure of the objective function will affect the behavior of the search algorithm. It is known, and quite intuitive, that incorporating problem-specific knowledge into a search algorithm can often substantially increase the performance of the algorithm. However, given multiple conflicting options for building such algorithms, comparatively little is known concerning the right choices, or even the right information necessary to make good decisions.

This work examines a set of tools developed to help gain insights into how various algorithms navigate complex multiobjective search spaces. Many of these tools have previously been described in relation to conventional optimization problems. In such cases, the implications of extending the tools into the multiobjective realm are carefully examined. In this work, we propose some methods by which such relevant information may be obtained and exploited, and we apply these methods to a pair of classes of assignment problems exhibiting markedly different types of structure.

2 Fitness Landscape Analysis

As more and more researchers have turned their attention to modeling search algorithm performance, a number of techniques have been proposed to classify fitness landscapes, generally corresponding to fundamental properties of a given search

V. Maniezzo, R. Battiti, and J.-P. Watson (Eds.): LION 2007 II, LNCS 5313, pp. 110–124, 2008.

space. While most commonly defined in terms of classical single objective optimization, many of these properties have straightforward generalizations to the multiobjective domain. However, multiobjective optimization introduces additional features which may be analyzed and exploited by search algorithms. By studying these features, one should be able to design metaheuristic algorithms to better take advantage of the peculiarities of multiobjective optimization of a given problem.

Often, multiobjective search algorithms are defined in terms of a simpler, single objective algorithm. The n objectives of the original problem are scalarized using a particular weight vector, and the resulting single objective problem is attacked using the component search method. In these cases, the multiobjective problem may be completely characterized, for the purpose of modeling the performance of such an algorithm, by a family of related fitness landscapes. Each landscape is a window on the problem as viewed through a specific weight vector.

Abstractly, one may thus consider the ruggedness of adjacent landscapes. In fact, this notion of the similarity between nearby landscapes is what determines in part the success of different types of multiobjective local search algorithms. Following convention established by analysis of single objective landscapes, we may characterize a multiobjective problem as "smooth" if small changes to the underlying weight vector impose small changes on the fitness landscape. Conversely, a "rugged" multiobjective problem is one in which making a small change to the weight vector drastically alters the resulting fitness landscape.

Given a single good solution to a multiobjective optimization problem, the difficulty in finding other good solutions is largely determined by the smoothness of the family of landscapes. It is somewhat intuitive that smoothness implies that a good solution on a particular landscape should be nearby to good solutions on nearby landscapes. This spatial locality makes algorithms which attempt to build from one solution to a multiobjective problem to find many others more attractive. On the other hand, as the family of landscapes becomes more rugged, the information gained by finding one good solution becomes less useful in finding others.

The following sections describe a number of potentially useful metrics by which fitness landscapes, both single and multiobjective, may be characterized. In general, many of the tools in common use for analysis of single objective landscapes have fairly straightforward generalizations to the multiobjective realm. In addition, multiobjective algorithms which consist entirely of a sequence of independent runs of some underlying single objective optimization method may be directly studied using the single objective tools. However, it is also true that multiobjective optimization provides additional opportunities to exploit problem knowledge, and much of the goal of this work is to study these opportunities and apply the resulting knowledge to the problem of designing more effective algorithms. The remainder of this chapter is focused on defining a number of tools by which we may obtain such useful information about a given multiobjective problem instance.

2.1 Distribution of Local and Pareto Optima

Intuitively, the number and distribution of local optima would seem to have a profound impact on the performance of a general purpose search algorithm.

A problem with but a single local optimum is by definition unimodal, and thus easily solved by any number of simple algorithms. As the number of local optima increase, the chances of becoming trapped in a local optimum are increased correspondingly. However, the distribution of local optima throughout the space is at least as important as the number of such optima. Classical notions such as deception are rooted entirely in the notion of unfortunate distributions of local optima.

One of the best-known examples of real-world problems with very different optima distributions is the comparison between the traveling salesman problem (TSP) and the quadratic assignment problem (QAP). In the TSP, problem instances exhibit what is commonly known as a "Big Valley" structure [1]. This term refers to the phenomena that almost all local optima are concentrated around a line that approaches the global optima, with the tour lengths of points along that line tending to increase as the distance to the global optimum increases.

In contrast, QAP instances tend to exhibit almost no structure when viewed in this same manner. The local optima for a typical QAP instance are very nearly uniformly distributed throughout the search space, with many, perhaps most, of all local optima lying close to the maximum possible distance from the global optimum. The difference in performance of a local search algorithm on TSP versus QAP is therefore quite dramatic.

From a multiobjective standpoint, many Pareto optimal solutions are also global optima of some single objective problem. Most commonly, given a Pareto front which is globally convex, there exists a weight vector for each Pareto optimal solution, a scalarization of the problem by which would result in the solution being the globally optimal solution of the resulting single objective problem. Thus, the distribution of local optima affects multiobjective problems just as it does for their single objective counterparts. However, multiobjective landscapes add an additional consideration, in that different Pareto optima are not generally local optima of the same single objective slice of the landscape. Therefore, the distribution of Pareto optima is in some sense a different aspect of the landscape than is the distribution of local optima leading to a single point on the Pareto front. In essense, you have to find both solutions, and the entire landscape can and often does change underneath you as you try to switch from one to the other.

2.2 Fitness Distance Correlation

Fitness distance correlation as a tool for modeling algorithm performance is based on the notion that good local optima should be near to the global optimum in terms of fitness as well. If this is the case, in principle there should be a clear trail from any local optimum to the global optimum in which each step requires only small changes to the current solution. If instead, large jumps are required to move from a local optimum to a better solution nearer to the global optimum, most search algorithms may be expected to suffer.

Primarily defined in terms of single-objective optimization, fitness distance correlation is the correlation coefficient between the distance in objective space and the distance in parameter space between a set of randomly distributed local optima and the respective nearest global optimum to each. The standard Pearson correlation coefficient may be used to describe the results, although some useful information is not captured by this single summary statistic. Instead, scatter plots of parameter space distance versus objective space distance are often reported.

In the context of multiobjective optimization, the basic distinction is that the set of global optima is taken to be the set of nondominated solutions. While conceptually a simple extension, each Pareto optimal solution may or may not be the optimum of some mono-objective problem associated with a particular set of weights. The novelty in such a formulation is that, because each solution is, in essence, the optimum of a different fitness function, the correlations between nondominated solutions need bear no resemblance to the correlation between different *local* optima of a single function. Thus, considering the correlation between nondominated solutions can provide very useful information concerning the relative difficulty of moving "along" the Pareto front.

There are possibly other, more useful, ways to generalize this concept to multiobjective landscapes. The basic restriction of FDC is that one needs to get a single number indicating the distance to each optima. While one can certainly treat the Euclidean distance between a fitness vectors as the requisite metric, one may also consider, for example, the angle between vectors. If we treat distance as being defined by the angle between fitness vectors, then the nearest optimum will be, in a sense, "aligned" with the solution, in that they will have their component fitness values in the nearest proportion with one another. The nearest Pareto optimum under this definition will be that which lies adjacent on the Pareto front. The impact of these different choices on the resulting analysis is still an open question, but it is worthwhile to keep in mind the various possibilities that arise when dealing with vector valued optimization.

2.3 Ruggedness

Ruggedness is a somewhat vague notion, and a number of attempts have been made to formalize it [2]. However, intuitively, a landscape is rugged if there are many local optima of highly varying fitness concentrated in any constrained region of the space. Thus, it would seem that any definition of ruggedness, or conversely, smoothness must take into account both the number and distribution of local optima.

One straightforward measure of smoothness is to consider the correlation between adjacent points in the search space, where adjacency is dependent on the specification of a suitable neighborhood operator [3]. This correlation function value provides vital insights into the structure of a given search space under the chosen operator. A high correlation coefficient implies that adjacent positions in the search space tend to have very similar fitness values. In this case, a local search algorithm might be expected to perform well, since it seems possible to exploit information gained by prior fitness function evaluations to effectively guide the choice of points to evaluate in the future.

As computing the true correlation requires exhaustive knowledge, researchers have often substituted autocorrelation instead. Autocorrelation typically arises in signal processing applications, and roughly speaking, measures the degree to which a time series is correlated with a time-shifted version of itself. In the current context of landscape analysis, one may construct a time series by setting out on a random walk and recording the fitness values at each point along the walk [3]. An uninformed random walk provides information about the overall ruggedness of the search space, which is certainly useful information for one seeking to design an effective algorithm.

However, if one knows the location of the global optima, or even of a set of high quality local optima, we can measure the autocorrelation of points along a path leading to these desirable solutions. For example, suppose we know the location of the global optimum for some problem. We can create a large number of random initial solutions, then send each one on a walk toward the global optimum. If the average autocorrelation is high, then we have reason to believe that other similar problem instances may be successfully attacked using fairly simple local improvement operators.

Multiobjective optimization provides another possible use for autocorrelation analysis. One vital piece of information concerns the relative difficulty of using previously located solutions to guide the search for additional Pareto optimal solutions versus performing a random restart to begin searching for other points. One way to attempt to answer this question is to look at the autocorrelation of random walks between known Pareto optimal solutions. If the path between Pareto optima is very rugged, it may be difficult for a two-phase algorithm to navigate the minefield of local optima. As a result, it may actually be beneficial to perform a restart to begin the search for additional Pareto optimal solutions.

2.4 Random Walk Analysis

Merz in [4] models crossover and mutation in hybrid evolutionary algorithms as random walks initiated from local optima. A series of mutations is represented by an undirected random walk starting from a local optimum. In contrast, crossover is modeled as a random walk starting at one local optimum and ending at another. Provided that the crossover operator is respectful [5,4], this random walk between local optima explores the space in which offspring will be produced.

In [6], this method was extened to multiobjective optimization by considering the impact of crossover and mutation operating on the current nondominated frontier at various points during the evolutionary process. Using the mQAP as an example, it was shown that the ability of crossover to generate points closer to the Pareto front accounts for some of the success of a very good memetic algorithm. On instances where crossover more closely resembled mutation, the memetic algorithm could not outperform a simpler local search metaheuristic.

Modeling genetic operators is not the only valid use of random walks, however. In the context of local search operators, random walks can help to provide estimates of many different properties of fitness landscapes. Watson et. al. [7,8] used random walks to estimate the depth of each basin of attraction in job shop scheduling problems, and using the insights provided, developed an algorithm called I-JAR (iterated jump and re-descend) which was competitive with highly refined tabu search algorithms on JSP instances. I-JAR works by simply running a local search algorithm to a local optimum, then taking a random walk of length n, where n was empirically determined for the JSP, followed by an additional local improvement stage. The result is a very efficient algorithm which finds a local optimum, then "jumps" the minimum distance required to escape the basin of attraction before continuing the local search.

In multiobjective optimization, the notion of depth in relation to a basin of attraction is not so straightforward. Depending on both the location of the basin itself, or more accurately, the local optimum at the extremum of the basin, and the direction in which the search algorithm chooses to attempt its escape, the depth may vary dramatically. In general, even true Pareto optimal points will usually have a non-zero escape probability given an arbitrary direction of escape. This makes the notion of a barrier or basin more complicated in multiobjective landscapes, and further study is required to better understand the impact of changing direction during the search.

3 The Generalized Assignment Problem

The generalized assignment problem (GAP) deals with a set of m agents and a set of n tasks. Each task must be completed by exactly one agent. Each agent is allocated a specific number of resource units, and each agent requires a particular number of units to complete each task. Additionally, each agent incurs a specified cost for each task. The resource requirements and costs for a given task may differ between agents. The overall goal is to assign all tasks such that no agent violates the capacity constraints and the total costs incurred are minimized.

Formally, we may introduce an m-dimensional vector $\mathbf{B}$, with b_j denoting the total capacity alloted to agent j. We further introduce $m \times n$ matrices $\mathbf{A}$ and $\mathbf{C}$, denoting the resource matrix and the cost matrix respectively. Finally, we introduce an $m \times n$ binary matrix $\mathbf{X}$, with $x_{ij} = 1$ only if task i assigned to agent j by a particular candidate solution. The goal is thus to find such a solution so that

$$\min_{\mathbf{X}} \sum_{i=1}^{m} \sum_{j=1}^{n} x_{ij} c_{ij}, \tag{1}$$

subject to

$$\sum_{i=1}^{m} x_{ij} a_{ij} \leq b_j \quad \forall j : 1 \leq j \leq n \tag{2}$$

and

$$\sum_{j=1}^{n} x_{ij} = 1 \quad \forall i : 1 \leq i \leq m. \tag{3}$$

(2) are known as the *capacity constraints*, and (3) are called the *semi-assignment constraints*.

The GAP is known to be $\mathcal{NP}$-hard [9], and exact algorithms have proven tractable only for problems of a few hundred tasks or less [10]. Thus, for large instances, heuristic and metaheuristic methods have received a great deal of attention, including tabu search approaches [11,12], variable depth search [13,14], ant colony optimization [15], evolutionary algorithms [16], and more recently, path relinking algorithms [17,18,19,20].

Most experimental studies of the GAP have chosen a benchmark set of randomly generated instances divided into five classes, imaginatively named A, B, C, D, and E. The A and B-type instances are not considered challenging to modern algorithms and are not considered in this work. The D and E-type instances considered more difficult than the C-type due to the fact that the costs are inversely related to the resource allocations in the former instances.

Type C:
$$a_{ij} = \mathrm{U}(5, 25)$$
$$c_{ij} = \mathrm{U}(10, 50)$$
$$b_i = 0.8 \sum_j \frac{a_{ij}}{m}$$

Type D:
$$a_{ij} = \mathrm{U}(1, 100)$$
$$c_{ij} = 111 - a_{ij} + \mathrm{U}(-10, 10)$$
$$b_i = 0.8 \sum_j \frac{a_{ij}}{m}$$

Type E:
$$a_{ij} = 1 - 10 \ln(\mathrm{U}(0, 1])$$
$$c_{ij} = \frac{1000}{a_{ij}} - 10\mathrm{U}[0, 1]$$
$$b_i = 0.8 \sum_j \frac{a_{ij}}{m}$$

3.1 The Multiobjective GAP

Like many other combinatorial optimization problems, the GAP is often applicable in situations calling for simultaneous optimization of more than one objective function. We propose a straightforward extension of the GAP to a multiobjective problem. The reasons for this are to provide an additional test for multiobjective algorithms, but also because we feel that a multiobjective benchmark problem featuring constraints and a quite different type of structure to the now well-known mQAP should be added to the suite of widely available benchmarks. In the remainder of this paper, the mGAP will be described and examined with particular attention to the structure of the resulting search space and its impact on MOGLS algorithm performance.

The most basic formulation of the mGAP is to augment the description in the previous section with additional cost matrices. Each task must then be assigned to a single agent in such a way as to minimize a vector of costs. Formally, the optimization problem is thus to find a solution s such that

$$\min_{s \in \Lambda} \sum_{i=1}^{M} C^{k}_{s_i, i}, \tag{4}$$

where C^k denotes the kth cost matrix and the min operation denotes vector minimization, or Pareto optimization. The constraints are unchanged from the single objective version of the problem.

Proceeding analogously to Knowles and Corne for the mQAP [21,22], the correlation between cost matrices would seem to be an important indicator of search algorithm performance. The analysis presented for the mQAP indicates that in some situations, a MOSA-like strategy could be quite effective. In such an algorithm, a single solution is optimized with respect to some weight vector. The weights are then modified slightly and a new optimization pass is initialized from the locally optimal solution found in the first stage. The process could continue until some termination criteria is satisfied, at which point the algorithm could terminate or choose to restart from a new randomly chosen starting point and a new weight vector.

However, unlike the mQAP, the mGAP is constrained. In the mGAP, the capacity constraints make it unlikely that large numbers of neighboring solutions are feasible. One may allow the search to visit infeasible regions of the space, provided some appropriate method were in place to guide the search back toward good feasible solutions. Alternately, the algorithm may be prohibited from exploration outside the feasible region. In the latter case, "neighbors" along the Pareto optimal front may be disconnected. This could have profound implications on the ability of the two-phase local search algorithms to adequately explore the Pareto optimal front.

A related issue in the description and formulation of the mGAP is the tightness of the constraints. Intuitively, if the given capacity constraints for each agent roughly equal the average resource usage times the average number of tasks per agent, then the problem would seem to be quite difficult. Simply finding a feasible solution may be very challenging.

Furthermore, the relationship between the resource usage and costs can also impact the performance of search algorithms. If the costs are inversely related to the resource usage, then the most cost effective solutions are likely to involve mappings of tasks to agents that incur large resource usage, thus moving ever nearer to the boundaries of the feasible regions of the space. This can be of particular importance to neighborhood-based search algorithms such as those considered in this work. Additionally, two of the three classes of random instances involved cost matrices which are negatively correlated with the resource matrices. As we have previously seen, correlation between cost matrices may be relevant in determining search algorithm performance. If, for example, one generates a mGAP instance with negatively correlated cost matrices, one of the objectives will then be positively correlated with the resource matrix. It might thus be worth exploring additional forms of random instances in the case of multiobjective optimization.

However, in generalizing the GAP to include multiple objectives, additional problems arise. Standard problem instances are designed to ensure that the feasible region is somewhat tight and the type D and E instances are designed to be more difficult by correlating their cost function with the resource requirements.

Generating multiple cost matrices based on these definitions yields problem instances in which essentially all Pareto optimal solutions lie in a very small feasible region, as the high correlation (or anti-correlation) means that the resulting costs matrices are very similar. Care must be taken in generating instances so that the resulting problem is not degenerate in some form.

3.2 Local Search Algorithms for the GAP

The GAP, and by extension, the mGAP admits more than one possible type of move in relation to a local search algorithm. Unlike the QAP, in which the swap is essentially universally used as the basis of local search methods, local search algorithms for the GAP tend to utilize both *shifts* and swaps. In a shift operation, a single task is reassigned to a different agent. Shifts can be effective when trying to move from an infeasible to a feasible solution, since they can take free up resources from one agent without adding other resources in return. Of course, if a solution is feasible, a shift may be unlikely to maintain feasibility for exactly the same reason.

Unlike shifts, swaps cannot alter the number of tasks assigned to any agents. This lack of explorative power would likely yield very poor performance. Therefore, most local search approaches to the GAP rely on some combination of shifts and swaps. In principle, the swap neighborhood is a strict subset of the neighborhood imposed by two subsequent shifts, and therefore might be viewed as extraneous. However, in practice, the ability to perform a true swap is useful enough to warrant special support.

There exist different approaches to utilizing shifts and swaps in the GAP neighborhood. Possibly the simplest effective method is to randomly select either a shift or swap at each iteration. That approach will be adopted here, largely for two reasons. First, a local search algorithm using such a neighborhood can be reasonably effective. However, the major reason is one of convenience. In order to gather information concerning the properties of the fitness landscape, it is necessary to be able to compute the number of moves which separate any two solutions. Using a full shift/swap neighborhood, it is relatively easy to compute the number of moves required to transform a solution into any other feasible solution. By contrast, in one of the most recent approaches [10], ejection chains are used to search the GAP space. Ejection chains are formed by a series of n successive shift operations. Because the neighborhood size grows exponentially with n, that work utilizes classical integer programming bounds to prune the neighborhood to a manageable number of solutions. While providing high performance, the highly dynamic neighborhoods prohibit the sort of analysis required by this work.

To compute the distance between any two candidate solutions under the shift/swap neighborhood, let us consider two candidate solutions A and B. Let A be considered the reference solution, and we want to compute the minimum number of moves necessary to reach A starting from B. Each locus of the two solutions is compared, and where they assign a given task to different agents, a move is deemed necesssary. To determine what type of move should be taken, a linear search of the remaining unexamined positions of B is performed, looking for a task in B assigned to the agent assigned to the current task in A. If such a task is found in B, then a swap of the two positions in B moves B two steps closer to A and is thus taken. If no such element is found, then any swap in B must by necessity move just one step closer, and thus is no more effective than a shift. When B is equal to A, the number of moves taken denotes the distance between the original two solutions.

3.3 Examining the mGAP Pareto Front

To gain some initial insights into the structure of mGAP instances, several sets of randomly generated toy problems were considered. The problem sizes were deliberately kept small so that the search spaces could be exhaustively searched in reasonable periods of time. Each problem was then enumerated and all Pareto optimal points recorded. Table 1 shows the number and types of instances examined along with some summary statistics on the resulting Pareto optimal sets.

In Table 1, *Cor* refers to the coefficient of correlation between the two cost matrices, $|PF|$ to the number of Pareto optimal solutions, and r to the Pearson coefficient of fitness-distance correlation. Two important conclusions can be drawn immediately from inspection of the Pareto fronts on these small problems. First, the FDC coefficient (denoted by r in the table) is much higher than is observed in the mQAP, where it is typically very close to zero. This implies that there should exhibit more structure among Pareto optimal solutions than we typically see in mQAP instances. This is promising for algorithms which attempt to utilize information from previously located efficient solutions to help generate additional efficient solutions.

Secondly, E-type problems with uncorrelated cost matrices seem to admit very few Pareto optimal solutions, often only one. Extrapolating to larger problems, the Pareto fronts for such problems would seem to be restricted to a very small region of the search space. However, with respect to this property, such extrapolation is fraught with danger. It seems equally likely that the underlying cause of this behavior is an increase in the tightness of the constraints. If this is true, then the trend toward small numbers of Pareto optimal solutions might continue as the problem size was increased. However, the actual solutions might be somewhat more distributed throughout the space, with large regions of infeasibility in between. More research is required to answer this question with any certainty.

A consequence of the positive fitness distance correlation coefficient is that there exist local optima at many different distances to the Pareto front. In contrast, in mQAP instances, it is usually the case that all local optima are fairly uniformly distributed with no two lying too close together. Although this structure in mGAP instances can be usefully exploited by search algorithms attempting to move along the Pareto front, the presence of constraints means that not all solutions on a path between two efficient solutions need be feasible. If these intermediate solutions lie deep within infeasible regions of the space, an

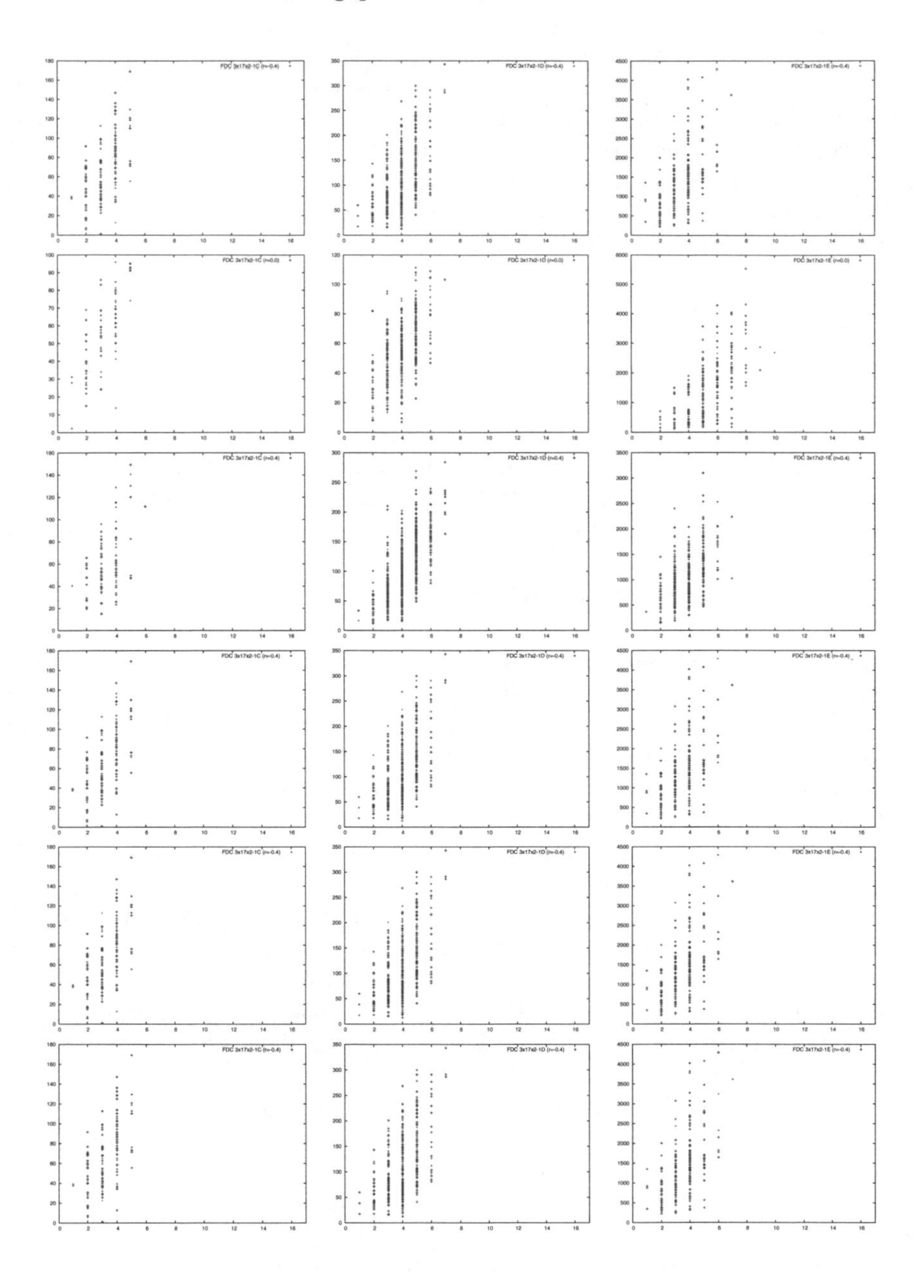

Fig. 1. Fitness distance correlation plots for the various instances considered. Columns left to right are r ={C, D, E} type instances. Row 1, 3x17 $r = -0.4$; Row 2, 3x17 $r = 0.0$; Row 3, 3x17 $r = 0.4$; Row 4, 5x12 $r = -0.4$; Row 5, 5x12 $r = 0.0$; Row 6, 5x12 $r = 0.4$.

Table 1. Toy mGAP instances examined. For each problem type, number of objectives, and correlation between cost matrices, the number of Pareto optimal solutions, the fitness distance correlation coefficient, and the ratio of infeasible to feasible solutions along each path to the front were recorded

Size	Type	Obj	Cor	$\|PF\|$	Pearson's r	$frac_{infeas}$
3x17	C	2	-0.4	14.43±4.83	0.48±0.11	0.86
3x17	D	2	-0.4	13.67±4.56	0.45±0.12	0.86
3x17	E	2	-0.4	17.90±8.19	0.59±0.09	0.82
3x17	C	2	0.0	9.23±5.39	0.53±0.13	0.88
3x17	D	2	0.0	7.16±2.29	0.44±0.08	0.87
3x17	E	2	0.0	1.13±0.43	0.52±0.09	0.88
3x17	C	2	0.4	7.30±3.27	0.48±0.28	0.89
3x17	D	2	0.4	9.73±3.93	0.44±0.10	0.87
3x17	E	2	0.4	8.80±4.43	0.66±0.08	0.83
5x12	C	2	-0.4	14.30±7.05	0.44±0.15	0.89
5x12	D	2	-0.4	12.27±5.00	0.46±0.11	0.85
5x12	E	2	-0.4	16.53±6.13	0.46±0.11	0.82
5x12	C	2	0.0	9.37±4.12	0.51±0.15	0.91
5x12	D	2	0.0	5.50±2.21	0.40±0.09	0.91
5x12	E	2	0.0	1.20±0.41	0.41±0.18	0.92
5x12	C	2	0.4	6.17±2.57	0.56±0.12	0.91
5x12	D	2	0.4	9.13±2.71	0.46±0.11	0.87
5x12	E	2	0.4	7.73±3.04	0.52±0.12	0.83

algorithm might need to take special measures to allow large constraint violations to occur during the search.

To test the extent to which constraint violations occur between efficient solutions, the number of infeasible solutions visited along the shortest path between each local optima and the nearest efficient solution was recorded during the FDC analysis described above. The percentage of the path to the nearest Pareto optimal solution is shown in Table 1 as $frac_{infeas}$. Across all types and sizes of instances, this percentage remains reasonably stable between about 82 and 92 percent. Looking at the FDC scatter plots shown in Figure 1, it is clear that while many of the local optima are a significant distance from the Pareto front, there is a moderately strong positive correlation and reasonably significant structure in the landscape. Despite this structure, the high percentage of infeasible solutions along the shortest path to the front implies that an algorithm seeking to approach the Pareto front from a random local optimum may need to move deep into the infeasible regions of the space.

In addition to the toy instances, a set of ten $20x100$ bi-objective GAP instances were also generated, and optimized using both a straightforward multiobjective tabu search algorithm as well as a two-phase tabu search method. The structure exhibited by the GAP would seem to provide the two-phase algorithm the ability to better exploit its prior work than would the much more random structure inherent in QAP instances, for example. Figure 2 shows the median empirical attainment surfaces over 30 trials of the two algorithms applied to one of the

Table 2. Performance of the MOTS and Two-Phase TS algorithms on a set of $20x100$ mGAP instances, measured as the hypervolume mean and standard deviation. All results are averaged over 30 trials.

Instance	MOTS	Two-Phase TS
1	1.5562e7 ± 2.26e5	1.6415e7 ± 1.91e5
2	1.3271e7 ± 1.71e5	1.3994e7 ± 1.70e5
3	1.5562e7 ± 2.26e5	1.6415e7 ± 1.91e5
4	1.3271e7 ± 1.71e5	1.3994e7 ± 1.70e5
5	1.5562e7 ± 2.26e5	1.6415e7 ± 1.91e5
6	1.5562e7 ± 2.26e5	1.6415e7 ± 1.91e5
7	1.3271e7 ± 1.71e5	1.3994e7 ± 1.70e5
8	1.5562e7 ± 2.26e5	1.6415e7 ± 1.91e5
9	1.3271e7 ± 1.71e5	1.3994e7 ± 1.70e5
10	1.5562e7 ± 2.26e5	1.6415e7 ± 1.91e5

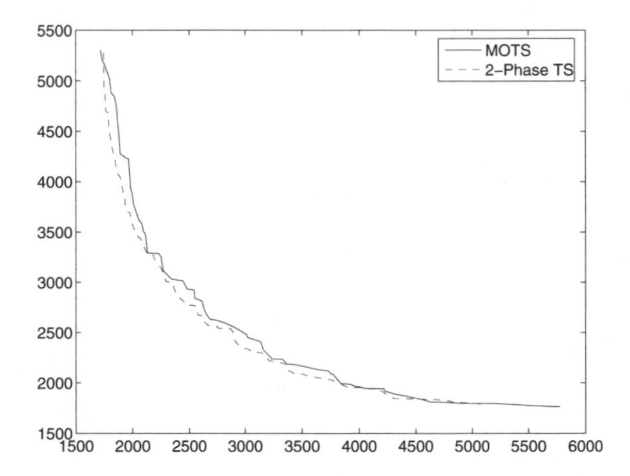

Fig. 2. Empirical attainment surfaces for a simple multiobjective tabu search and a simple two-phase variant for a sample 20x100 bi-objective generalized assignment problem instance

larger instances. The results are consistent across the ten instances, with the two phase algorithm showing a moderate, but consistent and statistically significant increase in performance, as shown in Table 2.

However, as shown in Table 1, a high percentage of the points leading to a new Pareto optimal solution are generally infeasible. This can adversely affect the ability of a simple two-phase local search algorithm to move along the Pareto front, once it has located a single solution. In multiobjective optimization, where it is important to be able to move between good nondominated solutions during the search, the effects of tight constraints are not yet well understood. An algorithm that could selectively and dynamically adjust the degree to which it allowed constraint violations might allow even greater improvements in multiobjective optimization via these types of two-phase heuristics.

4 Conclusions

It is known that no single algorithm is appropriate for all search and optimization tasks. Taking advantage of domain specific knowledge and/or known structural properties of a given search space is a must in designing a high-performance algorithm. However, in the multiobjective realm, comparatively little attention has been paid to the unique aspects of multiobjective landscapes. In particular, not all multiobjective algorithms approach the Pareto front via a single synthesized objective function which can be studied using existing landscape analysis tools. Many multiobjective evolutionary algorithms in particular fit this description, instead relying on a type of coevolution to move the population as a whole toward the Pareto front.

This paper has described some basic tools for examining the properties of multiobjective fitness landscapes. Many of these tools are straightforward generalizations of well-known ideas from the world of single-objective algorithms. However, the unique requirements of multiobjective optimization mean that there are additional types of information that may be exploited by effective algorithms, and there can be subtle differences in the interpretation of familiar techniques. Only by gaining a further understanding of multiobjective search spaces can we hope to systematically design faster and more effective algorithms.

References

1. Boese, K.D.: Cost versus distance in the traveling salesman problem. Technical Report TR-950018, University of California at Los Angeles (1995)
2. Hoos, H., Stützle, T.: Stochastic Local Search: Foundations and Applications. Elsevier, Amsterdam (2005)
3. Weinberger, E.D.: Correlated and uncorrelated fitness landscapes and how to tell the difference. Biological Cybernetics 63(5), 325–336 (1990)
4. Merz, P.: Advanced fitness landscape analysis and the performance of memetic algorithms. Evolutionary Computation 12(3), 303–325 (2004)
5. Radcliffe, N.: Forma analysis and random respectful recombination. In: Belew, R., Booker, L. (eds.) Proceedings of the Fourth International Conference on Genetic Algorithms. Morgan Kaufman, San Francisco (1991)
6. Garrett, J.D., Dasgupta, D.: Analyzing the performance of hybrid evolutionary algorithms on the multiobjective quadratic assignment problem. In: Proceedings of the IEEE Congress on Evolutionary Computation (2006)
7. Watson, J.P.: Empirical Modeling and Analysis of Local Search Algorithms for the Job-Shop Scheduling Problem. PhD thesis, Colorado State University, Fort Collins, CO (2003)
8. Watson, J.P., Howe, A., Whitley, L.D.: An analysis of iterated local search for job-shop scheduling. In: Proceedings of the Fifth Metaheuristics International Conference (2003)
9. Sahni, S., Gonzalez, T.: P-complete approximation problems. Journal of the ACM 23(3), 555–565 (1976)
10. Yagiura, M., Ibaraki, T., Glover, F.: An ejection chain approach for the generalized assignment problem. INFORMS Journal on Computing 16(2), 133–151 (2004)

11. Díaz, J.A., Fernández, E.: A tabu search heuristic for the generalized assignment problem. European Journal of Operational Research 132(1), 22–38 (2001)
12. Laguna, M., Kelly, J.P., González-Velarde, J.L., Glover, F.: Tabu search for the multilevel generalized assignment problem. European Journal of Operational Research 82(1) (1995)
13. Yagiura, M., Yamaguchi, T., Ibaraki, T.: A variable depth search algorithm with branching search for the generalized assignment problem. Optimization Methods and Software 10(3), 419–441 (1998)
14. Racer, M., Amini, M.M.: A robust heuristic for the generalized assignment problem. Annals of Operations Research 50(1), 487–503 (1994)
15. Lourenço, H.R., Serra, D.: Adaptive search heuristics for the generalized assignment problem. Mathware and Soft Computing 9(3), 209–234 (2002)
16. Chu, P.C., Beasley, J.E.: A Genetic Algorithm for the Generalized Assignment Problem. Computers and Operations Research 24(1), 17–23 (1997)
17. Alfandari, L., Plateau, A., Tolla, P.: A two-phase path relinking algorithm for the generalized assignment problem. In: Proceedings of the Fourth Metaheuristics International Conference, pp. 175–179 (2001)
18. Alfandari, L., Plateau, A., Tolla, P.: A path relinking algorithm for the generalized assignment problem. In: Resende, M.G.C., Sousa, J.P. (eds.) Metaheuristics: Computer Decision-Making, pp. 1–17. Kluwer Academic Publishers, Dordrecht (2004)
19. Yagiura, M., Ibaraki, T., Glover, F.: An effective metaheuristic algorithm for the generalized assignment problem. In: 2001 IEEE International Conference on Systems, Man, and Cybernetics, pp. 242–250 (2001)
20. Yagiura, M., Ibaraki, T., Glover, F.: A path relinking approach for the generalized assignment problem. In: Proceedings of the International Symposium on Scheduling, pp. 105–108 (2002)
21. Knowles, J., Corne, D.: Towards landscape analyses to inform the design of a hybrid local search for the multiobjective quadratic assignment problem. In: Abraham, A., del Solar, J.R., Koppen, M. (eds.) Soft Computing Systems: Design, Management and Applications, pp. 271–279. IOS Press, Amsterdam (2002)
22. Knowles, J., Corne, D.: Instance generators and test suites for the multiobjective quadratic assignment problem. In: Evolutionary Multi-Criterion Optimization (EMO 2003), Second International Conference, pp. 295–310 (2003)

Limited-Memory Techniques for Sensor Placement in Water Distribution Networks

William E. Hart, Jonathan W. Berry, Erik Boman, Cynthia A. Phillips, Lee Ann Riesen, and Jean-Paul Watson

Sandia National Laboratories
P.O. Box 5800
Albuquerque, NM 87185 USA
{wehart,jberry,egboman,lafisk,caphill,jwatson}@sandia.gov

Abstract. The practical utility of optimization technologies is often impacted by factors that reflect how these tools are used in practice, including whether various real-world constraints can be adequately modeled, the sophistication of the analysts applying the optimizer, and related environmental factors (e.g. whether a company is willing to trust predictions from computational models). Other features are less appreciated, but of equal importance in terms of dictating the successful use of optimization. These include the scale of problem instances, which in practice drives the development of approximate solution techniques, and constraints imposed by the target computing platforms. End-users often lack state-of-the-art computers, and thus runtime and memory limitations are often a significant, limiting factor in algorithm design. When coupled with large problem scale, the result is a significant technological challenge. We describe our experience developing and deploying both exact and heuristic algorithms for placing sensors in water distribution networks to mitigate against damage due intentional or accidental introduction of contaminants. The target computing platforms for this application have motivated limited-memory techniques that can optimize large-scale sensor placement problems.

1 Introduction

Real-world optimization problems are often complicated by factors that make them more challenging than problem formulations that are studied by academic researchers. For example, industrial scheduling problems differ significantly from academic scheduling models like job-shop and resource-constrained scheduling problems because of large numbers of company-specific constraints. Most researchers acknowledge the impact of such side constraints, but what is far less appreciated is the degree to which factors like problem size and target computational platform impact the practical solution of real-world optimization problems. In this paper we present a real-world case study that highlights how memory limitations impact a deployed solution technology.

This paper presents a case study that considers the protection of drinking water distribution systems, found in municipalities throughout the world. Public water

V. Maniezzo, R. Battiti, and J.-P. Watson (Eds.): LION 2007 II, LNCS 5313, pp. 125–137, 2008.

distribution systems are inherently vulnerable to accidental or intentional contamination because of their distributed geography. The use of on-line, real-time contaminant warning systems (CWSs) is a promising strategy for mitigating these risks. The general goal of a CWS is to identify a low-probability, high-impact contamination incident while allowing sufficient time for an appropriate response that mitigates adverse impacts. A CWS may complement conventional routine monitoring by quickly providing information on unusual threats to a water supply.

A key element of the design of an effective CWS is the strategic placement of sensors throughout the distribution network. We have recently demonstrated that a canonical sensor placement formulation is equivalent to the well-known p-median facility location problem. However, the p-median problems that arise in real-world sensor placement applications are much larger than typical facility location instances considered in the literature. For example, the largest p-median instances commonly investigated are limited to approximately 10,000 facilities and customers. In contrast, the p-median instances arising in real-world CWS design can involve as many as 50,000 facilities and hundreds of thousands of customers. The magnitude of these instances requires efficient computational techniques to deal with the increase in solution difficulty. Further, these large-scale instances have significant memory footprints that exceed the limits available in the computing platforms of most water utilities.

The goal of this paper is to summarize how memory limitations have influenced the development of sensor placement algorithms in the TEVA-SPOT Toolkit [1] (SPOT). SPOT provides a sensor placement framework that facilitates research in sensor placement optimization and enables the practical application of sensor placement solvers to real-world CWS design applications. SPOT contains algorithms for solving the integer programming formulation exactly (e.g., via CPLEX), heuristically via GRASP, and heuristically via Lagrangian relaxation. The details of these techniques are documented in other papers, but current goal is to provide an overview of how our focus on limited-memory techniques has led to algorithmic challenges that are motivated by a real-world application. The United States Environmental Protection Agency (USEPA) has funded the development of SPOT to support the analysis of US water distribution networks in the USEPA TEVA program [2]. SPOT's support of limited-memory sensor placement techniques has been crucial for the successful analysis of these large-scale networks.

This paper is organized as follows. We begin in Section 2 with a detailed description of the CWS problem and the corresponding integer programming formulation. We discuss limited-memory strategies for integer programming, GRASP and Lagrangian relaxation in Sections 3, 4, and 5 respectively. We conclude in Section 6 with a discussion of the implications of our results for real-world problem solving.

2 Background

Contamination warning systems (CWSs) have been proposed as a promising approach for detecting contamination incidents in drinking water distribution systems. The goal of a CWS is to detect contamination incidents early enough to allow for effective public health and/or water utility intervention to limit potential public health or economic impacts. There are many challenges to detecting contaminants in drinking water systems: municipal distribution systems are large, consisting of hundreds or thousands of miles of pipe; flow patterns are driven by time-varying demands placed on the system by customers; and distribution systems are looped, resulting in mixing and dilution of contaminants. The drinking water community has proposed that CWSs be designed to maximize the number of contaminants that can be detected in drinking water distribution systems by combining online sensors with public health surveillance systems, physical security monitoring, customer complaint surveillance, and routine sampling programs [3].

For CWS design, the general goal of sensor placement optimization is to place a limited number of sensors in a water distribution network such that the impact to public health of contaminant injection is minimized. However, there is no specific formulation of the problem that is widely accepted by the water resources management community. There are a wide range of important design objectives for sensor placements (e.g., minimizing the cost of sensor installation and maintenance, the response time to a contamination incident, and the extent of contamination), and researchers have developed different formulations when studying these objectives. Further, researchers have developed a variety of technical approaches for solving sensor placement problems including mixed-integer programming (MIP) models [4, 5, 6, 7, 8, 9], combinatorial heuristics [10, 11, 12], and general-purpose metaheuristics (e.g., [12]).

A common feature of most sensor placement formulations is that they rely either directly or indirectly on contaminant transport simulation models. Simulation tools, like EPANET [13], perform extended-period simulation of the hydraulic and water quality behavior within pressurized pipe networks. These models can be used to evaluate the expected flow in water distribution systems, and they can model the transport of contaminants and related chemical interactions. Thus, a water utility can assess risks to their distribution network by considering simulations of an ensemble of contamination incidents, which reflect the impact of contamination at different locations, times of the day, etc.

A key limitation of early sensor placement formulations is that they incorporate contamination transport simulation results indirectly. Consequently, the optimized value of the final solution may not accurately approximate a risk assessment performed with contaminant transport simulations. We have proposed a mixed-integer programming (MIP) model that resolves this difficulty by directly integrating contaminant transport simulation results [14, 4]. The MIP objective exactly captures water utilities' current risk metrics. Furthermore, this model can minimize a variety of different design objectives simply by integrating different statistics from the simulation results. This model assumes that a potentially large number of contamination incidents can be simulated, but these simulations are preprocessing steps that can be done in advance of the optimization process. Thus, the time needed for simulation does not impact the time spent performing sensor placement.

Our MIP formulation for sensor placement is:

$$\text{(SP) minimize} \sum_{a \in \mathcal{A}} \alpha_a \sum_{i \in \mathcal{L}_a} d_{ai} x_{ai}$$

$$\text{where} \begin{cases} \sum_{i \in \mathcal{L}_a} x_{ai} = 1 & \forall a \in \mathcal{A} \\ x_{ai} \leq s_i & \forall a \in \mathcal{A}, i \in \mathcal{L}_a \\ \sum_{i \in L} s_i \leq p & \\ s_i \in \{0, 1\} & \forall i \in L \\ 0 \leq x_{ai} \leq 1 & \forall a \in \mathcal{A}, i \in \mathcal{L}_a \end{cases}$$

This MIP minimizes the expected impact of a set of contamination incidents defined by $\mathcal{A}$. For each incident $a \in \mathcal{A}$, α_a is the weight of incident a, frequently a probability. The EPANET simulator reports contamination levels at a set of *locations*, denoted by L, where a location refers to network junction. For each incident a, $\mathcal{L}_a \subseteq L$ is the set of locations that can be contaminated by a. Thus a sensor at a location $i \in \mathcal{L}_a$ can detect contamination from incident a at the time contamination first arrives at location i. Each incident is *witnessed* by the first sensor to see it. For each incident $a \in \mathcal{A}$ and location $i \in \mathcal{L}_a$, d_{ai} defines the impact of the contamination incident a if it is witnessed by location i. This impact measure assumes that as soon as a sensor witnesses contamination, then any further contamination impacts are mitigated (perhaps after a suitable delay that accounts for the response time of the water utility). The s_i variables indicate where sensors are placed in the network, subject to a budget p, and the x_{ia} variables indicate whether incident a is witnessed by a sensor at location i.

We may not be able to witness all contamination incidents with a given set of sensors. To account for this, L contains a *dummy* location. This dummy location is in all subsets $\mathcal{L}_a$. The impact for this location is the impact of the contamination incident after the entire contaminant transport simulation has finished, which corresponds to the impact that would occur without an online CWS.

Remarkably, SP is identical to the well-known p-median facility location problem [15]. In the p-median problem, p facilities (e.g., central warehouses) are to be located on m potential sites such that the sum of distances d_{ai} between each of n customers (e.g., retail outlets) and the nearest facility i is minimized. In comparing SP and p-median problems, we observe equivalence between (1) sensors and facilities, (2) contamination incidents and customers, and (3) contamination impacts and distances. While SP allows placement of *at most* p sensors, p-median formulations generally enforce placement of all p facilities; in practice, the distinction is irrelevant unless p approaches the number of possible locations.

The flexibility of this sensor placement formulation is illustrated by the TEVA-SPOT Toolkit (SPOT) [1], which integrates a variety of sensor placement solvers developed by Sandia National Laboratories and the Environmental Protection Agency, along with many academic collaborators [5, 14, 4]. SPOT includes general-purpose heuristic solvers, integer programming heuristics, exact solvers, and linear-programming bounding techniques. SPOT can place sensors to minimize a variety of design objectives, including population-based public health measures, time to detection, extent of pipe contamination, volume consumed, and number of failed detections.

The size of the SP formulation is largely a function of the the total number of impacts, D. This is the dominant term in the number of constraints, the number of variables, and the number of nonzeros in the constraint matrix. We can compute a lower bound on SP by relaxing the integrality constraints on the variables and solving the resulting *linear program* (LP). Solving the LP involves linear algebra. Dense methods would require space proportional to D^2. Solvers use sparse methods as much as possible. However, the space requirements are generally superlinear in D. Because integer-programming solvers use LPs for bounding subproblems, they require at least as much space as LP asymptotically.

Water distribution networks analyzed in the TEVA program have 1,000s to 10,000s of pipes and junctions. Due to memory limitations, contamination incidents are typically restricted to a small number of times during a day (e.g. morning, afternoon, evening and night incidents). The number of locations contaminated by an incident can be highly variable; although many incidents impact a small number of locations, some large networks have many incidents that contaminate a large fraction of the network. Many of the SP analyses performed in the TEVA program have had millions of impact values. Very large sensor placement problems considered in the TEVA program have had over 40,000 potential sensor placement locations, 20,000 contamination incidents and close to 30 million impact values. Furthermore, real-world analyses will ultimately require the consideration of many more contamination incidents, for example to model changes in weekday vs. weekend demands, as well as seasonal changes in demands.

3 Integer Programming

The SP MIP model provides a generic approach for performing sensor placement with a variety of design objectives. However, the size of this MIP formulation can quickly become prohibitively large, especially for 32-bit computers (yielding a maximum of 4GB of RAM in the case of Unix systems, and in practice 3GB of RAM in the case of Windows systems). As noted in the previous section, SP can require millions of impact values for large water distribution systems.

In Berry et al. [16, 14], we note that for any given contamination incident a, there are often many impacts d_{ai} that have the same value. If the contaminant reaches two junctions at approximately the same time, then the impact for these two junctions would have the same impact values. For example, this occurs frequently when we use a coarse reporting time-step for the water quality simulation.

This observation led to a revised formulation that treats sensor placement locations as equivalent if their corresponding contamination impacts are the same for a given contamination incident. Let $\mathcal{L}_{ai}$ be a maximal set of locations in $\mathcal{A}$ that all have the same impact for incident a. Considering any witness in

$\mathcal{L}_{ai}$ equivalent reduces the set of effective witness "locations" to a new set $\hat{\mathcal{L}}_a$. The new MIP formulation is:

$$
\text{(waSP) minimize} \sum_{a\in\mathcal{A}} \alpha_a \sum_{i\in\hat{\mathcal{L}}_a} d_{ai}x_{ai}
$$

$$
\text{where} \begin{cases} \sum_{i\in\hat{\mathcal{L}}_a} x_{ai} = 1 & \forall a \in \mathcal{A} \\ x_{ai} \le \sum_{j\in\hat{\mathcal{L}}_{ai}} s_j & \forall a \in \mathcal{A}, i \in \hat{\mathcal{L}}_a \\ \sum_{i\in L} s_i \le p & \\ s_i \in \{0,1\} & \forall i \in L \\ 0 \le x_{ai} \le 1 & \forall a \in \mathcal{A}, i \in \hat{\mathcal{L}}_a \end{cases}
$$

This MIP selects both a group of sensors to witness an incident and an actual sensor from the group. The fundamental structure of this formulation changes only slightly from SP, but in practice this MIP often requires significantly less memory. Specifically a grouping of k equivalent locations removes $k-1$ entries from the the objective, $k-1$ variables, and $k-1$ constraints. Every feasible solution for SP has a corresponding solution in waSP with the same sensor placement. We can always map the selected observation variable to a real sensor with the same impact. Because the impact for each incident is the same, the objective value is the same, so we can use waSP to find optimal sensor placements.

The waSP model revises SP to exploit structure in SP that can make the MIP formulation smaller. We have developed two extensions of this idea in Berry et al. [16]: witness aggregation and incident aggregation. These aggregation strategies attempt to consolidate the impact values to create smaller MIP formulations for sensor placement that approximate SP.

3.1 Witness Aggregation

We can generalize the waSP formulation to consider location values as equivalent if their impact values are approximately equal. For each incident a, consider a list of locations in $\mathcal{L}_a$ sorted by impact. A *superlocation* is a contiguous sublist of this sorted list. Generally, we group locations into a superlocation if the difference in their impact values meets a given threshold. In Berry et al. [14], we describe two ways for creating superlocations: (1) the ratio of largest to the smallest impact in the superlocation is small, and (2) the difference between the largest and the smallest impact is small. Note that the locations grouped in a superlocation for an incident are not necessarily located physically close in the network even though the contamination for incident a reaches them at approximately the same time.

Let $\tilde{\mathcal{L}}_{ai} \subseteq \mathcal{L}_a$ be the locations in the ith *superlocation* for incident a. We denote the set of superlocations for incident a by $\tilde{\mathcal{L}}_a$. Let $\tilde{d}_{ai}$ be the largest impact value for incident a if witnessed by any location in $\tilde{\mathcal{L}}_{ai}$ (that is, $\tilde{d}_{ai} = \max_{i\in\tilde{\mathcal{L}}_{ai}} d_{ai}$). And let x_{ai} be a binary variable that is 1 if incident a is witnessed by some location in $\tilde{\mathcal{L}}_{ai}$. Then the MIP for general witness aggregation is the waSP formulation where we replace d_{ai} by $\tilde{d}_{ai}$, replace $\hat{\mathcal{L}}_{ai}$ by $\tilde{\mathcal{L}}_{ai}$, and replace $\hat{\mathcal{L}}_{ai}$ by $\tilde{\mathcal{L}}_{ai}$.

We have shown that the optimal solution to a problem with ratio aggregation is guaranteed to be an approximation for the original problem with quality proportional to the ratio. However, it is hard for a user to determine a good threshold without carefully exploring the data.

3.2 Incident Aggregation

In some cases, we can replace a pair or a group of contamination incidents with a single new incident that is equivalent. In Berry et al. [14], we describe one such strategy (called scenario aggregation in that paper for historical reasons). This aggregation strategy combines two incidents that impact the same locations in the same order, allowing for the possibility that one incident continues to impact other locations. For example, two contamination incidents should travel in the same pattern if they differ only in the nature of the contaminant, though one may decay more quickly than the other. Aggregated incidents can be combined by simply averaging the impacts that they observe and updating the corresponding incident weight α_a.

3.3 Impact

These aggregation techniques significantly improved our ability to apply MIP solvers to real-world sensor placement applications. The use of the waSP formulation is critical to solve large sensor placement problems, even on high-end workstations with large memory. For example, in Berry et al. [14] we show that aggregating witnesses with the same impacts can reduce the number of nonzeros in the MIP model by a factor of two, and it reduces the total runtime by a factor of four. Further, ratio witness aggregation and incident aggregation can be combined to formulate an approximate sensor placement formulation that reduces the number of nonzeros by a factor of 7 and the runtime by a factor of 200, while generating a solution that is within 5% of optimal.

4 The GRASP Heuristic

The MIP formulations described in the previous section cannot be solved to optimality for very large networks, even on high-end workstations with a lot of RAM. Thus, we have adapted heuristic algorithms for the p-median problem to solve SP. The current state-of-the-art heuristic for the p-median problem is the GRASP algorithm recently introduced by Resende and Werneck [17]. This GRASP heuristic is a three-phase search procedure. In the first phase, a set of high-quality solutions are generated using biased greedy construction techniques. Steepest-descent hill-climbing is then used to transform each of the resulting solutions into local optima. Finally, path relinking is used to further explore the set of solutions lying at the intersection of the resulting local optima. For a complete description of this heuristic, we refer the reader to [17].

On a series of wide-ranging tests, we observed that the GRASP heuristic was able to locate solutions to very large p-median instances (with over 10,000 facilities and 50,000 customers) in approximately ten minutes of run-time on a modern workstation-class computer [18]. This is approximately 5-10 times faster than CPLEX when solving the waSP MIP formulation introduced in Section 3. Further, the solutions obtained by GRASP were often optimal (as verified by comparison with exact solutions to the MIP formulation). The only drawback to the GRASP heuristic involved the memory requirements, which reached 16GB of RAM for the largest instance considered. This capacity is beyond the limits of what is available in most end-user environments for which CWS design is targeted; here, the typical platform is either a 32-bit workstation (with a maximum capacity of 4GB of RAM) or a Windows workstation, which is limited to 8GB of RAM even when running on 64-bit CPUs.

The GRASP heuristic creates a dense matrix of all customer-facility "distances", as the cost of determining the decrease in "cost" during a local search move is dictated by the lookup cost of specific d_{ai} impact values. The dense matrix approach replicates information, but in doing so yields constant-time lookup of the d_{ai} coefficients. An alternative "sparse" representation simply stores, for each $a \in \mathcal{A}$, a tree containing pairs (i, d_{ai}) for all i defined for the incident a. The resulting representation yields logarithmic (in the number of defined d_{aj} for a given a) lookup costs, necessarily slowing the execution of the GRASP heuristic. However, in practice the slow-down is less than 50%, while the memory requirements are reduced by a factor of four or more.

SPOT provides variants of the GRASP heuristic using the dense and sparse storage schemes for the d_{ai}, and this optimizer has been widely used in the USEPA TEVA program. However, even with the sparse representation the largest networks considered in the USEPA TEVA program are still too large for 32-bit workstations. Other avenues have been used to reduce the problem size further for these problems, such as restricting the number of locations for sensors. These strategies may preclude the optimal solution, but they provide a practical alternative for heuristic optimization.

5 Lagrangian Heuristic

In this section, we present a Lagrangian-based bounding procedure and approximation heuristic for sensor placement [19], which requires $O(n+D)$ space, where n is the number of sensor locations and D is the total number of impacts. This is an asymptotically optimal memory requirement for an in-core implementation. We use the Lagrangian-based lower-bounding method for the p-median problem described by Avella, Sassano, and Vasil'ev [20]. They give a Lagrangian model for which one can compute the optimal solution, given a set of Lagrangian multipliers, in linear space and near-linear time. Barahona and Chudak [21] give a Lagrangian formulation for the related unconstrained facility location problem, where one balances a facility opening cost with the service costs rather than limiting the number of facilities. Barahona and Chudak detail how to use subgradient search, specifically Barahona and Anbil's Volume algorithm [22], to find Lagrangian multipliers that produce progressively higher lower bounds. We adapted their method to the p-median problem. This search converges to a set of

Lagrangian multipliers for which the optimal solution to our relaxed problem is an optimal solution to the p-median LP relaxation. We then use our constrained rounding algorithm [23] to randomly select p sensor locations biased by the LP relaxation.

We now describe the Lagrangian relaxation model. As with all Lagrangian relaxation, we remove some of the constraints, leaving behind a problem that is easy to solve. We apply pressure to satisfy the constraints we have relaxed by adding penalties to the objective function. These penalties are proportional to the constraint violations. Thus there is no penalty if a constraint is met, a small penalty for a small violation, and a larger penalty for a larger violation.

We relax the first set of constraints in the SP formulation, those that require each incident is witnessed by some sensor; recall that this might be the *dummy* sensor that indicates a failure to detect the incident. This constraint is written as an equality, because that is a more efficient integer programming formulation. However, the difficult part of the constraint is insuring that at least one sensor witnesses each incident. The objective will prevent over-witnessing, so for the sake of the Lagrangian relaxation, we consider these constraints to be inequalities. For some incident a, this constraint is violated for a proposed setting of the s_i and x_{ai} variables if $\sum_{i \in \mathcal{L}_a} x_{ai} < 1$, giving a violation of $1 - \sum_{i \in \mathcal{L}_a} x_{ai}$. We weight each such violation with its own Lagrangian multiplier λ_a, which allows us to penalize some violations more than others. Adding a penalty term $\lambda_a - \lambda_a \sum_{i \in \mathcal{L}_a} x_{ai}$ to the objective for each incident a, the Lagrangian model becomes:

$$\text{(LAG) minimize} \sum_{a \in \mathcal{A}} \left(\alpha_a \sum_{i \in \hat{\mathcal{L}}_a} (d_{ai} - \lambda_a) x_{ai} \right) + \sum_{a \in \mathcal{A}} \alpha_a \lambda_a$$

$$\text{where} \begin{cases} x_{ai} \leq s_i & \forall a \in \mathcal{A}, i \in \hat{\mathcal{L}}_a \\ \sum_{i \in L} s_i \leq p & \\ s_i \in \{0,1\} & \forall i \in L \\ 0 \leq x_{ai} \leq 1 & \forall a \in \mathcal{A}, i \in \hat{\mathcal{L}}_a \end{cases}$$

For a fixed set of λ_a, we can compute the optimal value of LAG in linear space and near-linear time using a slight variation on the method described by Avella, Sassano, and Vasil'ev [20]. The optimal solution to LAG gives a valid lower bound on the value of an optimal solution to the p-median (SP) problem. This is because any feasible solution to the p-median problem is feasible for LAG. It has a zero violation for each of the lifted constraints and a value equal to the original p-median value.

Barahona and Chudak [21] describe the Volume subgradient method as applied to the unconstrained facility location problem. This method begins with $\lambda_a = 1$ for all incidents a, solves the relaxed problem, then iteratively updates the multipliers, increasing the multipliers in proportion to the violation. The updates require space and time linear in the number of variables. We modified the Vol unconstrained facility location code, available in the COIN-OR repository [24] for the p-median problem. This will converge to an optimal solution for the p-median problem.

Given a fractional solution to the p-median LP, we can treat the fractional values as probabilities and select sensors randomly according to this probability. However, one is unlikely to get precisely p sensors this way. We use the method of Berry and Phillips [23] for efficiently sampling over the "lucky" distribution where we select precisely k sensors. If necessary, we then select the dummy location.

In preliminary tests with a moderate[1] sized problem, the Lagrangian method required approximately 1/3 the space of the GRASP heuristic and usually found a solution almost as good while running up to 2.5 times longer. For example, on a problem with 3358 locations, 1621 incidents, and 5 sensors, considering four different types of objectives, the Lagrangian solver required 45Mb of memory while the GRASP heuristic required 154Mb of memory. The GRASP heuristic found the optimal solution in all four cases as verified by the MIP. The Lagrangian heuristic was within .5% of this for three out of the four objectives. Running times for GRASP ranged from 33.8 seconds to 44 seconds. The Lagrangian ran in less than 86 seconds for 3 out of 4 objectives. For the fourth objective, Lagrangian ran for 105 seconds and had a gap of 64%, showing that the Lagrangian behavior can be less stable than GRASP.

As we noted above, the Lagrangian method provides a lower bound on the value of an optimal solution to the p-median (SP) problem. Further, this lower bound is computed with less memory than an LP relaxation of SP. Thus, another practical motivation for applying the Lagrangian method is that it computes a valid lower bound on the value of solutions generated by GRASP!

We also consider the use of witness aggregation to further reduce the memory required for the Lagrangian method, particularly aggregation of locations that have the same impact values. However, we cannot embed the set-cover constraints (the second set of constraints in the waSP formulation) without altering the Lagrangian model. We can run the heuristic with the aggregated witnesses where the superlocations are not directly associated with their constituent locations. This creates a straight p-median problem for the Lagrangian solver that now no longer has the same optimal solution. Because there are fewer opportunities to witness incidents, this revised formulation has a higher optimal impact, and therefore the current Lagrangian solver does not give a valid lower bound. However, we can still compute a heuristic solution by solving this modified problem and mapping superlocations back to real locations.

We have developed a preliminary version of an aggregated Lagrangian heuristic that simply selects the first real location in a superlocation list. For a large-scale problem with 42,000 junctions, the Lagrangian heuristic required only 100Mb for the aggregation problem where we equated only witnesses of equal impact. This is a considerable reduction from the 1.8GB the Lagrangian method required with no witness aggregation, even of equal impact (the SP version). The GRASP heuristic required 17GB; there is no value for witness aggregation in the GRASP heuristic, so this is the memory requirement for the SP version. However, the objective of the Lagrangian solution is 60% worse than the solution

[1] This problem is the same size as those Avella et al. call "large-scale."

found by GRASP. The significant reduction in space motivates more work on this aggregated Lagrangian heuristic, and we expect more sophisticated techniques for mapping from supernodes to real location will improve its performance.

6 Discussion and Conclusion

In practice, a particular sensor placement problem must be solved numerous times, e.g., to generate sensor budget versus performance trade-off curves, or to guide search toward solutions of a specific form. Consequently, the design of SPOT was initially focused on execution speed, to facilitate maximal analysis throughput. For the smaller sensor placement problems examined in the early phases of this project, emphasis on run-time achieved this goal.

However, as larger and larger problems became available, our design focus rapidly shifted from minimizing run-time toward minimizing the memory footprint for large sensor placement problems. SPOT is intended for general use by water utilities throughout the United States, most of which do not possess high-end computing platforms, and limited-memory sensor placement strategies are needed for commonly available workstations. This change in emphasis focused algorithm design and development efforts in fundamentally new, unanticipated directions.

There is a broad lesson here: a strict focus on run-time can severely limit the applicability of algorithmic techniques to real-world problems. Non-algorithmic considerations can significantly impact the practicality of an algorithmic approach. End-users consider a wide range of factors when deciding to use a computational tool, such as likely acceptance in their organization, the background of users, and required computation resources. These factors can easily outweigh algorithmic considerations like run-time efficiency. This is not a new observation, but what is surprising is that most discrete optimization research appears to be driven strictly by run-time considerations, e.g., to obtain either new best-known solutions to benchmark problems or reduce the run-time required to obtain high-quality solutions. This project illustrates that focusing on other performance factors can lead to fundamentally new algorithmic challenges, and that assessments of algorithmic strategies should consider trade-offs between factors that impact their use in practice.

Acknowledgements

We have collaborated with Regan Murray (US EPA), Rob Janke (US EPA), Jim Uber (University of Cincinnati) and Tom Taxon (Argonne National Laboratory) on the development of the TEVA-SPOT Toolkit. Sandia is a multiprogram laboratory operated by Sandia Corporation, a Lockheed Martin Company, for the United States Department of Energy's National Nuclear Security Administration under Contract DE-AC04-94-AL85000.

Bibliography

[1] Hart, W.E., Berry, J., Murray, R., Phillips, C.A., Riesen, L.A., Watson, J.P.: SPOT: A sensor placement optimization toolkit for drinking water contaminant warning system design. Technical Report SAND2007-4393 C, Sandia National Laboratories (2007)
[2] Morley, K., Janke, R., Murray, R., Fox, K.: Drinking water contamination warning systems: Water utilities driving water research. Journal AWWA, 40–46 (2007)
[3] USEPA: WaterSentinel System Architecture. Technical report, U.S. Environmental Protection Agency (2005)
[4] Berry, J., Hart, W.E., Phillips, C.A., Uber, J.: A general integer-programming-based framework for sensor placement in municipal water networks. In: Proc. World Water and Environment Resources Conference (2004)
[5] Berry, J., Fleischer, L., Hart, W.E., Phillips, C.A., Watson, J.P.: Sensor placement in municipal water networks. J. Water Planning and Resources Management 131(3), 237–243 (2005)
[6] Lee, B.H., Deininger, R.A., Clark, R.M.: Locating monitoring stations in water distribution systems. Journal, Am. Water Works Assoc., 60–66 (1991)
[7] Lee, B.H., Deininger, R.A.: Optimal locations of monitoring stations in water distribution system. Journal of Environmental Engineering 118(1), 4–16 (1992)
[8] Propato, M., Piller, O., Uber, J.: A sensor location model to detect contaminations in water distribution networks. In: Proc. World Water and Environmental Resources Congress, American Society of Civil Engineers (2005)
[9] Watson, J.P., Greenberg, H.J., Hart, W.E.: A multiple-objective analysis of sensor placement optimization in water networks. In: Proc. World Water and Environment Resources Conference (2004)
[10] Kessler, A., Ostfeld, A., Sinai, G.: Detecting accidental contaminations in municipal water networks. Journal of Water Resources Planning and Management 124(4), 192–198 (1998)
[11] Kumar, A., Kansal, M.L., Arora, G.: Discussion of 'detecting accidental contaminations in municipal water networks'. Journal of Water Resources Planning and Management 125(4), 308–310 (1999)
[12] Ostfeld, A., Salomons, E.: Optimal layout of early warning detection stations for water distribution systems security. Journal of Water Resources Planning and Management 130(5), 377–385 (2004)
[13] Rossman, L.A.: The EPANET programmer's toolkit for analysis of water distribution systems. In: Proceedings of the Annual Water Resources Planning and Management Conference (1999),
`http://www.epanet.gov/ORD/NRMRL/wswrd/epanet.html`
[14] Berry, J., Hart, W.E., Phillips, C.E., Uber, J.G., Watson, J.P.: Sensor placement in municiple water networks with temporal integer prog ramming models. J. Water Resources Planning and Management 132(4), 218–224 (2006)
[15] Mirchandani, P., Francis, R. (eds.): Discrete Location Theory. John Wiley and Sons, Chichester (1990)
[16] Berry, J., Carr, R.D., Hart, W.E., Phillips, C.A.: Scalable water sensor placement via aggregation. In: Proc. Water Distribution System Symposium (2007)
[17] Resende, M., Werneck, R.: A hybrid heuristic for the p-median problem. Journal of Heuristics 10(1), 59–88 (2004)

[18] Ostfeld, A., Uber, J.G., Salomons, E., Berry, J.W., Hart, W.E., Phillips, C.A., Watson, J.P., Dorini, G., Jonkergouw, P., Kapelan, Z., di Pierro, F., Khu, S.T., Savic, D., Eliades, D., Polycarpou, M., Ghimire, S.R., Barkdoll, B.D., Gueli, R., Huang, J.J., McBean, E.A., James, W., Krause, A., Leskovec, J., Isovitsch, S., Xu, J., Guestrin, C., VanBriesen, J., Small, M., Fischbeck, P., Preis, A., Propato, M., Piller, O., Trachtman, G.B., Wu, Z.Y., Walski, T.: The battle of the water sensor networks (BWSN): A design challenge for engineers and algorithms. J. Water Resource Planning and Management (submitted, 2007)

[19] Berry, J.W., Boman, E., Phillips, C.A., Riesen, L.A.: Low-memory Lagrangian relaxation methods for sensor placement in municipal water networks. In: Proc. EWRI (to appear, 2008)

[20] Avella, P., Sassano, A., Vasil'ev, I.: Computational study of large-scale p-median problems. Mathematical Programming 109(1), 89–114 (2007)

[21] Barahona, F., Chudak, F.: Near-optimal solutions to large-scale facility location problems. Discrete Optimization 2, 35–50 (2005)

[22] Barahona, F., Anbil, R.: The volume algorithm: producing primal solutions with a subgradient method. Mathematical Programming 87(3), 385–399 (2000)

[23] Berry, J., Phillips, C.: Randomized rounding for sensor placement problems (preparation) (submitted, 2007)

[24] Computational INfrastructure for Operations Research home page (2008), `http://www.coin-or.org/`

A Hybrid Clustering Algorithm Based on Honey Bees Mating Optimization and Greedy Randomized Adaptive Search Procedure

Yannis Marinakis[1], Magdalene Marinaki[2], and Nikolaos Matsatsinis[1]

[1] Decision Support Systems Laboratory, Department of Production Engineering and Management, Technical University of Crete, 73100 Chania, Greece
marinakis@ergasya.tuc.gr, nikos@ergasya.tuc.gr

[2] Industrial Systems Control Laboratory, Department of Production Engineering and Management, Technical University of Crete, 73100 Chania, Greece
magda@dssl.tuc.gr

Abstract. This paper introduces a new hybrid algorithmic nature inspired approach based on the concepts of the Honey Bees Mating Optimization Algorithm (HBMO) and of the Greedy Randomized Adaptive Search Procedure (GRASP), for optimally clustering N objects into K clusters. The proposed algorithm for the Clustering Analysis, the Hybrid HBMO-GRASP, is a two phase algorithm which combines a HBMO algorithm for the solution of the feature selection problem and a GRASP for the solution of the clustering problem. This paper shows that the Honey Bees Mating Optimization can be used in hybrid synthesis with other metaheuristics for the solution of the clustering problem with remarkable results both to quality and computational efficiency. Its performance is compared with other popular stochastic/metaheuristic methods like particle swarm optimization, ant colony optimization, genetic algorithms and tabu search based on the results taken from the application of the methodology to data taken from the UCI Machine Learning Repository.

Keywords: Honey Bees Mating Optimization, Greedy Randomized Adaptive Search Procedure, Nature Inspired Intelligence, Clustering Analysis.

1 Introduction

During the last years, nature inspired intelligence becomes increasingly popular through the development and utilisation of intelligent methods in advanced information systems design and optimization. These methods are driven by concepts from nature and biology including advances in structural genomics, mapping of genes to proteins and proteins to genes, modelling of complete cell structures, functional genomics, self-organization of natural systems. For optimization within complex domains of data or information, the most popular nature inspired approaches are those representing successful animal and micro-organism team

V. Maniezzo, R. Battiti, and J.-P. Watson (Eds.): LION 2007 II, LNCS 5313, pp. 138–152, 2008.

behaviour. Thus, birds flocks or fish schools inspired Particle Swarm Optimization [28], the imitation of the biological immune systems led to the development of the artificial immune systems ([15], [14]), the ants foraging behaviours gave rise to Ant Colony Optimization [16] while the mating process of honey bees gave rise to the honey bees mating optimization algorithm ([1], [2]). Since then, the honey bees mating optimization algorithm has been used on different applications ([3], [17], [21]). The Honey Bees Mating Optimization algorithm simulates the mating process of the queen of the hive that begins when the queen flights away from the nest performing the mating flight during which the drones follow the queen and mate with her in the air.

In this paper, a new hybrid metaheuristic algorithm based on the Honey Bees Mating Optimization algorithm and on the Greedy Randomized Adaptive Search Procedure (GRASP) [18] is proposed for the clustering analysis. Clustering analysis is an important tool for data exploration and it has been applied in a wide variety of fields. The typical clustering analysis consists of four steps (i.e. feature selection or extraction, clustering algorithm design or selection, cluster validation and results interpretation) with feedback pathway. These steps are closely related to each other and affect the derived clusters. The proposed hybrid algorithm uses the Honey Bees Mating Optimization algorithm for the solution of the feature selection problem and the Greedy Randomized Adaptive Search Procedure for the clustering problem. It should be noted that such an algorithm that combines a nature inspired intelligence technique like HBMO and a stochastic metaheuristic like GRASP is applied for the first time for the solution of this kind of problems, at least to our knowledge. In order to assess the efficacy of the proposed algorithm, this methodology is evaluated on datasets from the UCI Machine Learning Repository. Also, the method is compared with the results of four other metaheuristic algorithms for clustering analysis that use a Tabu Search Based Algorithm [19], a Genetic Based Algorithm [20], an Ant Colony Optimization algorithm [16] and the Particle Swarm Optimization [28] for the solution of the feature selection problem [35]. All these algorithms use the GRASP for the clustering algorithm. Also, the efficiency of the proposed algorithm is compared with the results of a classical clustering approach, like k-means [43]. The rest of this paper is organized as follows: In the next section a description of the Clustering Analysis is presented. In the third section the proposed algorithm, the Honey Bees Mating Optimization Algorithm for the Clustering Analysis (Hybrid HBMO-GRASP) is presented and analyzed in detail. Computational results are presented and analyzed in the fourth section while in the last section conclusions and future research are given.

2 Clustering Analysis

Clustering analysis identifies clusters (groups) embedded in the data, where each cluster consists of objects that are similar to one another and dissimilar to objects in other clusters ([23], [42], [47]). As it has already been mentioned, the typical clustering analysis consists of four steps [47]:

- The ***basic feature selection problem (FSP)***, where the problem is to search through the space of feature subsets to identify the optimal or near-optimal one with respect to a performance measure. In the literature many successful feature selection algorithms have been proposed ([24], [35]). ***Feature extraction*** utilizes some transformations to generate useful and novel features from the original ones.
- The ***clustering algorithm design*** or ***selection*** step is usually combined with the selection of a corresponding proximity measure ([23], [42]) and the construction of a clustering criterion function which makes the partition of clusters a well defined optimization problem. The clustering objective functions are highly non-linear and multi-modal functions and, thus, the problem is NP-hard and as a consequence it is difficult to investigate the problem in an analytical approach. Many heuristic, metaheuristic and stochastic algorithms have been developed in order to find a near optimal solution in reasonable computational time. Analytical surveys of the clustering algorithms can be found in [23], [42], [47]. In [4], [12], [33] algorithms based on Tabu Search are presented. Simulated Annealing for clustering is used in [8], [10], [12] while in [9] a clustering algorithm based on Greedy Randomized Adaptive Search Procedure (GRASP) is applied. Genetic algorithms are used in [7], [13], [32], [37] while an analytical review of the use of neural networks in clustering is given in [31]. Clustering algorithms based on Ant Colony Optimization are used in [5], [6], [11], [22], [26], [37], [39], [44] while in [25], [27], [36], [39], [45] clustering algorithms based on Particle Swarm Optimization are applied. Clustering algorithms based on Artificial Immune Systems are presented in [30], [40] and, finally, a clustering algorithm based on Honey Bees Mating Optimization is presented in [17].
- ***Cluster validity*** analysis is the assessment of a clustering procedure's output. Effective evaluation standards and criteria are used in order to find the degree of confidence for the clustering results derived from the used algorithms. External indices, internal indices, and relative indices are used for cluster validity analysis ([23], [47]).
- In the ***results interpretation*** step, experts in the relevant fields interpret the data partition in order to guarantee the reliability of the extracted knowledge.

3 The Proposed Hybrid HBMO-GRASP Algorithm for Clustering

3.1 Introduction

The proposed algorithm (Hybrid HBMO-GRASP) for the solution of the clustering problem is a two phase algorithm which combines a Honey bees mating optimization (HBMO) algorithm for the solution of the feature selection problem and a Greedy Randomized Adaptive Search Procedure (GRASP) for the solution of the clustering problem. In this algorithm, the activated features are calculated by the Honey bees mating optimization (HBMO) algorithm (see 3.3) and the fitness (quality) of each particle is calculated by the clustering algorithm (see 3.4).

The problem of clustering N objects (patterns) into K clusters is considered. In particular the problem is stated as follows: Given N objects in R^n , allocate each object to one of K clusters such that the sum of squared Euclidean distances between each object and the center of its belonging cluster (which is also to be found) for every such allocated object is minimized. The clustering problem can be mathematically described as follows:

$$\text{Minimize } J(w,z) = \sum_{i=1}^{N}\sum_{j=1}^{K} w_{ij} \parallel x_i - z_j \parallel^2 \tag{1}$$

Subject to

$$\sum_{j=1}^{K} w_{ij} = 1, \qquad i = 1, ..., N \tag{2}$$

$$w_{ij} = 0 \text{ or } 1, \qquad i = 1, ..., N, j = 1, ..., K \tag{3}$$

where K is the number of clusters (given or unknown), N is the number of objects (given), $x_i \in R^n, (i = 1, ..., N)$ is the location of the ith pattern (given), $z_j \in R^n, (j = 1, ..., K)$ is the center of the jth cluster (to be found), ($z_j = \frac{1}{N_j}\sum_{i=1}^{N} w_{ij}x_i$, where N_j is the number of objects in the jth cluster), and w_{ij} is the association weight of pattern x_i with cluster j, (to be found), where w_{ij} is equal to 1 if pattern i is allocated to cluster j, $\forall i = 1, ..., N, j = 1, ..., K$ and is equal to 0, otherwise.

Initially in the first phase of the algorithm a number of features are activated, using the Honey Bees Mating Optimization Algorithm. In order to find the clustering of the samples (fitness or quality of the HBMO algorithm) a GRASP algorithm is used. The clustering algorithm has the possibility to solve the clustering problem with known or unknown number of clusters. When the number of clusters is known, the equation (1), denoted as SSE, is used in order to find the best clustering. In the case that the number of clusters is unknown, the selection of the best solution of the feature selection problem cannot be performed based on the sum of squared Euclidean distances because when the features are increased (or decreased) a number of terms are added (or subtracted) in equation (1) and the comparison of the solutions is not possible, using only the SSE measure. Thus the minimization of a validity index ([41], [46]) is used, given by:

$$validity = \frac{SSE}{SSC} \tag{4}$$

where $SSC = \sum_i^K \sum_j^K (\parallel z_i - z_j \parallel)^2$ is the distance between the centers of the clusters.

3.2 Honey Bees

Honey bees are social insects that work together in a highly structured social order and create hives. Each hive typically consists of a single queen, drones and workers [2]. The queen is fed with "royal jelly" (a milky-white colored, jellylike substance that makes her bigger than any other bee in the hive), can lay over 1500 eggs per day and will live five to six years [29]. Drones provide the queen with some sperm and in order to spot the queen when she is on her mating flight, drones have noticeably bigger eyes than those of the other castes [21]. At the end of the season the drones left in the hive will be driven out of the hive to die and, thus, drones live up to 6 months [3].

Workers are all sterile females and do all the different tasks needed to maintain and operate the hive, like comb construction, brood rearing (broods arise either from fertilized eggs representing potential queens or workers or unfertilized eggs representing prospective drones), tending the queen and drones, cleaning, temperature regulation and defending the hive when they are young and foraging outside the hive to gather nectar, pollen, water and certain sticky plant resins used in hive construction when they are older [21]. The workers that are born early in the season live about 6 weeks while those that are born in the fall live until the following spring [3].

In the marriage process, the queen mates during her mating flight far from the nest. During this flight the drones follow the queen and mate with her in the air [2]. The queen mates multiple times, but the drone only once as the insemination ends with the eventual death of the drone and the queen receiving the "mating sign". In each mating, sperm reaches the spermatheca and, thus, each time a queen lays fertilized eggs, she randomly retrieves a mixture of the sperm in her spermatheca to fertilize the egg [3].

3.3 Honey Bees Mating Optimization for the Feature Selection Problem

Feature selection is used as the first step of the clustering task in order to reduce the dimension of problem, decrease noise and improve the speed of the algorithm by the elimination of irrelevant or redundant features. In this paper, the Honey Bees Mating Optimization Algorithm is used for feature selection. Initially, we have to choose the population of the honey bees that will configure the initial hive. Each bee is randomly placed in the d-dimensional space as a candidate solution (in the feature selection problem d corresponds to the number of activated features). One of the key issues in designing a successful algorithm for Feature Selection Problem is to find a suitable mapping between Feature Selection Problem solutions and bees in Honey Bees Mating Optimization Algorithm. Every candidate feature in HBMO is mapped into a binary particle where the bit 1 denotes that the corresponding feature is selected and the bit 0 denotes that the feature is not selected. Afterwards the fitness of each idividual is calculated using the GRASP algorithm for clustering (see section 3.4) and the best member of the initial population of bees is selected as the queen of the hive while all the other members of the population are the drones.

Before the process of mating begins, the user has to define a number that corresponds to the queen's size of spermatheca. This number corresponds to the maximum number of mating of the queen in a single mating flight. Each time the queen succesfully mates with a drone the genotype of the drone is stored and a variable is increased by one until the size of spermatheca is reached. Another two parameters have to be defined, the number of queens and the number of broods that will be born by all queens. In this implementation of Honey Bees Mating Optimization (HBMO) Algorithm, the number of queens is set equal to one, because in the real life only one queen will survive in a hive, and the number of broods is set equal to the number corresponding to the queen's spermatheca size. Then, we are ready to begin the mating flight of the queen. At the start of the flight, the queen is initialized with some energy content (initially, the speed and the energy of the queen are generated at random) and returns to her nest when the energy is within some threshold from zero to full spermatheca [3]. A drone mates with a queen probabilistically using the following annealing function [1], [2]:

$$Prob(D) = e^{[\frac{-\Delta(f)}{Speed(t)}]} \tag{5}$$

where $Prob(D)$ is the probability of adding the sperm of drone D to the spermatheca of the queen (that is, the probability of a successful mating), $\Delta(f)$ is the absolute difference between the fitness of D and the fitness of the queen (for complete description of the calculation of the fitness function see 3.4) and $Speed(t)$ is the speed of the queen at time t. The probability of mating is high when the queen is still at the beginning of her mating flight, therefore her speed is high, or when the fitness of the drone is as good as the queen's. After each transition in space, the queen's speed and energy decays according to the following equations:

$$Speed(t+1) = \alpha \times Speed(t) \tag{6}$$

$$energy(t+1) = \alpha \times energy(t) \tag{7}$$

where α is a factor $\in (0, 1)$ and is the amount of speed and energy reduction after each transition and each step. A number of mating flights are realized. If the mating is successful (i.e., the drone passes the probabilistic decision rule), the drone's sperm is stored in the queen's spermatheca. By crossovering the drone's genotypes with the queen's, a new brood (trial solution) is formed which later can be improved, employing workers to conduct local search. One of the major differences of the Honey Bees Mating Optimization Algorithm from the classic evolutionary algorithms is that since the queen stores a number of different drone's sperm in her spermatheca she can use parts of the genotype of different drones to create a new solution which gives the possibility to have more fittest broods. In the crossover phase, we use a crossover procedure which initially identifies the common characteristics of the queen and the drone and, then, inherits them to the broods. Subsequently, a greedy procedure is applied to each brood in order to complete the solution.

In real life, the role of the workers is restricted to brood care and for this reason the workers are not separate members of the population but they are used as local search procedures in order to improve the broods produced by the mating flight of the queen. Each of the workers have different capabilities and the choise of two different workers may produce different solutions. Each of the worker have the possibility to activate or deactivate a number of different features. Each of the brood will choose, randomly, one worker to feed it with royal jelly (local search phase) having as a result the possibility of replacing the queen if the solution of the brood is better than the solution of the current queen. If the brood fails to replace the queen, then in the next mating flight of the queen this brood will be one of the drones.

3.4 Greedy Randomized Adaptive Search Procedure for the Clustering Problem

As it was mentioned earlier in the clustering phase of the proposed algorithm a **Greedy Randomized Adaptive Search Procedure (GRASP)** ([18], [34]) is used which is an iterative two phase search algorithm. Each iteration consists of two phases, a **construction phase** and a **local search phase**. In the construction phase, a randomized greedy function is used to built up an initial solution which is then exposed for improvement attempts in the local search phase. The final result is simply the best solution found over all iterations. In the first phase, a **randomized greedy technique** provides feasible solutions incorporating both greedy and random characteristics. This phase can be described as a process which stepwise adds one element at a time to the partial (incomplete) solution. The choice of the next element to be added is determined by ordering all elements in a candidate list (**Restricted Candidate List - RCL**) with respect to a greedy function. The heuristic is adaptive because the benefits associated with every element are updated during each iteration of the construction phase to reflect the changes brought on by the selection of the previous element. The probabilistic component of a **GRASP** is characterized by randomly choosing one of the best candidates in the list but not necessarily the top candidate. The greedy algorithm is a simple one pass procedure for solving the clustering problem. In the second phase, a **local search** is initialized from the solution of the first phase, and the final result is simply the best solution found over all searches. In the following the way the GRASP algorithm is applied for the solution of the clustering problem is analyzed in detail. An initial solution (i.e. an initial clustering of the samples in the clusters) is constructed step by step and, then, this solution is exposed for development in the local search phase of the algorithm. The first problem that we have to face was the selection of the number of the clusters. Thus, the algorithm works with two different ways.

If the number of clusters is known a priori, then a number of samples equal to the number of clusters are selected randomly as the initial clusters. In this case, as the iterations of GRASP increase the number of clusters do not change. In each iteration, different samples (equal to the number of clusters) are selected as initial clusters. Afterwards, the RCL is created. The RCL parameter determines the level of greediness or randomness in the construction. In our implementation, the best promising candidate samples are selected to create the RCL. The samples in the list are ordered taking into account the distance of each sample from all centers of the clusters and the ordering is from the smallest to the largest distance. From this list, the first E samples (E is a parameter of the problem) are selected in order to form the final RCL. The candidate sample for inclusion in the solution is selected randomly from the RCL using a random number generator. Finally, the RCL is readjusted in every iteration by recalculated all the distances based on the new centers and replacing the sample which has been included in the solution by another sample that does not belong to the RCL, namely the $(E+iter)$th sample where $iter$ is the number of the current iteration. When all the samples have been assigned to clusters two measures are calculated (see section 3.1) and a local search strategy is applied in order to improve the solution. The local search works as follows: For each sample the probability of its reassignment in a different cluster is examined by calculating the distance of the sample from the centers. If a sample is reassigned to a different cluster the new centers are calculated. The local search phase stops when in an iteration no sample is reassigned. If the number of clusters is unknown, then initially a number of samples are selected randomly as the initial clusters. Now, as the iterations of GRASP increase the number of clusters changes but cannot become less than two. In each iteration a different number of clusters can be found. The creation of the initial solutions and the local search phase work as in the previous case. The only difference compared to the previous case concerns the use of the validity measure in order to choose the best solution because as we have different number of clusters in each iteration the sum of squared Euclidean distances varies significantly for each solution.

4 Computational Results

4.1 Data and Parameter Description

The performance of the proposed methodology is tested on 9 benchmark instances taken from the UCI Machine Learning Repository. The datasets from the UCI Machine Learning Repository were chosen to include a wide range of domains and their characteristics are given in Table 1. The data varies in term of the number of observation from very small samples (Iris with 150 observations) up to larger data sets (Spambase with 4601 observations). Also, there are data sets with two and three clusters. In one case (Breast Cancer Wisconsin) the data set is appeared with different size of observations because in this data set there is a number of missing values. The problem of missing values was faced with two different ways. In the first way where all the observations are used we took the mean values of all the observations in the corresponding feature while in the second way where we have less values in the observations we did not take into account the observations that they had missing values. Some data sets involve only numerical features, and the remaining include both numerical and categorical features. For each data set, Table 1 reports the total number of features and the number of categorical features in parentheses. The parameter settings for

Table 1. Data Sets Characteristics

Data Sets	Observations	Features	Clusters
Australian Credit (AC)	690	14(8)	2
Breast Cancer Wisconsin 1 (BCW1)	699	9	2
Breast Cancer Wisconsin 2 (BCW2)	683	9	2
Heart Disease (HD)	270	13(7)	2
Hepatitis 1 (Hep1)	155	19 (13)	2
Ionosphere (Ion)	351	34	2
Spambase(spam)	4601	57	2
Iris	150	4	3
Wine	178	13	3

Hybrid HBMO-GRASP algorithm were selected after thorough empirical testing and they are: The number of queens is set equal to 1, the number of drones is set equal to 200, the number of mating flights (M) is set equal to 1000, the size of spermatheca is set equal to 50, number of broods is set equal to 50, α is set equal to 0.9, the number of workers (w) is set equal to 20 and, finally, the size of RCL is set equal to 50. The algorithm was implemented in Fortran 90 and was compiled using the Lahey f95 compiler on a Centrino Mobile Intel Pentium M 750 at 1.86 GHz, running Suse Linux 9.1.

4.2 Results of the Proposed Algorithm

The objective of the computational experiments is to show the performance of the proposed algorithm in searching for a reduced set of features with high clustering of the data. The purpose of feature variable selection is to find the smallest set of features that can result in satisfactory predictive performance. Because of the curse of dimensionality, it is often necessary and beneficial to limit the number of input features in order to have a good predictive and less computationally intensive model. In general there are $2^{numberoffeatures} - 1$ possible feature combinations and, thus, in our cases the problem with the fewest number of feature combinations is the Iris (namely $2^4 - 1$), while the most difficult problem is the Spambase where the number of feature combinations is $2^{57} - 1$.

A comparison with the classic k-means and other metaheuristic approaches for the solution of the clustering problem is presented in Table 2. In this Table, eight other algorithms are used for the solution of the feature subset selection problem and for the clustering problem. In the first one a Particle Swarm Optimization algorithm is used for the solution of the feature selection problem while a Greedy Randomized Adaptive Search Procedure is used in the clustering phase [36] (columns 4 and 5 of Table 2), while in the second an Ant Colony Optimization Algorithm is used in the feature selection problem with Greedy Randomized Adaptive Search Procedure in the clustering phase [37] (columns 6 and 7 of Table 2). Subsequently, in the second group of algorithms and columns 2 and 3 of Table 2 a genetic algorithm [20] is used in the first phase of the algorithm while a Greedy Randomized Adaptive Search Procedure is used in the second phase of the algorithm [38]. In the second group and in columns 4 and 5 of Table 2 a Tabu Search Algorithm [19] is used in the first phase and a GRASP algorithm is used in the second phase. Finally, classic k-means algorithm is used for the clustering problem using all features (columns 6 and 7 of Table 2 of the second group). In the third group of algorithms in columns 2 and 3 of Table 2 a Particle Swarm Optimization is used in the first phase and an Ant Colony Optimization algorithm is used in the second phase [39], in the second one in both phases (feature selection phase and clustering phase) an Ant Colony Optimization algorithm is used (columns 4 and 5 of Table 2 of the third group) while in the third one a Particle Swarm Optimization in both phases (feature selection phase and clustering phase) algorithm is used (columns 6 and 7 of Table 2 of the third group). The parameters and the implementation details of all of the algortihms presented in the comparisons are analyzed in papers [36], [37], [38], [39].

From this table, it can be observed that the Hybrid HBMO-GRASP algorithm performs better (has the largest number of correct clustered samples) than the other eight algorithms in all instances. It should be mentioned that in some instances the differences in the results between the Hybrid HBMO-GRASP algorithm and the other eight algorithms are very significant. Mainly, for the two data sets that have the largest number of features compared to the other data sets, i.e. in the Ionosphere data set the percentage of corrected clustered samples for the Hybrid HBMO-GRASP algorithm is 88.03%, for the Hybrid PSO-ACO algorithm is 86.03%, for the Hybrid PSO-GRASP algorithm is 85.47%, for the Hybrid ACO-GRASP algorithm is 82.90%, for the Genetic-GRASP algorithm is 75.78%, for the Tabu-GRASP algorithm is 74.92%, for the PSO algorithm is 74.35%, for the ACO algorithm is 73.50% and for the k-means is 70.65% and in the Spambase data set the percentage of corrected clustered samples for the Hybrid HBMO-GRASP algorithm is 87.54%, for the Hybrid PSO-ACO algorithm is 87.19%, for the Hybrid PSO-GRASP algorithm is 87.13%, for the Hybrid ACO-GRASP algorithm is 86.78%, for the ACO algorithm is 86.22%, for the PSO algorithm is 86.06%, for the k-means is 86.02%, for the Genetic-GRASP algorithm is 85.59% and for the Tabu-GRASP algorithm is 82.80%. It should, also, be noted that a hybridization algorithm performs always better than a no hybridized algorithm. More precisely, the only three algorithms that are competitive in almost all instances with the proposed Hybrid HBMO-GRASP algorithm are the Hybrid PSO - ACO, the Hybrid PSO - GRASP and the Hybrid ACO - GRASP algorithms. These results prove the significance of the solution of the feature selection problem in the clustering algorithm as when more sophisticated methods for the solution of this problem (Honey Bees Mating Optimization, Particle Swarm Optimization and Ant Colony Optimization) were used the performance of the clustering algorithm was improved. The significance of the solution of the feature selection problem using the Honey Bees Mating Optimization Algorithm is, also, proved by the fact that with this algorithm the best solution was found by using less features than the other algorithms used in the comparisons. More precisely, in the most difficult instance, the Spambase

Table 2. Results of the Algorithm

Instance	HBMO-GRASP		PSO-GRASP		ACO-GRASP	
	Selected Feat.	Correct Clustered	Selected Features	Correct Clustered	Selected Feat.	Correct Clustered
BCW2	5	664(97.21%)	5	662(96.92%)	5	662(96.92%)
Hep1	5	140(90.32%)	7	135(87.09%)	9	134(86.45%)
AC	8	604(87.53%)	8	604(87.53%)	8	603(87.39%)
BCW1	5	677(96.85%)	5	676(96.70%)	5	676(96.70%)
Ion	8	309 (88.03%)	11	300(85.47%)	2	291(82.90%)
spam	31	4028 (87.54%)	51	4009(87.13%)	56	3993(86.78%)
HD	8	237(87.77%)	9	232(85.92%)	9	232(85.92%)
Iris	3	146(97.33%)	3	145(96.67%)	3	145(96.67%)
Wine	7	176(98.87%)	7	176(98.87%)	8	176(98.87%)
Instance	Genetic-GRASP		Tabu-GRASP		k-Means	
	Sel. Feat.	Correct Clustered	Sel. Feat.	Correct Clustered	Sel Feat.	Correct Clustered
BCW2	5	662(96.92%)	6	661(96.77%)	9	654(95.74%)
Hep1	9	134(86.45%)	10	132(85.16%)	19	121(78.06%)
AC	8	602(87.24%)	9	599(86.81%)	14	580(84.05%)
BCW1	5	676(96.70%)	8	674(96.42%)	9	672(96.13%)
Ion	17	266(75.78%)	4	263(74.92%)	34	248(70.65%)
spam	56	3938(85.59%)	34	3810(82.80%)	57	3958(86.02%)
HD	7	231(85.55%)	9	227(84.07%)	13	220(81.48%)
Iris	4	145(96.67%)	3	145(96.67%)	4	144(96%)
Wine	7	175(98.31%)	7	174(97.75%)	13	172(96.92%)
Instance	PSO-ACO		ACO		PSO	
	Selected Features	Correct Clustered	Sel. Feat.	Correct Clustered	Sel. Feat.	Correct Clustered
BCW2	5	664(97.21%)	5	662(96.92%)	5	662(96.92%)
Hep1	6	139(89.67%)	9	133(85.80%)	10	132(85.16%)
AC	8	604(87.53%)	8	601(87.10%)	8	602(87.24%)
BCW1	5	677(96.85%)	8	674(96.42%)	8	674(96.42%)
Ion	7	302(86.03%)	16	258(73.50%)	12	261(74.35%)
spam	39	4012(87.19%)	41	3967(86.22%)	37	3960(86.06%)
HD	9	235(87.03%)	9	227(84.07%)	9	227(84.07%)
Iris	3	146(97.33%)	3	145(96.67%)	3	145(96.67%)
Wine	7	176(98.87%)	7	174(97.75%)	7	174(97.75%)

instance, the proposed algorithm needed 31 features in order to find the optimal solution, while the other seven algorithms (in the k-means the feature selection problem was not solved) the algorithms needed between 34 - 56 features to find their best solution. It should, also, be mentioned that the algorithm was tested with two options: with known and and unknown number of clusters. When the number of clusters was unknown and thus in each iteration of the algorithm different initial values of clusters were selected the algorithm always converged to the optimal number of clusters and with the same results as in the case that the number of clusters was known.

5 Conclusions and Future Research

In this paper a new metaheuristic algorithm, the Hybrid HBMO-GRASP, is proposed for solving the Clustering Problem. This algorithm is a two phase algorithm which combines a Honey Bees Mating Optimization (HBMO) algorithm for the solution of the feature selection problem and a Greedy Randomized Adaptive Search Procedure (GRASP) for the solution of the clustering problem. A number of metaheuristic algorithms and the classic k-means algorithm were also used for comparison purposes. The performance of the proposed algorithm was tested using various benchmark datasets from UCI Machine Learning Repository. The objective of the computational experiments, the desire to show the high performance of the proposed algorithms, was achieved as the algorithm gave very efficient results. The significance of the solution of the clustering problem by the proposed algorithm is proved by the fact that the percentage of the correct clustered samples is very high and in some instances is larger than 98%. Also, the focus in the significance of the solution of the feature selection problem is proved by the fact that the instances with the largest number of features gave better results when the HBMO algorithm was used. Future research is intended to be focused in using different algorithms both to the feature selection phase and to the clustering algorithm phase.

Acknowledgment

This work is a deliverable of the task KP_26 and is realized in the framework of the Operational Programme of Crete, and it is co-financed from the European Regional Development Fund (ERDF) and the Region of Crete with final recipient the General Secretariat for Research and Technology.

References

1. Abbass, H.A.: A monogenous MBO approach to satisfiability. In: Proceeding of the International Conference on Computational Intelligence for Modelling, Control and Automation, CIMCA 2001, Las Vegas, NV, USA (2001)
2. Abbass, H.A.: Marriage in honey-bee optimization (MBO): a haplometrosis polygynous swarming approach. In: The Congress on Evolutionary Computation (CEC 2001), Seoul, Korea, May 2001, pp. 207–214 (2001)
3. Afshar, A., Bozog Haddad, O., Marino, M.A., Adams, B.J.: Honey-bee mating optimization (HBMO) algorithm for optimal reservoir operation. Journal of the Franklin Institute 344, 452–462 (2007)
4. Al-Sultan, K.: A Tabu Search Approach to the Clustering Problem. Pattern Recognition 28(9), 1443–1451 (1995)
5. Azzag, H., Guinot, C.: Data and Text Mining with Hierarchical Clustering Ants. In: Abraham, A., Grosan, C., Ramos, V. (eds.) Swarm Intelligence in Data Mining, pp. 153–190 (2006)
6. Azzag, H., Venturini, G., Oliver, A., Gu, C.: A Hierarchical Ant Based Clustering Algorithm and its Use in Three Real-World Applications. European Journal of Operational Research 179, 906–922 (2007)

7. Babu, G., Murty, M.: A Near-Optimal Initial Seed Value Selection in K-means Algorithm Using a Genetic Algorithm. Pattern Recognition Letters 14(10), 763–769 (1993)
8. Brown, D., Huntley, C.: A Practical Application of Simulated Annealing to Clustering. Pattern Recognition 25(4), 401–412 (1992)
9. Cano, J.R., Cordón, O., Herrera, F., Sánchez, L.: A GRASP Algorithm for Clustering. In: Garijo, F.J., Riquelme, J.-C., Toro, M. (eds.) IBERAMIA 2002. LNCS (LNAI), vol. 2527, pp. 214–223. Springer, Heidelberg (2002)
10. Celeux, G., Govaert, G.: A Classification EM Algorithm for Clustering and Two Stochastic Versions. Computational Statistics and Data Analysis 14, 315–332 (1992)
11. Chen, L., Tu, L., Chen, H.: A Novel Ant Clustering Algorithm with Digraph. In: Wang, L., Chen, K., Ong, Y.S. (eds.) ICNC 2005. LNCS, vol. 3611, pp. 1218–1228. Springer, Heidelberg (2005)
12. Chu, S., Roddick, J.: A Clustering Algorithm Using the Tabu Search Approach with Simulated Annealing. In: Ebecken, N., Brebbia, C. (eds.) Data Mining II-Proceedings of Second International Conference on Data Mining Methods and Databases, Cambridge, U.K, pp. 515–523 (2000)
13. Cowgill, M., Harvey, R., Watson, L.: A Genetic Algorithm Approach to Cluster Analysis. Computers and Mathematics with Applications 37, 99–108 (1999)
14. de Castro, L.D., Timmis, J.: Artificial Immune Systems: A New Computational Intelligence Approach. Springer, Heidelberg (2002)
15. Dasgupta, D. (ed.): Artificial Immune Systems and their Application. Springer, Heidelberg (1998)
16. Dorigo, M., Stutzle, T.: Ant Colony Optimization. A Bradford Book/The MIT Press, Cambridge (2004)
17. Fathian, M., Amiri, B., Maroosi, A.: Application of Honey Bee Mating Optimization Algorithm on Clustering. Applied Mathematics and Computation 190(2), 1502–1513 (2007)
18. Feo, T.A., Resende, M.G.C.: Greedy randomized adaptive search procedure. Journal of Global Optimization 6, 109–133 (1995)
19. Glover, F.: Tabu Search I. ORSA Journal on Computing 1(3), 190–206 (1989)
20. Goldberg, D.E.: Genetic Algorithms in Search, Optimization, and Machine Learning. Addison-Wesley Publishing Company, INC., Massachussets (1989)
21. Haddad, O.B., Afshar, A., Marino, M.A.: Honey-Bees Mating Optimization (HBMO) Algorithm: A New Heuristic Approach for Water Resources Optimization. Water Resources Management 20, 661–680 (2006)
22. He, Y., Hui, S.C., Sim, Y.: A Novel Ant-Based Clustering Approach for Document Clustering. In: Ng, H.T., Leong, M.-K., Kan, M.-Y., Ji, D. (eds.) AIRS 2006. LNCS, vol. 4182, pp. 537–544. Springer, Heidelberg (2006)
23. Jain, A.K., Murty, M.N., Flynn, P.J.: Data Clustering: A Review. ACM Computing Surveys 31(3), 264–323 (1999)
24. Jain, A., Zongker, D.: Feature Selection: Evaluation, application, and Small Sample Performance. IEEE Transactions on Pattern Analysis and Machine Intelligence 19, 153–158 (1997)
25. Janson, S., Merkle, D.: A New Multi-objective Particle Swarm Optimization Algorithm Using Clustering Applied to Automated Docking. In: Blesa, M.J., Blum, C., Roli, A., Sampels, M. (eds.) HM 2005. LNCS, vol. 3636, pp. 128–141. Springer, Heidelberg (2005)

26. Kao, Y., Cheng, K.: An ACO-Based Clustering Algorithm. In: Dorigo, M., Gambardella, L.M., Birattari, M., Martinoli, A., Poli, R., Stützle, T. (eds.) ANTS 2006. LNCS, vol. 4150, pp. 340–347. Springer, Heidelberg (2006)
27. Kao, Y.-T., Zahara, E., Kao, I.-W.: A Hybridized Approach to Data Clustering. Expert Systems with Applications 34(3), 1754–1762 (2008)
28. Kennedy, J., Eberhart, R.: Particle swarm optimization. In: Proceedings of 1995 IEEE International Conference on Neural Networks, vol. 4, pp. 1942–1948 (1995)
29. Laidlaw, H.H., Page, R.E.: Mating designs. In: Rinderer, T.E. (ed.) Bee Genetics and Breeding, pp. 323–341. Academic Press Inc., NY (1986)
30. Li, Z., Tan, H.-Z.: A Combinational Clustering Method Based on Artificial Immune System and Support Vector Machine. In: Gabrys, B., Howlett, R.J., Jain, L.C. (eds.) KES 2006. LNCS, vol. 4251, pp. 153–162. Springer, Heidelberg (2006)
31. Liao, S.-H., Wen, C.-H.: Artificial Neural Networks Classification and Clustering of Methodologies and Applications - Literature Analysis from 1995 to 2005. Expert Systems with Applications 32, 1–11 (2007)
32. Liu, Y., Chen, K., Liao, X., Zhang, W.: W. Zhang A Genetic Clustering Method for Intrusion Detection. Pattern Recognition 37, 927–942 (2004)
33. Liu, Y., Liu, Y., Wang, L., Chen, K.: A Hybrid Tabu Search Based Clustering Algorithm. In: Khosla, R., Howlett, R.J., Jain, L.C. (eds.) KES 2005. LNCS, vol. 3682, pp. 186–192. Springer, Heidelberg (2005)
34. Marinakis, Y., Migdalas, A., Pardalos, P.M.: Expanding Neighborhood GRASP for the Traveling Salesman Problem. Computational Optimization and Applications 32, 231–257 (2005)
35. Marinakis, Y., Marinaki, M., Doumpos, M., Matsatsinis, N., Zopounidis, C.: Optimization of Nearest Neighbor Classifiers via Metaheuristic Algorithms for Credit Risk Assessment. Journal of Global Optimization 42, 279–293 (2008)
36. Marinakis, Y., Marinaki, M., Matsatsinis, N.: A Hybrid Particle Swarm Optimization Algorithm for Cluster Analysis. In: Song, I.-Y., Eder, J., Nguyen, T.M. (eds.) DaWaK 2007. LNCS, vol. 4654, pp. 241–250. Springer, Heidelberg (2007)
37. Marinakis, Y., Marinaki, M., Doumpos, M., Matsatsinis, N., Zopounidis, C.: A Hybrid ACO-GRASP Algorithm for Clustering Analysis. Annals of Operations Research (submitted, 2007)
38. Marinakis, Y., Marinaki, M., Doumpos, M., Matsatsinis, N., Zopounidis, C.: A Hybrid Stochastic Genetic - GRASP Algorithm for Clustering Analysis. Operational Research: An International Journal 8(1), 33–46 (2008)
39. Marinakis, Y., Marinaki, M., Matsatsinis, N.: A Stochastic Nature Inspired Metaheuristic for Clustering Analysis. International Journal of Business Intelligence and Clustering Analysis 3(1), 30–44 (2008)
40. Nasraoui, O., Gonzalez, F., Cardona, C., Rojas, C., Dasgupta, D.: A Scalable Artificial Immune System Model for Dynamic Unsupervised Learning. In: Cantú-Paz, E., Foster, J.A., Deb, K., Davis, L., Roy, R., O'Reilly, U.-M., Beyer, H.-G., Kendall, G., Wilson, S.W., Harman, M., Wegener, J., Dasgupta, D., Potter, M.A., Schultz, A., Dowsland, K.A., Jonoska, N., Miller, J., Standish, R.K. (eds.) GECCO 2003. LNCS, vol. 2723, pp. 219–230. Springer, Heidelberg (2003)
41. Ray, S., Turi, R.H.: Determination of Number of Clusters in K-means Clustering and Application in Colour Image Segmentation. In: Proceedings of the 4th International Conference on Advances in Pattern Recognition and Digital Techniques (ICAPRDT 1999), Calcutta, India (1999)
42. Rokach, L., Maimon, O.: Clustering Methods. In: Maimon, O., Rokach, L. (eds.) Data Mining and Knowledge Discovery Handbook, pp. 321–352. Springer, New York (2005)

43. Selim, S.Z., Ismail, M.A.: K-means-type algorithms: A generalized convergence theorem and characterization of local optimality. IEEE Transactions on Pattern Analysis and Machine Intelligence 6, 81–87 (1984)
44. Shelokar, P.S., Jayaraman, V.K., Kulkarni, B.D.: An Ant Colony Approach for Clustering. Analytica Chimica Acta 509, 187–195 (2004)
45. Shen, H.-Y., Peng, X.-Q., Wang, J.-N., Hu, Z.-K.: A Mountain Clustering Based on Improved PSO Algorithm. In: Wang, L., Chen, K., S. Ong, Y. (eds.) ICNC 2005. LNCS, vol. 3612, pp. 477–481. Springer, Heidelberg (2005)
46. Shen, J., Chang, S.I., Lee, E.S., Deng, Y., Brown, S.J.: Determination of Cluster Number in Clustering Microarray Data. Applied Mathematics and Computation 169, 1172–1185 (2005)
47. Xu, R., Wunsch II, D.: Survey of Clustering Algorithms. IEEE Transactions on Neural Networks 16(3), 645–678 (2005)

Ant Colony Optimization and the Minimum Spanning Tree Problem

Frank Neumann[1] and Carsten Witt[2,*]

[1] Max-Planck-Institut für Informatik, 66123 Saarbrücken, Germany
fne@mpi-inf.mpg.de
[2] LS Informatik 2, Technische Universität Dortmund, 44221 Dortmund, Germany
cw-lion@ls2.cs.uni-dortmund.de

Abstract. Ant Colony Optimization (ACO) is a kind of metaheuristic that has become very popular for solving problems from combinatorial optimization. Solutions for a given problem are constructed by a random walk on a so-called construction graph. This random walk can be influenced by heuristic information about the problem. In contrast to many successful applications, the theoretical foundation of this kind of metaheuristic is rather weak. Theoretical investigations with respect to the runtime behavior of ACO algorithms have been started only recently for the optimization of pseudo-Boolean functions.

We present the first comprehensive rigorous analysis of a simple ACO algorithm for a combinatorial optimization problem. In our investigations, we consider the minimum spanning tree problem and examine the effect of two construction graphs with respect to the runtime behavior. The choice of the construction graph in an ACO algorithm seems to be crucial for the success of such an algorithm. First, we take the input graph itself as the construction graph and analyze the use of a construction procedure that is similar to Broder's algorithm for choosing a spanning tree uniformly at random. After that, a more incremental construction procedure is analyzed. It turns out that this procedure is superior to the Broder-based algorithm and produces additionally in a constant number of iterations a minimum spanning tree if the influence of the heuristic information is large enough.

1 Introduction

Using Ant Colony Optimization (ACO) algorithms to obtain good solutions for combinatorial optimization problems has become very popular in recent years. In contrast to other kinds of randomized search heuristics such as Simulated Annealing or evolutionary algorithms, ACO algorithms have the ability to integrate knowledge about the problem into the construction of a new solution. In the case of a new combinatorial optimization problem, there is often some knowledge about the problem which can be incorporated into this kind of randomized

* This author was supported by the Deutsche Forschungsgemeinschaft (SFB) as a part of the Collaborative Research Center "Computational Intelligence" (SFB 531).

V. Maniezzo, R. Battiti, and J.-P. Watson (Eds.): LION 2007 II, LNCS 5313, pp. 153–166, 2008.

search heuristic. Therefore, the main application of ACO algorithms lies in the field of combinatorial optimization and the first problem to which this kind of heuristic has been applied was the traveling salesperson problem [6]. ACO is inspired by a colony of ants that search for a common source of food. It has been observed that ants are able to find a shortest path to such a source under certain circumstances by indirect communication. This communication is done by so-called pheromone values. The behavior of ants is put into an algorithmic framework to obtain solutions for a given problem. Solutions are constructed by random walks of artificial ants on a so-called construction graph, which has weights – the pheromone values – on the edges. Larger pheromone values lead to higher probability of the edges being traversed in the next walk. In addition, the random walk is usually influenced by heuristic information about the problem.

In contrast to successful applications, the theoretical foundation of the mentioned search heuristics is still in its infancy. A lot of applications show their practical evidence, but for a long time they were not analyzed with respect to their runtime or approximation qualities (see [5, 9] for an overview on different theoretical approaches including first steps to runtime analyses). We concentrate on the analysis of such heuristics with respect to their runtime behavior in a similar fashion to what is usually done for randomized algorithms. In this case, either the expected optimization time, which equals the number of constructed solutions until an optimal one has been obtained, or the success probability after a certain number of steps is analyzed.

The first results with respect to the runtime of a simple ACO algorithm have been obtained for the optimization of pseudo-Boolean functions [14, 4, 3, 11]. Many combinatorial optimization problems can be considered as the optimization of a specific pseudo-Boolean function. Especially in the case of polynomially solvable problems, we cannot hope that more or less general search heuristics outperform the best-known algorithms for a specific problem. Nevertheless, it is interesting to analyze them on such problems as this shows how the heuristics work and therefore improve the understanding of these, in practice successful, algorithms. A basic evolutionary algorithm called (1+1) EA has been considered for a wide class of combinatorial optimization problems in the context of optimizing a pseudo-Boolean function. All results with respect to the (1+1) EA transfer to a simple ACO algorithm called 1-ANT in this context [14]. This includes runtime bounds on some of the best-known polynomially solvable combinatorial optimization problems such as maximum matching, and the minimum spanning tree problem. In the case of NP-hard problems, the result of Witt [17] on the partition problem transfers to the 1-ANT.

In this paper, we conduct a first comprehensive runtime analysis of ACO algorithms on a combinatorial optimization problem. We have chosen the well-known minimum spanning tree (MST) problem as a promising starting point since different randomized search heuristics, in particular the (1+1) EA, have been studied w. r. t. this problem before, e. g., by Neumann and Wegener [13, 12] and Wegener [15]. Due to [14] and the result on the (1+1) EA in [13], the expected optimization time of the 1-ANT for the MST problem is $O(m^2(\log n + \log w_{\max}))$, where $w_{\max}$ is the largest weight of the input. In addition, a class of instances with polynomial weights has been presented in [13] where the expected time to obtain an optimal solution is $\Theta(n^4 \log n)$.

It is widely assumed and observed in experiments that the choice of the construction graph has a great effect on the runtime behavior of an ACO algorithm. The construction graph used in [14,4,3,11] is a general one for the optimization of pseudo-Boolean functions and does not take knowledge about the given problem into account. ACO algorithms have the advantage that more knowledge about the structure of a given problem can be incorporated into the construction of solutions. This is done by choosing an appropriate construction graph together with a procedure which allows to obtain feasible solutions. The choice of such a construction graph together with its procedure has been observed experimentally as a crucial point for the success of such an algorithm.

We examine ACO algorithms that work on a construction graphs which seem to be more suitable for MST problem. First, we consider a random walk on the input graph to construct solutions for the problem. It is well known how to choose a spanning tree of a given graph uniformly at random using random walk algorithms (see e. g. [1,16]). Our construction procedure produces solutions by a variant of Broder's algorithm [1]. We show a polynomial, but relatively large, upper bound for obtaining a minimum spanning tree by this procedure if no heuristic information influences the random walk. Using only heuristic information for constructing solutions, we show that the 1-ANT together with the Broder-based construction procedure with high probability does not find a minimum spanning tree or even does not present a feasible solution in polynomial time.

After that, we consider a more incremental construction procedure that follows a general approach proposed by Dorigo and Stützle [7] to obtain an ACO construction graph. We call this the Kruskal-based construction procedure as in each step an edge that does not create a cycle is chosen to be included into the solution. It turns out that the expected optimization time of the 1-ANT using the Kruskal-based construction procedure is $O(mn(\log n + \log w_{\max}))$. This beats the 1-ANT in the case that the minimum spanning tree problem is more generally modeled as an optimization problem of a special pseudo-Boolean function since then the above-mentioned lower bound $\Omega(n^4 \log n)$ of the (1+1) EA carries over. Using the 1-ANT together with the Kruskal-based construction procedure and a large influence of the heuristic information, the algorithm has even a constant expected optimization time. All our analyses show that and how ACO algorithms for combinatorial optimization can be analyzed rigorously using the toolbox from the analyses of randomized algorithms. In particular, we provide insight into the working principles of ACO algorithms by studying the effect of the (guided) random walks that these algorithms perform.

After having motivated our work, we introduce the model of the minimum spanning tree problem and the 1-ANT in Section 2. In Section 3, we consider a construction procedure which is influenced by Broder's algorithm and consider its effect with respect to the runtime behavior. Section 4 deals with the analysis of the 1-ANT using the Kruskal-based construction graph. We finish with conclusions and the discussion of some open problems.

2 Minimum Spanning Trees and the 1-ANT

Throughout the paper, we consider the well-known MST problem. Given an undirected connected graph $G = (V, E)$ with edge costs (weights) $w\colon E \to \mathbb{N}_{\geq 1}$, the goal is to find a spanning tree $E^* \subseteq E$ such that the total cost $\sum_{e \in E^*} w(e)$ becomes minimal. Denote $n := |V|$ and $m := |E|$ and assume w. l. o. g. that $E := \{1, \ldots, m\}$. Moreover, let $m \geq n$ since an existing spanning tree is unique if $m = n - 1$. The MST problem can be solved in time $O(m \log n)$ or $O(n^2)$ using the Greedy algorithms by Kruskal respectively Prim, see, e. g., [2].

We study the simple ACO algorithm called 1-ANT (see Algorithm 1), already analyzed in [14] for the optimization of pseudo-Boolean functions. In the 1-ANT, solutions are constructed iteratively by different construction procedures on a given directed construction graph $C = (X, A)$. In the initialization step, each edge $(u, v) \in A$ gets a pheromone value $\tau_{(u,v)} = 1/|A|$ such that the pheromone values sum up to 1. Afterwards, an initial solution x^* is produced by a random walk of an imaginary ant on the construction graph and the pheromone values are updated w. r. t. this walk. In each iteration, a new solution is constructed and the pheromone values are updated if this solution is not inferior (w. r. t. a fitness function f) to the best solution obtained so far.

Algorithm 1 (1-ANT)
1.) Set $\tau_{(u,v)} = 1/|A|$ *for all* $(u, v) \in A$.
2.) Compute a solution x *using a construction procedure.*
3.) Update the pheromone values and set $x^* := x$.
4.) Compute x *using a construction procedure.*
5.) If $f(x) \leq f(x^*)$, *update the pheromone values and set* $x^* := x$.
6.) Go to 4.).

We analyze the influence of different construction procedures on the runtime behavior of the 1-ANT algorithm. This is done by considering the expected number of solutions that are constructed by the algorithm until a minimum spanning tree has been obtained for the first time. We call this the *expected optimization time* of the 1-ANT.

3 Broder-Based Construction Graph

Since the MST problem is a graph problem, the first idea is to use the input graph G to the MST problem itself as the construction graph C of the 1-ANT. (Note that each undirected edge $\{u, v\}$ can be considered as two directed edges (u, v) and (v, u).) However, it is not obvious how a random walk of an ant on G is translated into a spanning tree. Interestingly, the famous algorithm of Broder [1], which chooses uniformly at random from all spanning trees of G, is a random walk algorithm.

We will use an ACO variant of Broder's algorithm as given in Algorithm 2. As usual in ACO algorithms, the construction procedure maintains pheromone values τ and heuristic information η for all edges of the construction graph G. Considering the MST problem, we assume that the heuristic information $\eta_{\{u,v\}}$ of an edge $\{u,v\}$ is the inverse of the weight of the edge $\{u,v\}$ in G. α and β are parameters that control the extent to which pheromone values respectively heuristic information is used.

Algorithm 2 (BroderConstruct(G, τ, η))

1.) *Choose an arbitrary node $s \in V$.*
2.) $u := s$, $T = \emptyset$
3.) *Let* $R := \sum_{\{u,v\} \in E} [\tau_{\{u,v\}}]^\alpha \cdot [\eta_{\{u,v\}}]^\beta$.
4.) *Choose neighbor v of u with probability* $\frac{[\tau_{\{u,v\}}]^\alpha \cdot [\eta_{\{u,v\}}]^\beta}{R}$.
5.) *If v has not been visited before, set $T := T \cup \{u,v\}$.*
6.) *Set $u := v$.*
7.) *If each node of G has been visited return T, otherwise go to 3.)*

Obviously, Algorithm 2 outputs a spanning tree T whose cost $f(T)$ is measured by the sum of the w-values of its edges. After a new solution has been accepted, the pheromone values τ are updated w. r. t. the constructed spanning tree T. We maintain upper and lower bounds on these values, which is a common measure to ensure convergence [5] and was also proposed in the previous runtime analysis of the 1-ANT [14]. We assume that after each update, the τ-value of each edge in the construction graph attains either the upper bound h or lower bound ℓ. Hence, for the new pheromone values τ' after an update, it holds that

$$\tau'_{\{u,v\}} = h \quad \text{if } \{u,v\} \in T \qquad \text{and} \qquad \tau'_{\{u,v\}} = \ell \quad \text{if } \{u,v\} \notin T.$$

So the last constructed solution is indirectly saved by the $n-1$ undirected edges that obtain the high pheromone value h. The ratio of the parameters ℓ and h is crucial since too large values of ℓ will lead to too large changes of the tree in subsequent steps whereas too large values of h will make changes of the tree too unlikely. We choose h and l such that $h = n^3\ell$ holds and will argue later on the optimality of this choice.

Note that choosing $\beta = 0$ or $\alpha = 0$ in Algorithm 2, only the pheromone value respectively the heuristic information influence the random walk. We examine the cases where one of these values is 0 to study the effect of the pheromone values respectively the heuristic information separately. First, we consider the case $\alpha = 1$ and $\beta = 0$ for the Broder-based construction graph. This has the following consequences. Let u be the current node of the random walk and denote by $R := \sum_{\{u,v\} \in E} \tau_{\{u,v\}}$ the sum over the pheromone values of all edges that are incident on u. Then the next node is chosen proportionally to the pheromone values on the corresponding edges, which means that a neighbor v of u is chosen with probability $\tau_{\{u,v\}}/R$.

For simplicity, we call the described setting of α, β, h and ℓ the *cubic update scheme*. To become acquainted therewith, we derive the following simple estimations on the probabilities of traversing edges depending on the pheromone values. Assume that a node v has k adjacent edges with value h and i adjacent edges with value ℓ. Note that $k + i \leq n - 1$ and $h = n^3\ell$. Then the probability of choosing an edge with value h is

$$\frac{kh}{kh + i\ell} = 1 - \frac{i}{kn^3 + i} \geq 1 - \frac{1}{n^2},$$

where among the edges with values h one edge is chosen uniformly at random. The probability of choosing a specific edge with value ℓ is at least

$$\frac{\ell}{\ell + (n-2)h} \geq \frac{\ell}{nh} \geq \frac{1}{n^4}.$$

This leads us to the following theorem, which shows that the 1-ANT in the described setting is able to construct MSTs in expected polynomial time provided $w_{\max}$, the largest weight of the edges, is not excessively large.

Theorem 1. *The expected optimization time of the 1-ANT using the procedure* BroderConstruct *with cubic update scheme is $O(n^6(\log n + \log w_{\max}))$. The expected number of traversed edges in a run of* BroderConstruct *is bounded above by $O(n^2)$ except for the initial run, where it is $O(n^3)$.*

Proof. We use the following idea for Theorem 2 in [13]. Suppose the spanning tree T^* was constructed in the last accepted solution. Let $T = T^* \setminus \{e\} \cup \{e'\}$ be any spanning tree that is obtained from T^* by including one edge e' and removing another edge e, and let $s(m,n)$ be a lower bound on the probability of producing T from T^* in the next step. Then the expected number of steps until a minimum spanning tree has been obtained is $O(s(m,n)^{-1}(\log n + \log w_{\max}))$. To prove the theorem, it therefore suffices to show that the probability of the 1-ANT producing T by the next constructed solution is $\Omega(1/n^6)$.

To simplify our argumentation, we first concentrate on the probability of rediscovering T^* in the next constructed solution. This happens if the ant traverses all edges of T^* in some arbitrary order and no other edges in between, which might require that an edge has to be taken more than once. (This is a pessimistic assumption since newly traversed edges are not necessarily included in the solution.) Hence, we are confronted with the cover time for the tree T^*. The cover time for trees on n nodes in general is bounded above by $2n^2$ [10], i. e., by Markov's inquality, it is at most $4n^2$ with probability at least $1/2$. We can apply this result if no so-called error occurs that an edge with pheromone value ℓ is taken. According to the above calculations, the probabilty of an error is bounded above by $1/n^2$ in a single step of the ant. Hence, there is no error in $O(n^2)$ steps with probability $\Omega(1)$. Therefore, the probability of rediscovering T^* in the next solution (using $O(n^2)$ steps of *BroderConstruct*) is at least $\Omega(1)$. Additionally taking into account the number of steps $O(n^3)$ for the initial solution [1], we have already bounded the expected number of traversed edges in a run of *BroderConstruct*.

To construct T instead of T^*, exactly one error is desired, namely e' has to be traversed instead of e. Consider the ant when it is for the first time on a node on which e' is incident. By the calculations above, the probability of including e' is $\Omega(1/n^4)$. Note that inserting e' into T^* closes a cycle c. Hence, when e' has been included, there may be at most $n-2$ edges of $\tilde{T} := T^* \setminus \{e\}$ left to traverse. We partition the edges of the forest $\tilde{T}$ into two subsets: The edges that belong to the cycle c are called critical and the remaining ones are called uncritical. The order of inclusion for the uncritical edges is irrelevant. However, all critical edges have to be included before the ant traverses edge e.

We are faced with the following problem: Let $v_1, \ldots, v_k, v_1$ describe the cycle c and suppose w.l.o.g. that $e' = \{v_1, v_k\}$. It holds that $e = \{v_i, v_{i+1}\}$ for some $1 \leq i \leq k-1$. Moreover, let v_s be the node of c that is visited first by the ant. W.l.o.g., $1 \leq s \leq i$. With probability $\Omega(1/n^4)$, the edge e' is traversed exactly once until a new solution has been constructed. Hence, after e' has been taken, the ant must visit the nodes $v_k, v_{k-1}, \ldots, v_{i+1}$ in the described order (unless an error other than including e' occurs), possibly traversing uncritical edges in between. To ensure that e is not traversed before, we would like the ant to visit all the nodes in $\{v_1, \ldots, v_i\}$, without visiting nodes in $\{v_{i+1}, \ldots, v_k\}$, before visiting v_k by traversing e'. We apply results on the Gambler's Ruin Problem [8]. The probability of going from v_s to v_i before visiting v_k is at least $\Omega(1/n)$. The same lower bound holds on the probability of going from v_i to v_1 before visiting v_{i+1}. These random walks are still completed in expected time $O(n^2)$. Hence, in total, the probability of constructing T is $\Omega((1/n^4) \cdot (1/n) \cdot (1/n)) = \Omega(1/n^6)$ as suggested. □

We see that the ratio $h/\ell = n^3$ leads to relatively high exponents in the expected optimization time. However, this ratio seems to be necessary for our argumentation. Consider the complete graph on n nodes where the spanning tree T^* equals a path of length $n-1$. The cover time for this special tree T^* is bounded below by $\Omega(n^2)$. To each node of the path, at most 2 edges with value h and at least $n-3$ edges with value ℓ are incident. Hence, the ratio is required to obtain an error probability of $O(1/n^2)$. It is much more difficult to improve the upper bound of Theorem 1 or to come up with a matching lower bound. The reasons are twofold. First, we cannot control the effects of steps where the ant traverses edges to nodes that have been visited before in the construction step. These steps might reduce the time until certain edges of T^* are reached. Second, our argumentation concerning the cycle $v_1, \ldots, v_k, v_1$ makes a worst-case assumption on the starting node v_s. It seems more likely that v_s is uniform over the path, which could improve the upper bound of the theorem by a factor $\Omega(n)$. However, a formal proof of this is open.

ACO algorithms often use heuristic information to direct the search process. In the following, we set $\alpha = 0$ and examine the effect of heuristic information for the MST problem. Recall that the heuristic information for an edge e is given by $\eta(e) = 1/w(e)$. Interestingly, for the obvious Broder-based graph, heuristic information alone does not help to find MSTs in reasonable time regardless of β. On the following example graph G^*, either the runtime of *BroderConstruct* explodes or MSTs are found only with exponentially small probability. W.l.o.g., $n = 4k+1$. Then G^*, a connected graph on the nodes $\{1, \ldots, n\}$, consists of

k triangles with weights $(1, 1, 2)$ and two paths of length k with exponentially increasing weights along the path. More precisely, let

$$T^* := \bigcup_{i=1}^{k} \{\{1, 2i\}, \{1, 2i+1\}, \{2i, 2i+1\}\},$$

where $w(\{1, 2i\}) = w(\{2i, 2i+1\}) := 1$ and $w(\{1, 2i+1\}) := 2$. Moreover, denote

$$P_1^* := \{1, 2k+2\} \cup \bigcup_{i=2}^{k} \{2k+i, 2k+i+1\},$$

where $w(\{1, 2k+2\}) := 2$ and $w(\{2k+i, 2k+i+1\}) := 2^i$, and, similarly,

$$P_2^* := \{1, 3k+2\} \cup \bigcup_{i=2}^{k} \{3k+i, 3k+i+1\},$$

where $w(\{1, 3k+2\}) := 2$ and $w(\{3k+i, 3k+i+1\}) := 2^i$. Finally, the edge set of G^* is $T^* \cup P_1^* \cup P_2^*$. Hence, all triangles and one end of each path are glued by node 1.

Theorem 2. *Choosing $\alpha = 0$ and β arbitrarily, the probability that the 1-ANT using* BroderConstruct *finds an MST for G^*, or the probability of termination within polynomial time is $2^{-\Omega(n)}$.*

Proof. Regardless of the ant's starting point, at least one path, w. l. o. g. P_1^*, must be traversed from 1 to its other end, and for least $k-1$ triangles, both nodes $2i$ and $2i+1$ must be visited through node 1. For each of these initially undiscovered triangles, the first move into the triangle must go from 1 to $2i$, otherwise the resulting tree will not be minimal. If the triangle is entered at node $2i$, we consider it a success, otherwise (entrance at $2i+1$) an error. The proof idea is to show that for too small β, i. e., when the influence of heuristic information is low, with overwhelming probability at least one triangle contains an error. If, on the other hand, β is too large, the ant with overwhelming probability will not be able to traverse P_1^* in polynomial time due to its exponentially increasing edge weights.

We study the success probabilities for the triangles and the path P_1. Given that the ant moves from 1 to either $2i$ or $2i+1$, the probability of going to $2i$ equals

$$\frac{(\eta(\{1, 2i\}))^\beta}{(\eta(\{1, 2i\}))^\beta + (\eta(\{1, 2i+1\}))^\beta} = \frac{1}{1 + 2^{-\beta}}$$

since $\eta(e) = 1/w(e)$. Therefore, the probability of $k-1$ successes equals, due to independence, $(1 + 2^{-\beta})^{-k+1}$. This probability increases with β. However, for $\beta \le 1$, it is still bounded above by $(2/3)^{k-1} = 2^{-\Omega(n)}$.

Considering the path P_1^*, we are faced with the Gambler's Ruin Problem. At each of the nodes $2k+i$, $2 \le i \le k-1$, the probability of going to a lower-numbered node and the probability of going to a higher-numbered have the same ratio of $r := (2^{-i+1})^\beta/(2^i)^\beta = 2^\beta$. Hence, starting in $2k+2$, the probability of reaching $3k+1$ before returning to 1 equals $\frac{r}{r^k-1} = \frac{2^\beta}{2^{k\beta}-1}$ (see [8]). This probability decreases with β. However, for $\beta \geq 1$, it is still bounded above by $2/(2^k-1) = 2^{-\Omega(n)}$. Then the probability of reaching the end in a polynomial number of trials is also $2^{-\Omega(n)}$. □

4 A Kruskal-Based Construction Procedure

Dorigo and Stützle [7] state a general approach how to obtain an ACO construction graph from any combinatorial optimization algorithm. The idea is to identify the so-called components of the problem, which may be objects, binary variables etc., with nodes of the construction graph and to allow the ant to choose from these components by moving to the corresponding nodes. In our setting, the components to choose from are the edges from the edge set $\{1,\dots,m\}$ of the input graph G. Hence, the canonical construction graph $C(G)$ for the MST problem is a directed graph on the $m+1$ nodes $\{0,1,\dots,m\}$ with the designated start node $s := 0$. Its edge set A of cardinality m^2 is given by

$$A := \{(i,j) \mid 0 \leq i \leq m,\ 1 \leq j \leq m,\ i \neq j\},$$

i. e., $C(G)$ is obtained from the complete directed graph by removing all self-loops and the edges pointing to s. When the 1-ANT visits node e in the construction graph $C(G)$, this corresponds to choosing the edge e for a spanning tree. To ensure that a walk of the 1-ANT actually constructs a tree, we define the feasible neigborhood $N(v_k)$ of node v_k depending on the nodes $v_1,\dots,v_k$ visited so far:

$$N(v_k) := \big(E \setminus \{v_1,\dots,v_k\}\big) \setminus \ \big\{e \in E \mid \big(V, \{v_1,\dots,v_k,e\}\big) \text{ contains a cycle}\big\}.$$

Note that the feasible neighborhood depends on the memory of the ant about the path followed so far, which is very common in ACO algorithms, see, e. g., [7].

A new solution is constructed using Algorithm 3. Again, the random walk of an ant is controlled by the pheromone values τ and the heuristic information η on the edges. Similarly to the Broder-based construction graph, we assume that the $\eta_{(u,v)}$-value of an edge (u,v) is the inverse of the weight of the edge of G corresponding to the node v in $C(G)$.

Algorithm 3 (Construct($C(G), \tau, \eta$))
1.) $v_0 := s$; $k := 0$.
2.) While $N(v_k)$ *is nonempty:*
a.) Let $R := \sum_{y \in N(v_k)} [\tau_{(v_k,y)}]^\alpha \cdot [\eta_{(v_k,y)}]^\beta$.
b.) Choose neighbor $v_{k+1} \in N(v_k)$ *with probability* $\frac{[\tau_{(v_k,v_{k+1})}]^\alpha \cdot [\eta_{(v_k,v_{k+1})}]^\beta}{R}$.
c.) Set $k := k+1$ *and go to 2.).*
3.) Return the path $p = (v_0,\dots,v_k)$ *constructed by this procedure.*

A run of Algorithm 3 returns a sequence of $k+1$ nodes of $C(G)$. It is easy to see that $k := n-1$ after the run, hence the number of steps is bounded above by n, and that $v_1, \ldots, v_{n-1}$ is a sequence of edges that form a spanning tree for G. Accordingly, we measure the fitness $f(p)$ of a path $p = (v_0, \ldots, v_{n-1})$ simply by $w(v_1) + \cdots + w(v_{n-1})$, i.e., the cost of the corresponding spanning tree. It remains to specify the update scheme for the pheromone values. As in the case of the Broder-based construction procedure, we only consider two different values h and ℓ. To allow the ant to rediscover the edges of the previous spanning tree equiprobably in each order, we reward all edges pointing to nodes from p except s, i.e., we reward $(m+1)(n-1)$ edges. Hence, the τ'-values are

$$\tau'_{(u,v)} = h \quad \text{if } v \in p \text{ and } v \neq s \quad \text{and} \quad \tau'_{(u,v)} = \ell \quad \text{otherwise.}$$

We choose h and ℓ such that $h = (m - n + 1)(\log n)\ell$ holds. In this case, the probability of taking a rewarded edge (if applicable) is always at least $1 - 1/\log n$.

We consider the case where the random walk to construct solutions is only influenced by the pheromone values on the edges of $C(G)$. The following result can be obtained by showing that the probability of obtaining from the current tree T^* a tree $T = T^* \setminus \{e\} \cup \{e'\}$ is lower bounded by $\Omega(1/(mn))$. The proof can be carried out in a similar fashion as done for Theorem 1.

Theorem 3. *Choosing $\alpha = 1$ and $\beta = 0$, the expected optimization time of the 1-ANT with construction graph $C(G)$ is bounded by $O(mn(\log n + \log w_{\max}))$.*

Proof. Let $e_1, \ldots, e_{n-1}$ be the edges of T^* and suppose w.l.o.g. that the edges of T are $e_1, \ldots, e_{n-2}, e'$ where $e' \neq e_i$ for $1 \leq i \leq n-1$. With probability $\Omega(1)$, exactly $n-2$ (but not $n-1$) out of the $n-1$ nodes visited by the 1-ANT in $C(G)$ form a uniformly random subset of $\{e_1, \ldots, e_{n-1}\}$. Hence, e_{n-1} is missing with probability $1/(n-1)$. Furthermore, the probability of visiting e' rather than e_{n-1} as the missing node has probability at least $\Omega(1/m)$. Hence, in total, T is constructed with probability $\Omega(1/(nm))$. Again we use the proof idea for Theorem 2 in [13]. It suffices to show the following claim. Suppose the 1-ANT has constructed the spanning tree T^* in the last accepted solution. Let $T = T^* \setminus \{e\} \cup \{e'\}$ be any spanning tree that is obtained from T^* by including one edge e' and removing another edge e. Then the probability of producing T by the next constructed solution is $\Omega(1/(nm))$.

Let $e_1, \ldots, e_{n-1}$ be the edges of T^* and suppose w.l.o.g. that the edges of T are $e_1, \ldots, e_{n-2}, e'$ where $e' \neq e_i$ for $1 \leq i \leq n-1$. We show that with probability $\Omega(1)$, exactly $n-2$ (but not $n-1$) out of the $n-1$ nodes visited by the 1-ANT in $C(G)$ form a uniformly random subset of $\{e_1, \ldots, e_{n-1}\}$. Hence, e_{n-1} is missing with probability $1/(n-1)$. Furthermore, we will show that the probability of visiting e' rather than e_{n-1} as the missing node has probability at least $\Omega(1/m)$. Hence, in total, T is constructed with probability $\Omega(1/(nm))$.

We still have to prove the statements on the probabilities in detail. We study the events E_i, $1 \leq i \leq n-1$, defined as follows. E_i occurs iff the first $i-1$ and the last $n-i-1$ nodes visited by the 1-ANT (excluding s) correspond to edges of T^* whereas the i-th one does not. Edges in $C(G)$ pointing to nodes of T^* have pheromone value h and all remaining edges have value ℓ. Hence, if $j-1$ edges of T^* have been found, the probability of not choosing another edge of T^* by the next node visited in $C(G)$ is at most

$$\frac{(m-(n-1))\ell}{((n-1)-(j-1))h} = \frac{1}{(n-j)\log n}.$$

Therefore, the first $i-1$ and last $n-i-1$ nodes (excluding s) visited correspond to edges of T^* with probability at least

$$1-\sum_{j=1}^{n-2}\frac{1}{(n-j)\log n} \geq 1-\frac{(\ln(n-1)+1)}{\log n}+\frac{1}{\log n} \geq 1-\frac{\ln n}{\log n} = \Omega(1)$$

(estimating the $(n-1)$-th Harmonic number by $\ln(n-1)+1$) and, due to the symmetry of the update scheme, each subset of T^* of size $n-2$ is equally likely, i. e., has probability $\Omega(1/n)$. Additionally, the probability of choosing by the i-th visited node an edge e' not contained in T^* equals

$$\frac{\ell}{(n-i)h+k\ell} \geq \frac{1}{(n-i+1)(m-n+1)\log n},$$

where k is the number of edges outside T^* that can still be chosen; note that $k\ell \leq h$. Hence, with probability at least $c/((n-i+1)mn\log n)$ for some small enough constant c (and large enough n), E_i occurs and the tree T is constructed. Since the E_i are mutually disjoint events, T is constructed instead of T^* with probability at least

$$\sum_{i=1}^{n-1}\frac{c}{(n-i+1)mn\log n} = \Omega(1/(mn))$$

as suggested. □

In the following, we examine the use of heuristic information for the Kruskal-based construction graph. Here it can be proven that strong heuristic information helps the 1-ANT mimicking the greedy algorithm by Kruskal.

Theorem 4. *Choosing $\alpha = 0$ and $\beta \geq 6w_{\max}\log n$, the expected optimization time of the 1-ANT using the construction graph $C(G)$ is constant.*

Proof. We show that the next solution that the 1-ANT constructs is with probability at least $1/e$ a minimum spanning tree, where e is Euler's number. This implies that the expected number of solutions that have to be constructed until a minimum spanning tree has been computed is bounded above by e.

Let $(w_1, w_2, \ldots, w_{n-1})$ the weights of edges of a minimum spanning tree. Let $w_i \leq w_{i+1}$, $1 \leq i \leq n-2$ and assume that the ant has already included $i-1$ edges that have weights $w_1, \ldots, w_{i-1}$ and consider the probability of choosing an edge of weight w_i in the next step. Let $M = \{e_1, \ldots, e_r\}$ be the set of edges that can be included without creating a cycle and denote by $M_i = \{e_1, \ldots, e_s\}$ the subset of M that includes all edges of weight w_i. W. l. o. g., we assume $w(e_i) \leq w(e_{i+1})$, $1 \leq i \leq r-1$.

The probability of choosing an edge of M_i in the next step is given by

$$\frac{\sum_{k=1}^{s}(\eta(e_k))^\beta}{\sum_{l=1}^{r}(\eta(e_l))^\beta} = \frac{\sum_{k=1}^{s}(\eta(e_k))^\beta}{\sum_{l=1}^{s}(\eta(e_l))^\beta + \sum_{l=s+1}^{r}(\eta(e_l))^\beta},$$

where $\eta(e_j) = 1/w(e_j)$ holds. Let $a = \sum_{k=1}^{s}(\eta(e_i))^\beta = \sum_{k=1}^{s}(1/w_i)^\beta$ and $b = \sum_{l=s+1}^{r}(\eta(e_l))^\beta$. The probability of choosing an edge of weight w_i is $a/(a+b)$, which is at least $1-1/n$ if $b \leq a/n$. The number of edges in $M \setminus M_i$ is bounded above by m, and the weight of such an edge is at least w_i+1. Hence, $b \leq m \cdot (1/(w_i+1))^\beta$.

We would like $m \cdot (1/(w_i+1))^\beta \leq s \cdot (1/w_i)^\beta/n$ to hold. This can be achieved by choosing

$$\beta \geq \frac{\log(mn/s)}{\log((w_i+1)/w_i)} = \frac{\log(mn/s)}{\log(1+1/w_i)},$$

which is at most

$$(\log(mn/s))/(w_i/2) \leq 6w_{\max}\log n$$

since $mn \leq n^3$ and $e^x \leq 1+2x$ for $0 \leq x \leq 1$. Due to our choices, the ant traverses the edge with weight w_i with probability at least $1-1/n$. Therefore, the probability that in every step i such an edge is taken is at least $(1-1/n)^{n-1} \geq 1/e$ as suggested. □

The result of Theorem 4 does not necessarily improve upon Kruskal's algorithm since the computational efforts in a run of the construction algorithm and for initializing suitable random number generators (both of which are assumed constant in our cost measure for the optimization time) must not be neglected. With a careful implementation of the 1-ANT, however, the expected computational effort w. r. t. the well-known uniform cost measure could be at least bounded above by the runtime $O(m \log m)$ of Kruskal's algorithm.

5 Conclusions

ACO algorithms have in particular shown to be successful in solving problems from combinatorial optimization. In contrast to many applications, first theoretical estimations of the runtime of such algorithms for the optimization of pseudo-Boolean functions have been obtained only recently. In the case of combinatorial optimization problems, the construction graphs used are more related to the problem at hand. For the first time, the effect of such graphs have been investigated by rigorous runtime analyses. We have considered a simple ACO algorithm 1-ANT for the well-known minimum spanning tree problem. In the case of the Broder-based construction procedure a polynomial, but relatively large, upper bound has been proven. In addition, it has been shown that heuristic information can mislead the algorithm such that an optimal solution with high probability is not found within a polynomial number of steps. In the case of the Kruskal-based construction procedure, the upper bound obtained shows that this construction graph leads to a better optimization process than the 1-ANT and simple evolutionary algorithms in the context of the optimization of pseudo-Boolean functions. In addition, a large influence of heuristic information makes the algorithm mimic Kruskal's algorithm for the minimum spanning tree problem. All analyses provide insight into the guided random walks that the 1-ANT performs in order to create solutions of our problem.

There are several interesting open questions concerning ACO algorithms. First, it would be desirable to obtain the expected optimization time for the considered algorithms asymptotically exactly. For the Broder-based construction graph, we have argued why we expect relatively large lower bounds. Nevertheless, a formal proof for that is open. On the other hand, the influence of the pheromone values and the heuristic information has been analyzed only separately. The same bounds should also hold if the effect of one of these parameters is low compared to the other one. It would be interesting to also consider cases where both have a large influence.

References

1. Broder, A.: Generating random spanning trees. In: Proc. of FOCS 1989, pp. 442–447. IEEE Press, Los Alamitos (1989)
2. Cormen, T.H., Leiserson, C.E., Rivest, R.L., Stein, C.: Introduction to Algorithms, 2nd edn. MIT Press, Cambridge (2001)
3. Doerr, B., Johannsen, D.: Refined runtime analysis of a basic ant colony optimization algorithm. In: Proc. of CEC 2007, pp. 501–507. IEEE Press, Los Alamitos (2007)
4. Doerr, B., Neumann, F., Sudholt, D., Witt, C., Witt, C.: On the runtime analysis of the 1-ANT ACO algorithm. In: Proc. of GECCO 2007, pp. 33–40. ACM Press, New York (2007)
5. Dorigo, M., Blum, C.: Ant colony optimization theory: A survey. Theoretical Computer Science 344, 243–278 (2005)
6. Dorigo, M., Maniezzo, V., Colorni, A.: The ant system: An autocatalytic optimizing process. Technical Report 91-016 Revised, Politecnico di Milano (1991)
7. Dorigo, M., Stützle, T.: Ant Colony Optimization. MIT Press, Cambridge (2004)
8. Feller, W.: An Introduction to Probability Theory and Its Applications, 3rd edn., vol. 1. Wiley, Chichester (1968)
9. Gutjahr, W.J.: Mathematical runtime analysis of ACO algorithms: Survey on an emerging issue. Swarm Intelligence 1, 59–79 (2007)
10. Motwani, R., Raghavan, P.: Randomized Algorithms. Cambridge University Press, Cambridge (1995)
11. Neumann, F., Sudholt, D., Witt, C.: Comparing variants of MMAS ACO algorithms on pseudo-boolean functions. In: Stützle, T., Birattari, M., H. Hoos, H. (eds.) SLS 2007. LNCS, vol. 4638, pp. 61–75. Springer, Heidelberg (2007)
12. Neumann, F., Wegener, I.: Minimum spanning trees made easier via multi-objective optimization. Natural Computing 5(3), 305–319 (2006)
13. Neumann, F., Wegener, I.: Randomized local search, evolutionary algorithms, and the minimum spanning tree problem. Theoretical Computer Science 378(1), 32–40 (2007)

14. Neumann, F., Witt, C.: Runtime analysis of a simple ant colony optimization algorithm. In: Asano, T. (ed.) ISAAC 2006. LNCS, vol. 4288, pp. 618–627. Springer, Heidelberg (2006)
15. Wegener, I.: Simulated annealing beats Metropolis in combinatorial optimization. In: Caires, L., Italiano, G.F., Monteiro, L., Palamidessi, C., Yung, M. (eds.) ICALP 2005. LNCS, vol. 3580, pp. 589–601. Springer, Heidelberg (2005)
16. Wilson, D.B.: Generating random spanning trees more quickly than the cover time. In: Proc. of STOC 1996, pp. 296–303. ACM Press, New York (1996)
17. Witt, C.: Worst-case and average-case approximations by simple randomized search heuristics. In: Diekert, V., Durand, B. (eds.) STACS 2005. LNCS, vol. 3404, pp. 44–56. Springer, Heidelberg (2005)

A Vector Assignment Approach for the Graph Coloring Problem

Takao Ono, Mutsunori Yagiura, and Tomio Hirata

Graduate School of Information Science, Nagoya University, Japan
ono@is.nagoya-u.ac.jp,
yagiura@nagoya-u.jp,
hirata@is.nagoya-u.ac.jp

Abstract. We consider the relationship between the graph coloring problem (GCP) and the vector assignment problem (VAP). Given an undirected graph, VAP asks to assign a vector to each vertex so as to maximize the minimum angle between the vectors corresponding to adjacent vertices. We show that any solution to the VAP in the 2-dimensional space, which we call the 2-dimensional VAP (2VAP), gives a feasible coloring, and that such transformation can be computed efficiently. We also show that any optimal solution to 2VAP gives an optimal coloring for GCP. Based on this fact, we propose a heuristic algorithm for GCP, whose search space is the set of solutions for 2VAP. The algorithm is quite simple and can be considered as a variant of the threshold accepting. The experiments show that our algorithm works well for graphs with relatively low degree.

1 Introduction

A vertex coloring for graph $G = (V, E)$ with $n = |V|$ vertices and $m = |E|$ edges is an assignment $c\colon V \to \mathbb{N}$ such that $c(i) \neq c(j)$ if $(i, j) \in E$. A k-coloring is a vertex coloring whose range is $\{1, 2, \ldots, k\}$. The graph coloring problem (GCP) is the problem of finding, for a given graph G, a k-coloring of G with minimum k. Such k is called the chromatic number of G and is denoted by $\chi(G)$. The graph coloring problem is one of the representative combinatorial optimization problems and is known to be NP-hard. It has many applications in such fields as scheduling, time-tabling and so forth. Hence much effort has been devoted to designing exact and heuristic algorithms for GCP.

Included among metaheuristic algorithms for GCP are tabu search [7], genetic algorithm [6], GRASP [13], variable neighborhood search [1], an so on. Kochenberger et. al. proposed a general solver to the unconstrained quadratic binary programming problem and applied it to GCP [12]. As for exact algorithms, Caramia and Dell'Olmo have proposed the *multistage* branch-and-bound method and solved instances with about 500 vertices [4].

Inspired by the theoretical success of Karger, Motwani, and Sudan [10], this paper considers the vector assignment problem (VAP). In this problem, we are given an undirected graph and asked to assign a vector to each vertex so as

V. Maniezzo, R. Battiti, and J.-P. Watson (Eds.): LION 2007 II, LNCS 5313, pp. 167–176, 2008.

to maximize the minimum angle between the vectors corresponding to adjacent vertices. When vectors are restricted in the plane, we call the resulting problem the 2-dimensional VAP (2VAP). A similar problem was considered e.g. in [3], in which the relationship with the maximum cut problem was discussed. In this paper, we show that any feasible solution to 2VAP can be transformed into a feasible coloring of G in time $O(n \log n + m)$ via the piercing set problem on circular-arc graphs [11]. Moreover, this procedure transforms any optimal solution for 2VAP to an optimal coloring for GCP.

Based on these facts, we propose a heuristic algorithm for GCP, whose search space is the set of feasible solutions to 2VAP. We call this algorithm VAP-COL. The algorithm is quite simple because our objective is to observe the fundamental behavior of algorithms based on such search space. Algorithm VAP-COL repeatedly modifies the position of a vector in such a way that (basically) expands the smallest angle between the vector in consideration and those corresponding to the adjacent vertices. The algorithm allows accepting moves that reduce the smallest angle slightly; this rule can be considered as a variant of the threshold accepting [5]. The experiments show that algorithm VAP-COL works well for graphs with relatively low degree.

The idea of using more than one search space alternately is confirmed to be promising [8,14]. Such a strategy is called *variable space search* or *formulation space search*, which can be considered as a generalization of variable neighborhood search. One of the contributions of our paper is that it proposes a new search space that can be exploited to devise more sophisticated algorithms for GCP by using the idea of variable space search.

The rest of the paper is structured as follows. In Sect. 2, we give the formulation of 2VAP and show how to obtain a coloring from the given solution to 2VAP. Section 3 describes the heuristic algorithm VAP-COL. We provide our experimental results in Sect. 4. The concluding remarks are in Sect. 5.

2 The Vector Assignment Problem

2.1 Definition

Assume that the given graph $G = (V, E)$ has $n = |V|$ vertices and $m = |E|$ edges. We also assume that $V = \{1, 2, \ldots, n\}$.

The vector assignment problem (VAP) for G is the problem of assigning a unit vector $\boldsymbol{v}_i$ for each vertex $i \in V$ so that the maximum of the inner products $\boldsymbol{v}_i \cdot \boldsymbol{v}_j$ for all $(i, j) \in E$ is minimized. This problem is formalized as follows:

$$\begin{aligned} \text{minimize} \quad & \alpha \\ \text{subject to} \quad & \alpha \geq \boldsymbol{v}_i \cdot \boldsymbol{v}_j \quad \text{for all } (i,j) \in E, \\ & \|\boldsymbol{v}_i\|^2 = 1 \quad \text{for all } i \in V. \end{aligned} \tag{1}$$

If all the vectors are restricted to be d-dimensional, we call the problem d-dimensional VAP (dVAP). We will consider 2VAP in the following, while Karger, Motwani, and Sudan used nVAP [10].

Because a 2-dimensional unit vector $\boldsymbol{v}$ is characterized by its angle θ as $\boldsymbol{v} = (\cos\theta, \sin\theta)$, we can rewrite 2VAP of formulation (1) as the following simpler form:

$$\begin{array}{lll} \text{maximize} & \beta & \\ \text{subject to} & \beta \leq \text{diff}(\theta_i, \theta_j) & \text{for all } (i,j) \in E, \\ & \theta_i \in [0, 2\pi) & \text{for all } i \in V, \end{array} \quad (2)$$

where $\text{diff}(\phi, \theta) = \min\{|\theta - \phi|, |\phi - \theta|\}$ denotes the difference (between 0 and π) of the angles ϕ and θ.

In the following, we will use the formulation (2) unless otherwise stated. We also assume that angles are in the range $[0, 2\pi)$.

2.2 Relationship with the Graph Coloring Problem

If the graph G is k-colorable, then the corresponding 2VAP defined by (2) has a feasible solution with objective value $2\pi/k$. To see this, let us fix a k-coloring $c\colon V \to \{1, \ldots, k\}$ of G. For $t = 1, \ldots, k$, let $\phi_t = 2t\pi/k$. Now we assign $\theta_i := \phi_{c(i)}$ for all $i \in V$. With this assignment, it is easy to see that $\text{diff}(\theta_i, \theta_j) \geq 2\pi/k$ holds for any adjacent pair $(i,j) \in E$ of vertices.

We then consider the opposite direction. That is, given a graph G and a solution $(\theta_1, \ldots, \theta_n)$ to (2) with objective value β, we show how to generate a valid coloring of G.

The first method, simple rounding (SR), is quite simple. Because we are given a solution $(\theta_1, \ldots, \theta_n)$ to (2), no vertices i and j are adjacent provided that $\text{diff}(\theta_i, \theta_j) < \beta$. In other words, we can give the same color to such vertices. Thus we can construct a k-coloring of G for $k = \lceil 2\pi/\beta \rceil$ by the following rule. Divide the unit circle into k equi-sized sectors $S_1, \ldots, S_k$, that is, $S_t = \{\theta \mid 2\pi(t-1)/k \leq \theta < 2\pi t/k\}$. Now we associate the color t to the sector S_t and give the color t to all vertices i with $\theta_i \in S_t$. This coloring is valid for the following reason. Assume that vertices i and j are adjacent. Their corresponding angles θ_i and θ_j should satisfy the inequality $\text{diff}(\theta_i, \theta_j) \geq \beta$. This implies that θ_i and θ_j are not in the same sector, because each sector has center angle $2\pi/k \leq \beta$. Thus, for any adjacent pair $(i,j) \in E$ of vertices, i and j have different colors.

We note that this method may produce a coloring with more colors than necessary. Suppose that the graph G is 3-partite. As we use a heuristic method to solve 2VAP, an obtained solution to (2) for such a graph can have $\beta = 2\pi/3 - \epsilon$ for a small ϵ. In this case, the above rounding method may require four colors. We will overcome this problem by a more sophisticated rounding method.

The second method, circular arc rounding (CAR), comes from the following observation. What we want to find is the division of unit circle into k sectors S_1, $\ldots$, S_k so that no two adjacent vertices i and j correspond to the angles θ_i and θ_j lying in the same sector.

To minimize the number of sectors, we further convert this problem. For two adjacent vertices i and j, we consider two counterclockwise circular arcs,

$\text{arc}(\theta_j, \theta_i)$ from θ_j to θ_i and $\text{arc}(\theta_i, \theta_j)$ from θ_i to θ_j, where $\text{arc}(\phi, \theta)$ is the set of angles defined by

$$\text{arc}(\phi, \theta) = \begin{cases} \{\psi \mid \phi < \psi < \theta\} & \theta \geq \phi, \\ \{\psi \mid \phi < \psi < 2\pi\} \cup \{\psi \mid 0 \leq \psi < \theta\} & \theta < \phi. \end{cases}$$

We assume that these arcs does not include their end points. Let $\mathcal{S}$ be the set of such arcs for all pairs of adjacent vertices. From the above observation, what we need is to find the set $\Phi = \{\phi_1, \dots, \phi_k\}$ of angles such that every arc in $\mathcal{S}$ includes at least one angle in Φ. Such a set is called a *piercing set* of $\mathcal{S}$, and the problem of finding the minimum cardinality piercing set is called the minimum piercing set problem for arcs. Katz, Nielsen and Segal [11] proposed an algorithm for the minimum piercing set problem, running in $O(s \log s)$ time for the given set of s arcs. Thus we can find an optimal coloring according to the solution $(\theta_1, \dots, \theta_n)$ as follows:

1. Construct the set of arcs $\mathcal{S}$.
2. Solve the minimum piercing set problem for $\mathcal{S}$. Let $\Phi = \{\phi_1, \dots, \phi_k\}$ be the minimum piercing set for $\mathcal{S}$, where we assume that $0 \leq \phi_1 < \phi_2 < \cdots < \phi_k < 2\pi$.
3. For each vertex $i \in V$, assign color t ($< k$) if $\theta_i \in \text{arc}(\phi_t, \phi_{t+1})$, and assign color k if $\theta_i \in \text{arc}(\phi_k, \phi_1)$.

It is obvious that no two adjacent vertices have the same color. The running time is $O(m \log m)$ because step 2 dominates the running time and $|\mathcal{S}| = 2m$ means that step 2 requires $O(m \log m)$ time.

We note that the running time can be reduced further for the following two reasons. One reason is that $\mathcal{S}$ has many unnecessary arcs. For example, assume that arc $\boldsymbol{a}$ completely contains another arc $\boldsymbol{b}$. In this case, any point in $\boldsymbol{b}$ is also in $\boldsymbol{a}$ and thus $\boldsymbol{a}$ is redundant. This means we need only two arcs for each vertex. Now the cardinality of $|\mathcal{S}|$ is reduced to $2n$. The other reason is that, the circular-arc graph $G_{\mathcal{S}}$ of $\mathcal{S}$ cannot be complete for $n \geq 4$. Therefore a clique cover $C_{\mathcal{S}}$ of $G_{\mathcal{S}}$ with s cliques provides a piercing set of $\mathcal{S}$ with s points, each corresponding to distinct clique. A minimum clique cover of circular-arc graph $G_{\mathcal{S}}$ can be found in linear time if the end points of all the arcs in $\mathcal{S}$ are sorted [9]. Thus we can accelerate the rounding method. The improved computation time is $O(m + n \log n)$.

3 A Heuristic Algorithm for the Graph Coloring Problem

We now propose a heuristic algorithm VAP-COL for GCP based on 2VAP. For a given graph G, this algorithm first solves 2VAP defined by (2) for G by a procedure described in the following and then rounds the feasible solution by the second rounding method in Sect. 2.2 to find a valid coloring for G.

We will use the difference $\text{cdiff}(\phi, \theta)$ of ϕ and θ for $0 \leq \theta$, $\phi < 2\pi$ defined as

$$\text{cdiff}(\phi, \theta) = \begin{cases} \theta - \phi & \text{if } \theta \geq \phi \\ 2\pi + \theta - \phi & \text{if } \theta < \phi. \end{cases}$$

That is, $x = \mathrm{cdiff}(\phi, \theta)$ means that the vector $(\cos\theta, \sin\theta)$ is obtained from the vector $(\cos\phi, \sin\phi)$ by the counterclockwise rotation with the angle x.

We now define the middle angle $\mathrm{mid}(\phi, \theta)$ of $\mathrm{arc}(\phi, \theta)$ by

$$\mathrm{mid}(\phi, \theta) = \begin{cases} (\phi + \theta)/2 & \phi \leq \theta \\ (\phi + \theta)/2 + \pi & \theta < \phi \text{ and } \theta + \phi < 2\pi \\ (\phi + \theta)/2 - \pi & \theta < \phi \text{ and } \theta + \phi \geq 2\pi \end{cases}$$

Our algorithm for 2VAP, which we call "threshold accepting vector assignment" (TAVA), is as follows:

1. Randomly initialize the angle $\theta_i \in [0, 2\pi)$ for each vertex $i \in V$.
2. Repeat l times:
3. For each vertex $i \in V$, do:
4. Let $\psi_1, \ldots, \psi_d$ be the vectors assigned to the vertices adjacent to i. We assume that these are sorted in the increasing order of $\mathrm{cdiff}(\theta_i, \psi_j)$. Currently $\theta_i \in \{\psi_d\} \cup \mathrm{arc}(\psi_d, \psi_1)$. Let us call $\mathrm{arc}(\psi_d, \psi_1)$ the current arc of θ_i.
5. Let $w = \mathrm{cdiff}(\psi_d, \psi_1)$ be the width of the current arc of θ_i.
6. Let $\mathrm{arc}(\psi_s, \psi_{s+1})$ be the widest arc among $\mathrm{arc}(\psi_j, \psi_{j+1})$ for $j = 1, \ldots, d-1$, and $w' = \mathrm{cdiff}(\psi_s, \psi_{s+1})$ be its width.
7. If $w' \geq (1 - f)w$ (f is a parameter) holds, then set $\boldsymbol{v}_i$ to $\mathrm{mid}(\psi_s, \psi_{s+1})$. Otherwise, set $\boldsymbol{v}_i$ to $\mathrm{mid}(\psi_d, \psi_1)$.
8. Output $\theta_1, \ldots, \theta_n$ as a feasible solution.

If the parameter $f = 0$, then this algorithm is the same as the hill-climbing heuristics, because at the line 7 the algorithm selects the widest range for θ_i in this case. The experiment in Sect. 4 shows that small positive f works better for random k partite graphs.

Note that we do not need to sort in line 4: for each vertex we can maintain the sorted order. We also note that $O(n + m)$ time is enough to check all the vertices in line 3.

For a graph G, the entire algorithm VAP-COL for GCP is described as follows:

1. Solve 2VAP for G by TAVA.
2. Apply CAR to the final solution $(\theta_1, \ldots, \theta_n)$ to find a coloring of G.

4 Experimental Results

We have conducted two numerical experiments to compare the performance of VAP-COL with others. In the first experiment, we use random k-partite graphs to observe the influence of the relaxation factor to the number of colors used. We also compare this result with other heuristic algorithms. In the second experiment, we use some graphs from the Second DIMACS Challenge.

We used the following heuristic algorithms for comparison: a tabu search of Hertz and de Werra (HW) [7] and DSATUR [2]. There are two reasons for this choice: both are simple heuristic algorithms and the codes are publicly available[1]. For HW, tabu tenure is set to 20 and execution is terminated after 10,000 iterations.

The experiments are done on a PC with PentiumD 3.4GHz and 2GB of RAM.

4.1 Experiment on Random k-Partite Graphs

We generate random k-partite graphs for $k = 5$, 9 as follows: Let V_i, $i = 1, \ldots, k$, be sets of 100 vertices each. Thus the entire graph has $100k$ vertices. For any pair of vertices $v \in V_i$ and $u \in V_j$ with $i \neq j$, we give edge (v, u) with probability p, which specifies the edge density. For the values of p, we consider $p = 0.05, 0.10, \ldots, 0.95$. For each k and probability p, we generate ten graphs.

In order to observe the influence of the parameter f, we run VAP-COL with $f = 0.00$, $f = 0.15$, $f = 0.35$ and $f = 0.50$ on each of the graph. The algorithm stops after $l = 1000$ iterations.

The results, the average number of colors used, are depicted in Figs 1 and 2. It seems that the factor of 0.35 or more is too big; in such cases, the performance is worse than the simple hill-climbing for most values of the edge density. The setting of $f = 0.00$ is a good choice for dense graphs, but it performs poorly for sparse graphs. For both $k = 5$ and 9, the setting $f = 0.15$ gives fairly good performance for all edge probabilities and thus we use $f = 0.15$ in the rest of the experiments.

The results of VAP-COL for $f = 0.00$ and $f = 0.15$, with the results of DSATUR and HW are in Figs 3 and 4. HW showed best performance for sparse graphs with $k = 5$, but for other graphs VAP-COL with $f = 0.15$ works fine.

4.2 Experiment on DIMACS Challenge Benchmarks

In Experiment 2, we select some benchmark graphs from the Second DIMACS Challenge to compare the performance of VAP-COL with other heuristics.

Table 1 shows the experimental results. In each entry, the minimum and the maximum numbers of colors used in ten independent runs are shown. The columns $f = 0.15$ and $f = 0.00$ are for VAP-COL with the corresponding factors. Entries in other columns are from the literatures.

From the Table 1, we can observe the effectiveness of allowing moves to non-improving solutions; compared to the case with $f = 0$, the setting of $f = 0.15$ improves the performance of VAP-COL. Moreover, VAP-COL with $f = 0.15$ obtains solutions of competitive (or sometimes better) quality compared to other heuristics. However, VAP-COL requires more computation time than HW and DSATUR. For example, on `flat1000_60_0`, HW and DSATUR uses one or two seconds, while VAP-COL needs about eighty seconds. Improving the speed of VAP-COL is one of the important directions of our future research.

[1] The program codes of HW and DSATUR are obtained from Culberson's Web site, `http://www.cs.ualberta.ca/~joe/Coloring/Colorsrc/color.tar.gz`

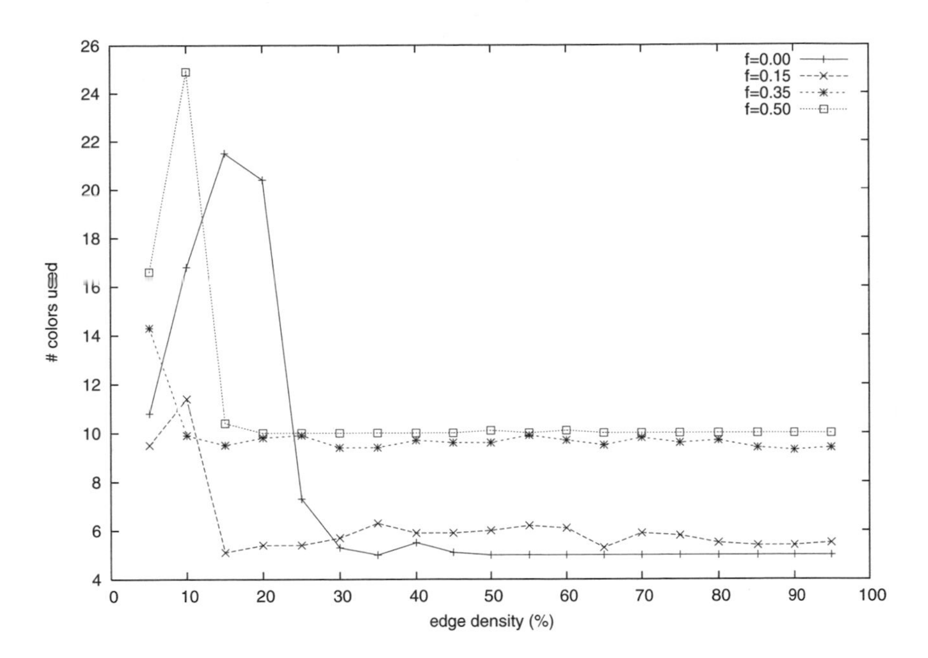

Fig. 1. Resuls for 5-partite graphs with various f

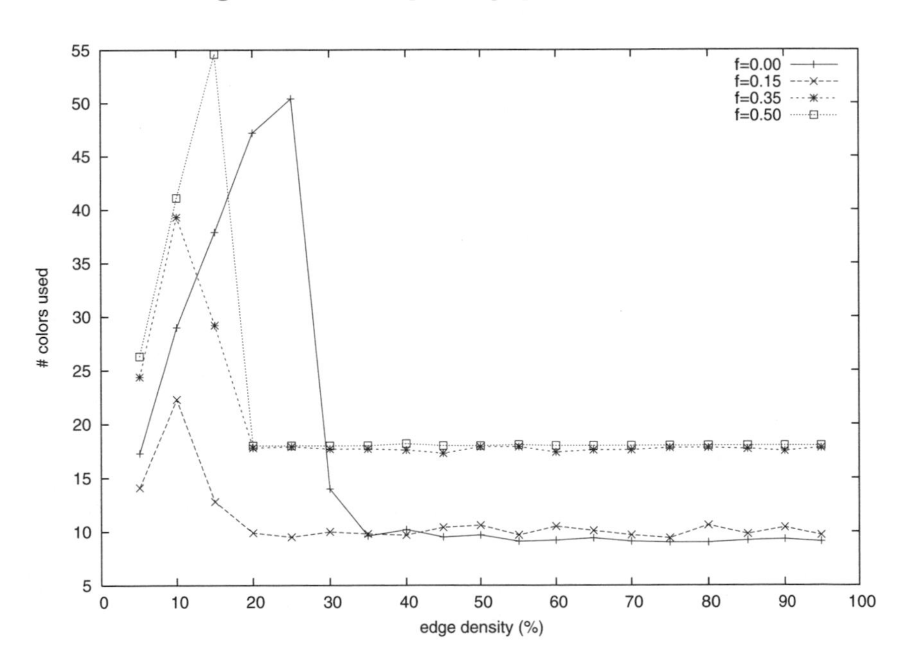

Fig. 2. Resuls for 9-partite graphs with various f

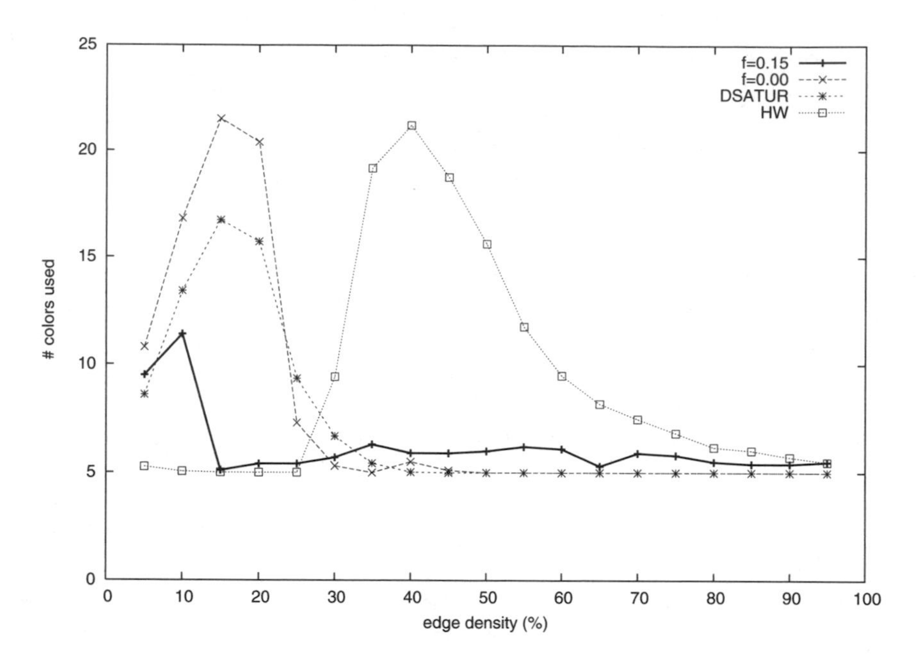

Fig. 3. Results for 5-partite graphs with other heuristics

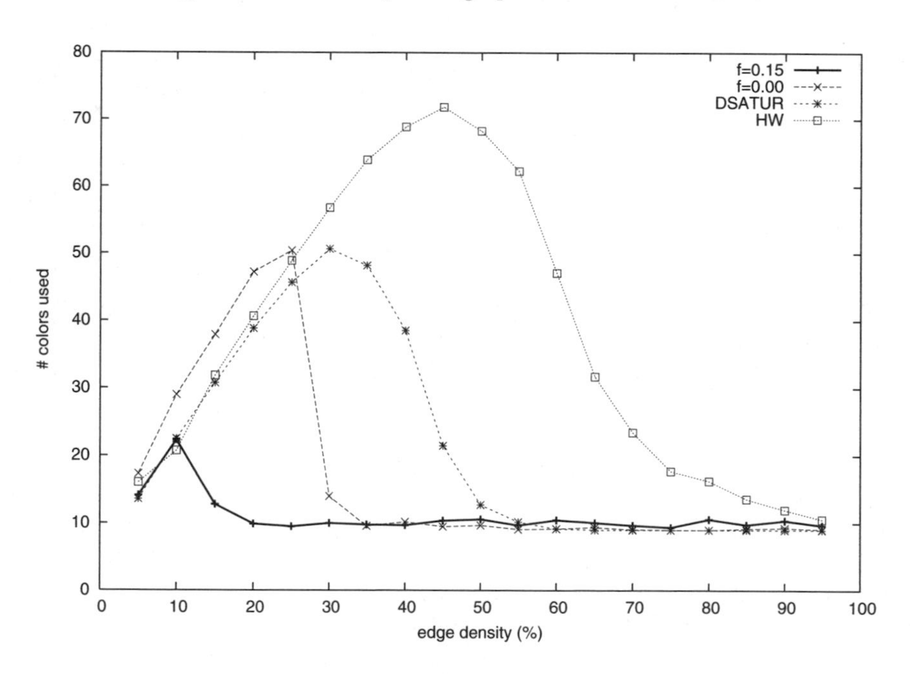

Fig. 4. Results for 9-partite graphs with other heuristics

Table 1. Results on the DIMACS benchmark instances

Instance	$\|V\|$	$\|E\|$	$f = 0.15$	$f = 0.00$	HW[7]	DSATUR[2]
DSJC250.5	250	31 336	34–37	39–46	41–43	37–39
DSJC500.5	500	125 248	59–63	70–80	71–74	65–68
DSJC1000.5	1 000	499 652	106–109	122–136	125–129	114–119
flat300_20_0	300	21 375	38–41	43–52	45–48	41–43
flat300_26_0	300	21 633	39–41	43–52	45–48	41–43
flat300_28_0	300	21 695	38–41	43–51	45–48	41–45
flat1000_50_0	1 000	245 000	104–107	121–133	124–127	114–117
flat1000_60_0	1 000	245 830	105–109	120–133	122–126	114–117
latin_square_10	900	307 350	123–126	137–156	146–156	126–134
le450_15a	450	8 168	20–21	21–25	17–20	16–18
le450_15b	450	8 169	19–22	21–25	17–20	16–17
le450_15c	450	16 680	25–28	31–36	29–30	24–25
le450_15d	450	16 750	27–28	31–36	28–31	24–26
mulsol.i.1	197	3 925	49–50	49–49	49–49	49–49
school1	385	19 095	15–30	29–45	14–15	14–29
school1_nsh	352	14 612	18–29	34–45	14–15	14–25

5 Conclusion

In this paper, we proposed the problem 2VAP and a new heuristic algorithm VAP-COL based on the fact that GCP can be reduced to 2VAP and the transformation is efficient. Thus 2VAP provides a new search space for the graph coloring problem. The experimental results show that VAP-COL works well for random k-partite graphs with relatively low degree. On the DIMACS benchmark instances, VAP-COL obtains solutions of competitive quality with DSATUR and tabu search though it requires more computation time. An important observation is that the graph types for which an algorithm performs better than others are quite different if the algorithm is different. This indicates that it is worth trying to incorporate the new search space proposed in this paper to other metaheuristic algorithms, e.g., by using the idea of variable space search.

References

1. Avanthay, C., Hertz, A., Zufferey, N.: A variable neighborhood search for graph coloring. European Journal of Operational Research 151, 379–388 (2003)
2. Brélaz, D.: New methods to color vertices of a graph. Communications of the ACM 22, 251–256 (1979)
3. Burer, S., Monteiro, R.D.C., Zhang, Y.: Rank-two relaxation heuristics for max-cut and other binary quadratic programs. SIAM Journal on Optimization 12(2), 503–521 (2002)
4. Caramia, M., Dell'Olmo, P.: Bounding vertex coloring by trancated *Multistage* branch and bound. Networks 44(4), 231–242 (2004)
5. Dueck, G., Scheuer, T.: Threshold accepting: A generel purpose optimization algorithm appearing superior to simulated annealing. Journal of Computational Physics 90, 161–175 (1990)

6. Eiben, A., van der Hauw, J.: Grpah coloring with adaptive genetic algorithms. Technical Report TR96-11, Leiden University (August 1996)
7. Hertz, A., de Werra, D.: Using tabu search techniques for graph coloring. Computing 39, 345–351 (1987)
8. Hertz, A., Plumettaz, M., Zufferey, N.: Variable space search for graph coloring, Working Paper (2007)
9. Hsu, W.L., Tsai, K.H.: Linear time algorithms on circular-arc graphs. Information Processing Letters 40, 123–129 (1991)
10. Karger, D., Motwani, R., Sudan, M.: Approximate graph coloring by semidefinite programming. Journal of the ACM 45(2), 246–265, 3 (1998)
11. Katz, M.J., Nielsen, F., Segal, M.: Maintenance of a piercing set for intervals with applications. Algorithmica 36(1), 59–73 (2003)
12. Kochenberger, G.A., Glover, F., Alidaee, B., Rego, C.: An unconstrained quadratic binary programming approach to the vertex coloring problem. Annals of Operations Research 139(1), 229–241 (2005)
13. Laguna, M., Martí, R.: A GRASP for coloring sparse graphs. Computational Optimization and Applications 19(2), 165–178 (2001)
14. Mladenović, N., Plastria, F., Urošević, D.: Formulation space search for circle packing problems. In: Stützle, T., Birattari, M., Hoos, H.H. (eds.) SLS 2007. LNCS, vol. 4638, pp. 212–216. Springer, Heidelberg (2007)

Rule Extraction from Neural Networks Via Ant Colony Algorithm for Data Mining Applications

Lale Özbakır[1], Adil Baykasoğlu[2], and Sinem Kulluk[1]

[1] Erciyes University, Faculty of Engineering, Department of Industrial Engineering, Kayseri, Turkey
[2] University of Gaziantep, Faculty of Engineering, Department of Industrial Engineering, Gaziantep, Turkey
lozbakir@erciyes.edu.tr, baykasoglu@gantep.edu.tr, skulluk@erciyes.edu.tr

Abstract. A common problem in Data Mining (DM) is the presence of noise in the data being mined. Artificial neural networks (ANN) are robust and have a good tolerance to noise, which makes them suitable for mining very noisy data. Although they may achieve high classification accuracy, they have the well-known disadvantage of having black-box nature and not discovering any high-level rule that can be used as a support for human understanding. The main challenge in using ANN in DM applications is to get explicit knowledge from these models. For this purpose, a study on knowledge acquirement from trained ANNs for classification problems is presented. The proposed method uses Touring Ant Colony Optimization (TACO) algorithm for extracting accurate and comprehensible rules from databases via trained artificial neural networks. The suggested algorithm is experimentally evaluated on different benchmark data sets. Results show that the proposed approach has a potential to generate accurate and concise rules.

Keywords: Artificial Neural Networks, Ant Colony Optimization, Rule Extraction.

1 Introduction

There has been a great interest in the area of data mining in which the general goal is to discover knowledge that is not only correct, but also comprehensible and interesting for the user. Data mining has been defined as the nontrivial extraction of implicit, previously unknown and potentially useful information from data [9]. DM encompasses a number of different technical approaches such as clustering, data summarization, learning classification rules, finding dependency networks and detecting anomalies [3].

Classification rule discovery is characterized by a concern for finding highly predictive rules, often by using heuristic techniques. Classification is the process of finding a set of models (or functions) which describe and distinguish data classes or concepts, for the purpose of being able to use the model to predict the class of objects whose class label is unknown [10]. The derived model is based on the analysis of a set

V. Maniezzo, R. Battiti, and J.-P. Watson (Eds.): LION 2007 II, LNCS 5313, pp. 177–191, 2008.

of training data. A rule generally represents a decision in the form of "**IF** ... **THEN** ..." proposition. The main goal of rule extraction is to discover hidden knowledge and explain it understandably, to extract previously unknown relations and to ensure reasoning and defining capability.

ANN is one of the most widely used techniques in classification. An ANN is a mathematical or computational model based on biological neural networks. It consists of an interconnected group of artificial neurons and processes information using a connectionist approach to computation. In most cases an ANN is an adaptive system that changes its structure based on external or internal information that flows through the network during the learning process. ANN can be used to model complex relationships between inputs and outputs or to find patterns in the data.

The NN method is highly accurate in classification and prediction of output. However classification and function approximation concepts of the NN are usually incomprehensible to the human user. This is because typical NN solutions consist of a large number of interacting non-linear elements, characterized by large sets of real-valued parameters that are hard to interpret. Distributed internal representations make it even harder to understand what exactly a network has learned and where it will fail to generate the correct answer [15]. Because of that, many researchers tend to develop new algorithms for rule extraction from NNs.

Hruschka and Ebecken [12] suggested a clustering-based approach for extracting rules from multilayer perceptrons in classification problems. Their rule extraction algorithm basically consists of two steps. First, a clustering genetic algorithm is applied to find clusters of hidden unit activation values. Then classification rules describing these clusters, in relation to the inputs, are generated.

Tokinaga et al [21] concentrated on the use of NN rule extraction techniques based on genetic programming (GP) to build intelligent and explanatory evaluation systems. They utilized GP to automate the rule extraction process in the trained neural networks where the statements changed into a binary classification.

Markowska-Kaczmar [16] described the experimental study of the influence of parameters on the final results of the rule extraction method from NN for classification problems. The method is based on evolutionary approach, where for each class evolves separate population. He also presented a method for rule extraction from a NN based on the genetic approach with Pareto optimization [17]. They described the idea of Pareto optimization and shown the details of the developed method.

Elalfi et al [7] presented a new algorithm for extracting accurate and comprehensible rules from databases via trained NN using a genetic algorithm (GA). Their algorithm does not depend on the NN training algorithms and does not modify the training results. The GA is used to find the optimal values of input attributes, which maximize the output function of output nodes. They decoded the optimal chromosome and used to get a rule which belongs to target class.

Santos et al [18] presented a method for extracting accurate, comprehensible rules from NNs. They proposed a method which uses GA to find a good NN topology. This topology is then passed to a rule extraction algorithm, and the quality of the extracted rules is then fed back to the GA.

Arbatlı and Akın [2] proposed an algorithm for rule extraction from trained NNs using GAs. The idea behind their approach is to use GA to optimize a NN topology and then extract relevant inputs from the input domain using the optimized topology, and finally to extract conjunctive rules using these relevant inputs.

Despite the successful applications of Ant Colony Optimization (ACO) algorithm in many optimization problems, it's observed that there is an absence of ACO application on rule extraction from trained NN. In this study, a Touring-ACO algorithm is developed and implemented for extracting rules from NNs. The TACO algorithm is used to find the optimal values of input attributes which maximize the output function of NN. Experiments on test data sets show that the proposed method is able to produce accurate and effective results.

2 Classification Rule Extraction from Trained NN Via TACO

NN is a powerful data modeling tool that is able to capture and represent complex input/output relationships. NNs resemble the human brain in two aspects; firstly a NN acquires knowledge through learning and secondly a NN's knowledge is stored within inter-neuron connection strengths known as synaptic weights.

The knowledge acquired by a NN is codified on its connection weights, which in turn are associated to both its architecture and activation functions [1]. In this context, the process of knowledge acquisition from NNs usually implies the use of algorithms based on the values of either connection weights or hidden unit activations. The algorithms designed to perform such task are generally called algorithms for rule extraction from NNs [12].

In this study, multi-layer perceptron (MLP) which is one of the most widely used NN is considered. MLP is especially useful for special problems such as classification, recognition and generalization. Multi-layer perceptron uses *sigmoid* or *tanh* functions in general. In this study L-36 Taguchi Design is carried out for determining effective structural parameters of MLP and improving convergence rates for all datasets.

Elements of the data set must be decoded in binary form in order to extract classification rules from the trained NN via the proposed TACO algorithm. A sample binary coding, with six input attributes and two output classes for "monks data set" is shown in Table 1. Attributes must be discretized before transformed into binary form for data sets which contain continuous attributes.

NN is trained on the encoded vectors of the input attributes and the corresponding vectors of the output classes until the convergence rate between actual and the desired output is achieved. The general methodology of rule extraction from NNs by TACO is shown in Figure 1.

Table 1. Coding the database in the binary format (R: rectangular, S: square, O: octagon, Y: yes, N: no, Sw: sword, F: flag, B: balloon, Re: red, Y: yellow, G: green, Bl: blue)

Head Shape			*Body Shape*			*Is smiling*		*Holding*			*Jacket Color*				*Has Tie*		*Target Class*	
R	S	O	R	S	O	Y	N	Sw	F	B	Re	Y	G	Bl	Y	N	Y	N
x_1	x_2	x_3	x_4	x_5	x_6	x_7	x_8	x_9	x_{10}	x_{11}	x_{12}	x_{13}	x_{14}	x_{15}	x_{16}	x_{17}	c_1	c_2
1	0	0	1	0	0	1	0	0	1	0	0	0	1	0	1	0	1	0

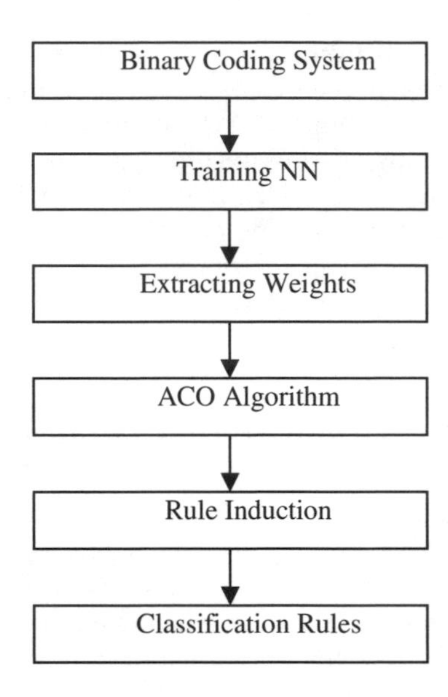

Fig. 1. General flowchart for the proposed methodology

After training the MLP, by using sigmoid activation function and one hidden layer as an example, two groups of weights can be obtained. The first group (WG1) consists of the weights from input layer to hidden layer and the second group (WG2) consists of the weights from hidden layer to output layer. Using these weights, final value of the kth output node, χ_k is given by:

$$\chi_k = \frac{1}{1 + e^{-\left[\sum_{j=1}^{J} (WG\,2)_{j,k} (1/1+e^{-\left[\sum_{i=1}^{I} x_i (WG\,1)_{i,j}\right]}\right]}} \tag{1}$$

The function χ_k is an exponential function and its maximum output value is equal to 1. Finally, for extracting rules between input attributes and related classes, it is necessary to find the input vector, which maximizes the function χ_k. This is an optimization problem and can be stated as:

$$\text{Maximizc} \quad \chi_k(x_i) = \frac{1}{1 + e^{-\left[\sum_{j=1}^{J} (WG\,2)_{j,k} (1/1+e^{-\left[\sum_{i=1}^{I} x_i (WG\,1)_{i,j}\right]}\right]}} \tag{2}$$

$$x_i = 0 \ or \ 1$$

Objective function χ_k is nonlinear and decision variables are binary. So this is a nonlinear integer programming problem and ACO algorithm can be used to solve it.

ACO algorithm is a meta-heuristic algorithm which was developed by simulating the behavior of real ants [6]. In this study, a version of Touring ACO algorithm which was first designed by Hiroyosu et al [11] for handling continuous variables in optimization problems is employed. In TACO algorithm, each solution is represented by a vector of parameters of which each is coded with a string of binary bits. At the decision stage for the value of a bit, ants use only the pheromone information. Once a solution is produced and quality value of the solution is calculated, an artificial pheromone to be attached to the sub-paths forming the solution is computed. After all ants in the colony have produced their solutions and the pheromone amount belonging to each solution has been calculated, the pheromones of sub-paths between the bits are updated [14].

As an example when calculating the probability of being preferred of a sub-path between 0 and 1 $(0 \rightarrow 1)$ the following equation is used:

$$p_{01} = \frac{\tau_{01}}{\tau_{01} + \tau_{00}} \tag{3}$$

Where; p_{01} is the probability associated with the sub-path $(0 \rightarrow 1)$, and τ_{00} and τ_{01} are the artificial pheromones of the sub-paths $(0 \rightarrow 0)$ and $(0 \rightarrow 1)$. Artificial pheromone is computed by using equation 4:

$$\Delta \tau_{01}^{k}(t, t+1) = \begin{cases} Q * F_k & \textit{if the ant } k \textit{ passes the subpath } (0 \rightarrow 1) \\ 0 & \textit{otherwise} \end{cases} \tag{4}$$

where $\Delta\tau_{01}^{k}$ is the pheromone attached to the sub-path $(0 \rightarrow 1)$ by the artificial ant k, F_k is the objective function value which is calculated by using the solution found by the ant k and Q is a positive constant. After N ants in the colony complete the search process and produce their solutions, the pheromone amount to be attached to the sub-path $(0 \rightarrow 1)$ between time t and $(t+1)$ is computed by using equation 5:

$$\Delta \tau_{01}(t, t+1) = \sum_{k=1}^{N} \Delta \tau_{01}^{k}(t, t+1) \tag{5}$$

The pheromone amount of the same sub-path at the time $(t+1)$ is updated by using equation 6:

$$\tau_{01}(t+1) = \rho\tau_{01}(t) + \Delta\tau_{01}(t, t+1) \tag{6}$$

Here, ρ is a coefficient called the evaporation rate of which the value is in between 0 and 1 $(0 \le \rho < 1)$.

The basic TACO algorithm faces with the premature convergence problem. This is due to the fact that it employs a pure pheromone-based direction selection strategy. In order to overcome this problem, the memory feature of Tabu Search (TS) algorithm can be incorporated into the TACO algorithm [13, 14].

The main feature of TS algorithm is that it has an explicit memory. The memory stores information about the past steps of search and new moves are produced in a certain neighborhood according to this memory. In TACO, the frequency based memory stores information about how often a sub-path is followed by ants. If ants sometimes choose their directions depending on this principle, they can follow different paths and get out of a local optimum, and hence be able to find the global optimum. The probability based on the frequency memory is calculated by using equation 7:

$$p_{01}(t) = \begin{cases} 1 & if \;\; (f * f_{01} < f_{00}), \\ \dfrac{\tau_{01}}{\tau_{01} + \tau_{00}} & otherwise \;\; , \end{cases} \tag{7}$$

Here, f is a frequency factor ($(f \geq 1)$. If the condition $(f * f_{01} < f_{00})$ is satisfied, then the path $(0 \rightarrow 1)$ is directly chosen; else the pheromone-based direction selection strategy is employed.

In the proposed TACO algorithm, solution strings are produced according to the χ_k quality function and rules whose quality values are bigger than a user specified threshold value are stored for rule induction procedure. In rule induction process, rules are selected based on the fitness values of the solution strings.

When a rule is used to classify a given training instance, one of the four possible concepts can be observed [19]: true positive (*tp*), false positive (*fp*), true negative (*tn*) and false negative (*fn*). The true positive and true negative are correct classifications, while false positive and false negative are incorrect classifications.

- True positive (*tp*): The rule predicts that the class is yes (positive) and the class of given instance is indeed yes (positive);
- True negative (*tn*): The rule predicts that the class is no (negative) and the class of given instance is indeed no (negative);
- False positive (*fp*): The rule predicts that the class is yes (positive) but the class of the given instance is no (negative);
- False negative (*fn*): The rule predicts that the class is no (negative) but the class of the given instance is yes (positive).

Using these concepts, the fitness function is defined as follows [3];

$$fitness = \frac{tp}{tp + fn} * \frac{tn}{tn + fp} \tag{8}$$

The main steps of the TACO algorithm which is used to find the best solution string by maximizing the χ_k is shown in Figure 2.

Step 1: Data coding in binary format
Step 2: Train the NN with binary inputs and outputs
Step 3: Extract weight groups from the NN.
Step 4: Initialize parameters, pheromone and frequency amounts. Set χ_k as quality function.
Step 5: Set fitness function as in (8).
Step 6: Repeat the following steps until stopping criteria is satisfied.
Step 6.1: Generate artificial ways for all artificial ants according to, pheromone and frequency amount

- Based on the probability value given in equation (7), choose a path for moving
- Repeat until all ants completed their ways.

Step 6.2: Compute the χ_k quality and fitness values for all ants.

Step 6.3: Update the pheromone and frequency attached on the ways based on χ_k, according to following formulas:

$$\Delta \tau_{ij}^{k}(t,t+1) = \begin{cases} Q * \chi_k & if \quad path \quad i-j \; is \; used \\ 0 & otherwise \end{cases}$$

$$\Delta \tau_{ij}(t,t+1) = \sum_{k=1}^{M} \Delta \tau_{ij}^{k}(t,t+1) \qquad \tau_{ij}(t+1) = \rho\tau_{ij}(t) + \Delta\tau_{ij}(t,t+1)$$

$$\Delta F_{ij}^{k}(t,t+1) = \begin{cases} 1 & if \quad path \quad i-j \; is \; used \\ 0 & otherwise \end{cases}$$

$$\Delta F_{ij}(t,t+1) = \sum_{k=1}^{M} \Delta F_{ij}^{k}(t,t+1) \qquad F_{ij}(t+1) = F_{ij}(t) + \Delta F_{ij}(t,t+1)$$

Step 7: Keep the ways (solutions) whose χ_k values are bigger than a user specified threshold value
Step 8: Rule induction: Initialize the rule list as empty. Do the following steps until a user specified training accuracy.
Step 8.1: From the solutions produced in step 7 find the solution whose fitness is the biggest.
Step 8.2: Remove the satisfied cases from the training data
Step 8.3: Keep the solution as a rule and add it to the rule list.
Step 9: Apply the rules to the testing data and find the accuracy of the rule set.
Step 10: Transform the solutions into linguistic rules.

Fig. 2. The main steps of the proposed algorithm

Table 2. Rule refinement and transformation to linguistic rule

Head Shape			*Body Shape*			*Is smiling*		*Holding*			*Jacket Color*				*Has Tie*		*Target Class*	
R	S	O	R	S	O	Y	N	Sw	F	B	Re	Y	G	Bl	Y	N	Y	N
x_1	x_2	x_3	x_4	x_5	x_6	x_7	x_8	x_9	x_{10}	x_{11}	x_{12}	x_{13}	x_{14}	x_{15}	x_{16}	x_{17}	c_1	c_2
1	0	0	1	0	0	1	1	0	1	0	1	1	1	1	0	0	1	0
Extracted Rule: If head shape is round **and** body shape is round **and** is smiling **or** not smiling **and** holding flag **and** jacket color is red **or** yellow **or** green **or** blue **then** robot is in target class																		
Refined Rule: If head shape is round **and** body shape is round **and** holding flag **then** robot is in target class.																		

The solution strings extracted by the proposed TACO algorithm are refined and transformed into linguistic rules. A sample transformation of rules and the refinement procedure is shown in Table 2.

3 Evaluation of the Proposed Algorithm

The proposed rule extraction algorithm is tested on five data sets namely Monks-2, Tic-Tac-Toe, Iris, Crx and Nursery from the UCI (University of California at Irvine) Machine Learning Repository, *http://mlearn.ics.edu/MLRepository.html.* Monks-2, Tic-Tac-Toe and Crx data sets have binary classes and Iris and Nursery data sets are n-ary classification problems.

The main characteristics of the data sets are summarized in Table 3. Table 3 shows the data set name, number of cases, number of categorical attributes, number of continuous attributes and number of classes in the data set.

Table 3. Main characteristics of the data sets

Data set	# Cases	# Categorical attributes	# Continuous attributes	Missing attribute values	# Classes
Monks-2	432	6	-	No	2
Tic-Tac-Toe	958	9	-	No	2
Iris	150	-	4	No	3
Crx	690	9	6	Yes	2
Nursery	12.960	8	-	No	5

The proposed rule extraction algorithm requires that data being mined have binary values. However, three of the data sets used in our experiments have categorical attributes, one of them has continuous attributes and the other one has categorical and continuous attributes. Categorical attributes are directly transformed into binary attributes as shown in Table 1. Continuous attributes in Iris and Crx data sets are primarily discretized by using Fayyad and Irani's [8] entropy based discretization method (MDL) and then transformed into binary attributes as a pre-processing step before evolving the neural network.

After coding the data sets in the binary form, related binary input attributes and output classes are obtained. Table 4 shows the attributes and corresponding binary variables of the data sets.

In Iris, Tic-Tac-Toe, Crx and Nursery data sets predictive accuracy is measured by well-known methodology of cross validation, with a cross-validation factor of ten. In other words, each data set is partitioned into ten data subsets and the rule extraction algorithm is run ten times. In each run a distinct data subset is used as the test set and the remaining nine partitions are used as the training set. The maximum, minimum and average predictive accuracies on the test set of the ten runs are presented. Also standard deviations of the corresponding predictive accuracies are calculated. Because of Monks-2 data set is originally partitioned into single train and test set, ten-fold cross validation is not applied to this data set.

Table 4. Attributes and corresponding binary variables of data sets

Binary variables (17)	Monks-2 variables
x_1, x_2, x_3	Head shape: X_1 (1, 2, 3) {round, square, octagon}
x_4, x_5, x_6	Body shape: X_2 (1, 2, 3) {round, square, octagon}
x_7, x_8	Is smiling : X_3 (1, 2) {yes, no}
x_9, x_{10}, x_{11}	Holding: X_4 (1, 2, 3) {sword, balloon, flag}
x_{12}, x_{13}, x_{14}, x_{15}	Jacket color: X_5 (1, 2, 3, 4) {red, yellow, green, blue}
x_{16}, x_{17}	Has tie: X_6 (1, 2) {yes,no}

Binary variables (54)	Crx variables
x_1, x_2	A1: X_1 (1, 2) {a,b}
x_3, x_4	A2: X_2 (1, 2) {(-; 38.96], (38.96; -)}
x_5, x_6	A3: X_3 (1, 2) {(-; 4.2075], (4.2075; -)}
x_7, x_8, x_9, x_{10}	A4: X_4 (1, 2, 3, 4) {u, y, l, t}
x_{11}, x_{12}, x_{13}	A5: X_5 (1, 2, 3) {g, p, gg}
x_{14},x_{15},…., x_{26},x_{27}	A6: X_6 (1,2,3,….,13,14) {w,q,m,r,cc,k,c,d,x,I,e,aa,ff,j}
x_{28},x_{29},…..,x_{35},x_{36}	A7: X_7 (1,2,3,4,5,6,7,8,9) {v,h,bb,ff,j,z,o,dd,n}
x_{37}, x_{38}	A8: X_8 (1, 2) {(-; 1.02] , (1.02; -)}
x_{39}, x_{40}	A9: X_9 (1,2) {t, f}
x_{41}, x_{42}	A10: X_{10} (1, 2) {t, f}
x_{43}, x_{44}, x_{45}	A11: X_{11} (1, 2, 3) {(-; 0.5], (0.5; 2.5], (2.5; -)}
x_{46}, x_{47}	A12: X_{12} (1, 2) {t, f}
x_{48}, x_{49}, x_{50}	A13: X_{13} (1, 2, 3) {g, s, p}
x_{51}, x_{52}	A14: X_{14} (1,2) {(-; 105], (105; -)}
x_{53}, x_{54}	A15: X_{15} (1,2) {(-; 492], (492, -)}

Binary variables (27)	Tic-Tac-Toe variables
x_1, x_2, x_3	Top-left-square: X_1 (1,2,3) {x,o,b}
x_4, x_5, x_6	Top-middle-square: X_2 (1,2,3) {x,o,b}
x_7, x_8, x_9	Top-right-square: X_3 (1,2,3) {x,o,b}
x_{10}, x_{11}, x_{12}	Middle-left-square: X_4 (1,2,3) {x,o,b}
x_{13}, x_{14}, x_{15}	Middle-middle-square: X_5 (1,2,3) {x,o,b}
x_{16}, x_{17}, x_{18}	Middle-right-square: X_6 (1,2,3) {x,o,b}
x_{19}, x_{20}, x_{21}	Bottom-left-square: X_7 (1,2,3) {x,o,b}
x_{22}, x_{23}, x_{24}	Bottom-middle-square:X_8 (1,2,3) {x,o,b}
x_{25}, x_{26}, x_{27}	Bottom-right-square: X_9 (1,2,3) {x,o,b}

Binary variables (12)	Iris variables
x_1, x_2, x_3	Sepal length: X_1 (1, 2, 3) {(-;5.55],(5.55;6.15],(6.15;-)}
x_4, x_5, x_6	Sepal width: X_2 (1, 2, 3) {(-;2.95],(2.95;3.35],(3.35;-)}
x_7, x_8, x_9	Petal length: X_3 (1, 2, 3) {(-;2.45],(2.45;4.75],(4.75;-)}
x_{10}, x_{11}, x_{12}	Petal width: X_4 (1, 2, 3) {(-;0.8],(0.8;1.75],(1.75;-)}

Binary variables (27)	Nursery variables
x_1, x_2, x_3	parents: X_1 (1,2,3) {usual, pretentious, great_pret}
x_4,x_5,x_6,x_7,x_8	has_nurs: X_2 (1,2,3,4,5) {prop,less_prop,improp,crit,very_crit}
x_9, x_{10}, x_{11},x_{12}	form: X_3 (1,2,3,4) {complete,completed,incomplete,foster}
x_{13},x_{14},x_{15},x_{16}	children: X_4 (1,2,3,4) {1, 2, 3, more}
x_{17}, x_{18}, x_{19}	housing: X_5 (1,2,3) {convenient, less_conv, critical}
x_{20}, x_{21}	finance: X_6 (1,2) {convenient, inconv}
x_{22}, x_{23}, x_{24}	social: X_7 (1,2,3) {non-prob,slightly-prob,problematic}
x_{25}, x_{26}, x_{27}	health:X_8 (1,2,3) {recommended,priority, not_recom}

In this study, L-36 (2**4 3*1) Taguchi Design is created using Minitab 14 statistical software for determining the best parameter settings of MLP. Taguchi Design factors and their levels are shown in Table 5.

Table 5. Taguchi Design factors and factor levels (*n*: number of input vectors)

Factor Name / Level Values	1	2	3
Number of hidden layer	1	2	
Processing elements in hidden layer(s)	n/3	(n/3) ± 4	
Transfer function	sigmoidaxon	tanhaxon	
Max epoch	10000	20000	
Learning rule	momentum	conjugategradient	quickprob

The NN is trained on the binary input attribute vectors and the corresponding output class vectors with different Taguchi design parameter levels by using NeuroSolutions 5 software for all data sets and mean square errors (MSE) of neural networks are achieved. Best parameter settings of MLPs are determined according to the MSEs (response factor). Due to the space limitations, only the best parameter setting for all data sets are included in Table 6.

Table 6. L-36 Taguchi Design analysis results

Data set	Number of hidden layer	Processing elements in hidden layer(s)	Transfer function	Max. epoch	Learning rule
Monks-2	1	8	tanhaxon	10000	conjugategradient
Tic-Tac-Toe	2	8	tanhaxon	20000	conjugategradient
Iris	2	9	sigmoidaxon	20000	momentum
Crx	2	18	sigmoidaxon	20000	conjugategradient
Nursery	2	13	sigmoidaxon	20000	conjugategradient

NNs are trained on the best parameter settings of Taguchi design for all data sets and weights are extracted. After getting the weights from trained NNs, TACO algorithm is applied to solve the equation χ_k . Table 7 shows the parameter setting of the proposed TACO algorithm.

Table 7. Parameter setting of TACO algorithm

No. of Ants (M)	*No. of Iterations (T)*	*Frequency Factor (f)*	*Evaporation Parameter (ρ)*	*Constant Q*
100	1000	2	0.8	5

Solutions whose χ_k values are bigger than 0.9 (predefined threshold value) are stored for rule induction procedure. Table 8 summarizes the 10-fold cross-validation results for each of the five data sets. Average, maximum and minimum predictive accuracies ((tp+tn)/(tp+tn+fp+fn)) on the test data sets, standard deviations and average number of rules are presented. Since Monks-2 dataset is divided into two parts, accuracies are obtained by evaluating ten different executions of the algorithm on the same training and test data sets.

Table 8. Predictive accuracies of proposed algorithm

Dataset	Max (%)	Average (%)	Min (%)	S. D. (%)	Average number of rules
Monks-2	100	98.99	94.67	1.74	7
Tic-Tac-Toe	100	99.37	97.92	0.73	16
Iris	100	98.67	93.33	2.81	3
Crx	98.55	92.61	88.41	3.83	11
Nursery	99.15	97.16	94.60	1.60	22

Table 9. The best rule sets, quality and fitness value of each rule

Data sets	Quality (χ_k)	Fitness	Best rule sets
Monk-2	0,9013	0,3192	If X_5=2 or 3 then class 1
	0,9121	0,3076	If X_1=2 or 3 and X_5=1 or 2 or 3 then class 1
	0,9554	0,3071	If X_3=1 and X_6=2 then class 1
	0,9184	0,2827	If X_4=3 then class 1
	0,9926	0,2833	If X_5=1 or 2 then class 1
	0,9822	0,2499	If X_1=1 then class 1
	0,9681	0,2499	If X_1=2 or 3 and X_6=1 then class 1
	0,9245	0,2430	If X_1=1 or 2 and X_6=2 then class 1
	0,9963	0,2188	If X_1=1 or 3 then class 1
			Else class 2.
Tic-Tac-Toe	0,9002	0,3439	If X_2=1 or 2 and X_5=2 or 3 and X_8=1 or 2 then class 2
	0,9003	0,3349	If X_1=1 or 2 and X_5=2 or 3 and X_8=1 or 2 then class 2
	0,9002	0,2906	If X_1=2 or 3 and X_2=1 or 3 and X_5=2 or 3 then class 2
	0,9000	0,2814	If X_3=2 or 3 and X_4=1 or 2 then class 2
	0,9002	0,2510	If X_1=1 or 2 and X_4=1 or 2 and X_7=2 then class 2
	0,9002	0,2384	If X_2=1 or 3 and X_5=2 and X_7=2 or 3 then class 2
	0,9002	0,2354	If X_2=1 or 2 and X_5=1 or 3 and X_8=1 or 3 and X_9=2 or 3 then class 2
	0,9001	0,1931	If X_1=2 or 3 and X_2=1 or 2 and X_5=2 or 3 and X_6=1 or 3 and X_8=1 or 3 then class 2
	0,9000	0,1910	If X_2=1 or 2 and X_4=1 or 2 and X_5=2 or 3 and X_7=1 or 2 and X_9=1 or 3 then class 2
	0,9000	0,1864	If X_2=1 or 2 and X_5=1 or 2 and X_6=1 and X_9=2 or 3 then class 2
	0,9003	0,1859	If X_1=1 or 2 and X_2=2 or 3 and X_4=1 or 3 and X_5=1 or 2 and X_8=1 or 2 then class 2
	0,9006	0,1821	If X_3=1 or 2 and X_5=1 or 2 and X_6=2 or 3 and X_7=2 or 3 then class 2
	0,9001	0,1684	If X_1=1 or 2 and X_2=2 or 3 and X_4=1 or 2 and X_9=1 or 3 then class 2
	0,9001	0,1551	If X_2=1 or 2 and X_5=1 or 2 and X_6=1 or 3 and X_7=2 or 3 then class 2
	0,9002	0,1490	If X_1=2 or 3 and X_2=2 or 3 and X_3=1 or 2 and X_6=1 or 3 and X_8=1 or 2 then class 2
	0,9004	0,1422	If X_1=2 and X_2=2 or 3 and X_9=1 or 3 then class 2
	0,9001	0,1380	If X_1=1 or 2 and X_2=2 or 3 and X_3=2 or 3 and X_4=1 or 3 and X_7=1 or 2 then class 2
	0,9002	0,1267	If X_1=2 or 3 and X_2=2 or 3 and X_3=1 or 2 and X_4=1 or 3 and X_5=1 or 2 and X_8=1 or 2 then class 2

Table 9. (*continued*)

	0,9000	0,1229	If X_1=2 or 3 and X_2=1 and X_3=2 or 3 and X_5=1 or 2 and X_6=1 or 2 and X_7=1 or 2 then class 2
	0,9002	0,0638	If X_1=1 or 3 and X_2=1 or 2 and X_3=2 or 3 and X_4=1 or 3 and X_6=2 then class 2
			Else class 1.
Iris	0,9074	1,0000	If X_3= 1 then class 1
	0,9104	0,9237	If X_3= 2 or 3 and X_4=2 then class 2
			Else class 3.
Crx	0,9999	0,5079	If X_4=1 or 2 and X_6=1 or 3 or 5 or 6 or 7 or 9 or 10 or 11 and X_7=1 or 2 or 5 or 6 or 8 and X_9=1 and X_{13}=1 or 3 then class 1
	0,9999	0,3307	If X_4=1 or 3 and X_6=1 or 2 or 3 or 6 or 7 or 8 or 11 or 12 or 14 and X_7=1 or 2 or 3 or 4 or 7 or 8 and X_{11}=2 or 3 and X_{13}=1 or 2 then class 1
	0,9999	0,3301	If X_6=2 or 5 or 6 or 7 or 9 or 10 or 11 or 12 and X_7=1 or 2 or 5 or 6 or 7 and X_{13}=1 or 3 then class 1
	0,9999	0,2727	If X_4=1 or 2 and X_6=2 or 5 or 7 or 8 or 9 or 11 and X_7=1 or 3 or 5 or 6 or 8 and X_{13}=1 or 3 then class 1
	0,9950	0,2499	If X_5=1 and X_6=1 or 2 or 4 or 5 or 6 or 7 or 9 or 10 or 11 or 12 or 14 and X_7=1 or 2 or 4 or 5 or 6 or 9 and X_{11}=1 or 3 and X_{15}=2 then class 1
	0,9900	0,2492	If X_4=1 or 3 and X_5=1 and X_6=1 or 2 or 4 or 5 or 6 or 7 or 9 or 10 or 11 or 13 or 14 and X_7=1 or 2 or 4 or 5 or 6 or 9 and X_{11}=1 or 3 and X_{13}=1 or 3 and X_{15}=2 then class 1
	0,9942	0,2377	If X_3= 2 and X4=1 or 2 and X_5=1 or 3 and X_6=1 or 2 or 5 or 6 or 7 or 10 or 11 or 13 and X_7=1 or 2 or 3 or 5 or 6 or 7 or 9 and X_8=2 and X_{13}=1 or 2 then class 1
	0,9995	0,2356	If X_3= 1 and X4=1 or 3 and X_5=1 and X_6=1 or 2 or 3 or 5 or 6 or 7 or 9 or 10 or 11 or 12 or 14 and X_7=1 or 2 or 3 or 7 and X_9=1 and X_{11}=1 or 3 and X_{13}=1 then class 1
	0,9719	0,2113	If X4=1 or 2 and X_5=1 or 2 and X_6=1 or 2 or 3 or 4 or 5 or 6 or 7 or 8 or 13 or 14 and X_7=1 or 2 or 3 or 8 and X_9=1 and X_{11}= 1 or 3 and X_{13}=1 or 2 and X_{15}=1 then class 1
			Else class 2.
Nursery	0,9474	0,4808	If X_2=2 or 3 or 4 or 5 and X_8=1 or 3 then class 1
	0,9509	0,4696	If X_1=1 or 2 and X_4=2 or 3 or 4 and X_8=1 or 2 then class 4
	0,9509	0,4546	If X_1=1 or 2 and X_2=1 or 2 or 3 or 5 and X_3=2 or 3 or 4 and X_8=1 or 2 then class 4
	0,9503	0,4015	If X_2=1 or 2 and X_3=1 or 2 and X_4=1 or 2 and X_7=1 or 2 then class 3
	0,9519	0,3950	If X_5=1 or 2 and X_7=1 or 2 and X_8=1 or 2 then class 4
	0,9508	0,3703	If X_2=1 or 2 or 3 or 4 and X_4=2 or 3 or 4 and X_7=2 or 3 and X_8=1 or 2 then class 4
	0,9508	0,3693	If X_2=1 or 2 or 3 or 5 and X_4=1 or 3 or 4 and X_5=1 or 3 and X_8=1 or 2 then class 4
	0,9521	0,3605	If X_2=2 or 3 or 4 and X_3=2 or 3 or 4 and X_8=1 or 2 then class 4
	0,9506	0,3597	If X_2=1 or 2 or 3 or 4 and X_4=2 or 3 or 4 and X_5=2 or 3 and X_8=1 or 2 then class 4
	0,9509	0,3590	If X_1=1 or 2 and X_2=2 or 3 or 4 and X_3=1 or 2 or 3 and X_8=1 or 2 then class 4
	0,9533	0,3508	If X_1=1 or 2 and X_2=1 or 2 or 3 or 5 and X_3=1 or 2 or 3 and X_4=1 or 2 and X_7=2 and X_8=1 then class 3
	0,9338	0,3505	If X_2=1 or 2 or 3 or 5 and X_3=1 or 2 or 4 and X_4=1 or 2 or 3 and X_8=1 or 3 then class 1

Table 9. *(continued)*

0,9499	0,3412	If X_2=1 or 2 or 5 and X_3=1 or 3 or 4 and X_4=1 or 2 or 4 and X_8=3 then class 1
0,9503	0,3388	If X_2=1 or 2 or 5 and X_3=1 or 2 or 4 and X_4=1 or 2 or 4 and X_8=3 then class 1
0,9317	0,3299	If X_3=1 or 3 or 4 and X_4=1 or 2 or 4 and X_5=1 or 2 X_8=1 or 2 then class 4
0,9506	0,3188	If X_2=1 or 2 or 3 or 4 and X_3=1 or 3 or 4 and X_4=1 or 3 or 4 and X_5=1 or 3 and X_8=1 or 2 then class 4
0,9345	0,3104	If X_1=1 and X_2=1 or 2 or 3 and X_3=1 or 2 and X_4=1 or 2 and X_7=1 or 2 then class 3
0,9502	0,2694	If X_1=2 or 3 and X_2=1 or 3 or 4 or 5 and X_4=2 or 3 and X_8=3 then class 1
0,9312	0,2548	If X_2=1 or 2 or 3 or 5 and X_3=1 or 2 or 3 and X_4=2 or 3 X_8=1 or 3 then class 1
0,9507	0,2316	If X_2=1 or 2 or 3 or 4 and X_4=1or 3 or 4 and X_7=1 or 3 and X_8=1 then class 4
0,9518	0,2236	If X_2=1 or 2 or 5 and X_3=2 or 3 or 4 and X_4=1 or 2 or 3 and X_7=1 or 3 and X_8=1 or 2 then class 4
		Else class 5.

Table 10. Comparison results of proposed algorithm with other rule based classifiers

Dataset	*Algorithm*	*Mean Accuracy (%)*	*Average Number of Rules*
Monks-2	Tan et al [19]	71,5	6
	Chen et al [4]	67,13	**2**
	TACO Algorithm	**98,36**	7
Tic-Tac-Toe	Baykasoğlu and Özbakır [3]	94,47	**2**
	Thabtah and Cowling [20]	**100**	26
	Chen et al [4]	**100**	26
	TACO Algorithm	97,59	16
Iris	Tan et al [19]	92,81	4
	Thabtah and Cowling [20]	93,87	15
	Santos et al [18]	93,33	5
	Chen et al [4]	94	7
	TACO Algorithm	**98**	**3**
Crx	Baykasoğlu and Özbakır [3]	**96,96**	**2**
	Tan et al [19]	85,48	5
	Chen et al [4]	82,5	21
	TACO Algorithm	92,61	11
Nursery	Baykasoğlu and Özbakır [3]	95,83	**5**
	Dehuri and Mall [5]	76,65	7
	TACO Algorithm	**97,16**	22

Table 9 shows the extracted rules for Monks2, Tic-Tac-Toe, Iris, Crx and Nursery problems. As it can be seen from the Tables 8 and 9 proposed method is able to extract rules with high predictive accuracy and comprehensibility.

Accuracy and comprehensibility of extracted rules in comparison to some other rule based classifiers which were presented in the literature [3, 4, 5, 18, 19, 20] are shown in Table 10. As shown in Table 10, the performance of the proposed algorithm is competitive to the compared rule-based classifiers from the literature. For Monks-2, Iris and Nursery data sets the proposed algorithm outperforms the other algorithms with better accuracies. If the average number of extracted rules is considered, ACO algorithm finds shorter rule sets on Iris data set. The proposed ACO algorithm has also very good predictive accuracy in comparison to compared algorithms.

4 Conclusion

ACO algorithm is a global optimization technique with certain advantages. In this study, a new framework for extracting comprehensible and accurate classification rules from trained artificial neural networks by using Touring ACO is presented. The methodology does not make any approximation to the activation function; it only uses weights to extract rules belonging to certain classes. Performance of the proposed algorithm is tested on the five benchmark data sets. The computational results have shown that in all of the five data sets, the proposed ACO based algorithm extracted comprehensible rules with high accuracy rates. The presented applications are studied as an introductory stage of this research. Different fitness functions, addition of the fitness function into TACO probability calculation as the visibility measure and different approaches to avoid entrapment into local optimum instead of frequency factor are scheduled as future works.

Acknowledgements

Prof. Dr. Adil Baykasoğlu is grateful to Turkish Academy of Sciences (TÜBA) for supporting his scientific studies.

References

1. Andrews, R., Diederich, J., Tickle, A.B.: A Survey, Critique of Techniques for Extracting Rules from Trained Artificial Neural Networks. Knowledge Based Sys. 8(6), 373–389 (1995)
2. Arbatlı, A.D., Akın, H.L.: Rule Extraction from Trained Neural Networks Using Genetic Algorithms. Nonlinear Analysis, Theory, Methods & Applications 30(3), 1639–1648 (1997)
3. Baykasoğlu, A., Özbakır, L.: MEPAR-miner: Multi-Expression Programming for Classification Rule Mining. European Journal of Operational Research 183(2), 767–784 (2007)
4. Chen, G., Liu, H., Yu, L., Wei, Q., Zhang, X.: A New Approach to Classification Based on Association Rule Mining. Decision Support Systems 42, 674–689 (2006)
5. Dehuri, S., Mall, R.: Predictive and Comprehensible Rule Discovery Using A Multi-Objective Genetic Algorithm. Knowledge-Based Systems 19, 413–421 (2006)
6. Dorigo, M., Maniezzo, V., Colorni, A.: Positive Feedback as A Search Strategy, Technical Report, N. 91-016, Politecnico di Milano (1991)
7. Elalfi, E., Haque, R., Elalami, M.E.: Extracting Rules from Trained Neural Network Using GA for Managing E-business. Applied Soft Computing 4, 65–77 (2004)
8. Fayyad, U., Irani, K.B.: Multi-interval Discretization of Continuous-valued Attributes for Classification Learning. In: Proceedings of 13th International Joint Conference on Artificial Intelligence, pp. 1022–1027 (1993)

9. Frawley, W., Piatetsky-Shapiro, G., Matheus, C.W.: Knowledge Discovery in Databases: An Overview. AI Magazine, 213–228 (1992)
10. Han, J., Kamber, M.: Data Mining: Concepts and Techniques, p. 24. Academic Press, London (2001)
11. Hiroyasu, T., Miki, M., Ono, Y., Minami, Y.: Ant Colony for Continuous Functions, The Science and Engineering Doshisha University XX(Y) (2000)
12. Hruschka, E.R., Ebecken, N.F.F.: Extracting Rules from Multilayer Perceptrons in Classification Problems: A Clustering-based Approach. Neurocomputing 70, 384–397 (2006)
13. Kalinli, A., Karaboga, N., Karaboga, D.: A Modified Touring Ant Colony Optimisation Algorithm For Continuous Functions. In: The 16th International Symposium on Computer and Information Sciences (ISCIS XVI), pp. 437–444. Isik University, Antalya (2001)
14. Karaboga, N., Kalinli, A., Karaboga, D.: Designing Digital IIR Filters Using Ant Colony Optimisation Algorithm. Engineering Applications of Artificial Intelligence 17, 301–309 (2004)
15. Kuttuyil, A.S.: Survey of Rule Extraction Methods, Master of Science Thesis, 1 (2004)
16. Markowska-Kaczmar, U.: The Influence of Prameters in Evolutionary Based Rule Extraction Method from Neural Network. In: Proceedings of the 2005 5th International Conference on Intelligent Systems Design and Applications, ISDA 2005 (2005)
17. Markowska-Kaczmar, U., Wnuk-Lipinski, P.: Rule Extraction from Neural Network by Genetic Algorithm with Pareto Optimization. In: Rutkowski, L., Siekmann, J.H., Tadeusiewicz, R., Zadeh, L.A. (eds.) ICAISC 2004. LNCS, vol. 3070, pp. 450–455. Springer, Heidelberg (2004)
18. Santos, R.T., Nievola, J.C., Freitas, A.A.: Extracting Comprehensible Rules From Neural Network via Genetic Algorithms. In: Proc. 2000 IEEE Symp. On Combinations of Evolutionary Computation and Neural Networks (ECNN 2000), San Antonio, TX, USA, pp. 130–139 (2000)
19. Tan, K.C., Yu, Q., Ang, J.H.: A Dual-Objective Evolutionary Algorithm for Rules Extraction in Data Mining. Computational Optimization and Applications 34, 273–294 (2006)
20. Thabtah, F.A., Cowling, P.I.: A Greedy Classification Algorithm Based on Association Rule. Applied Soft Computing 7, 1102–1111 (2007)
21. Tokinaga, S., Lu, J., Ikeda, Y.: Neural Network Rule Extraction by Using the Genetic Programming and Its Applications to Explanatory Classifications. IECE Trans. Fundamentas E88-A(10), 2627–2635 (2005)

Tuning Local Search by Average-Reward Reinforcement Learning

Steven Prestwich

Cork Constraint Computation Centre
Department of Computer Science, University College, Cork, Ireland
s.prestwich@cs.ucc.ie

Abstract. Reinforcement Learning and local search have been combined in a variety of ways, in order to learn how to solve combinatorial problems more efficiently. Most approaches optimise the total reward, where the reward at each action is the change in objective function. We argue that it is more appropriate to optimise the average reward. We use R-learning to dynamically tune noise in standard SAT local search algorithms on single instances. Experiments show that noise can be successfully automated in this way.

1 Introduction

Local search is used to solve many combinatorial problems, but to be effective it often requires at least one (and sometimes several) runtime parameters to be tuned manually by the user. Automatic parameter tuning and algorithm selection are active areas of research, and machine learning is a promising approach. In this paper we focus on Reinforcement Learning (RL) applied to the task of tuning local search noise level on single SAT instances. First we motivate the approach.

1.1 Local Optima and Long-Term Rewards

Local search performs iterative descent to find local minima, and applies *noise* to escape local minima in the hope of finding deeper ones.[1] Noise is the tendency of local search to make moves that appear counter-productive, but are necessary in order to escape from local optima. This may be a simple probability, a *temperature* as used in Simulated Annealing, or some other parameter used to increase *diversification* and/or decrease *intensification*.

Different noise levels might be best in different regions of the search space. For example consider the minimisation problem in Figure 1. For high values of the objective function it takes a "big valley" form, which is common in optimisation problems [1]. A good strategy for finding minima in such a space is low-noise local search [14], which is able to escape the small local minima while moving

[1] We assume without loss of generality that optimisation problems are minimisation, rather than maximisation.

V. Maniezzo, R. Battiti, and J.-P. Watson (Eds.): LION 2007 II, LNCS 5313, pp. 192–205, 2008.

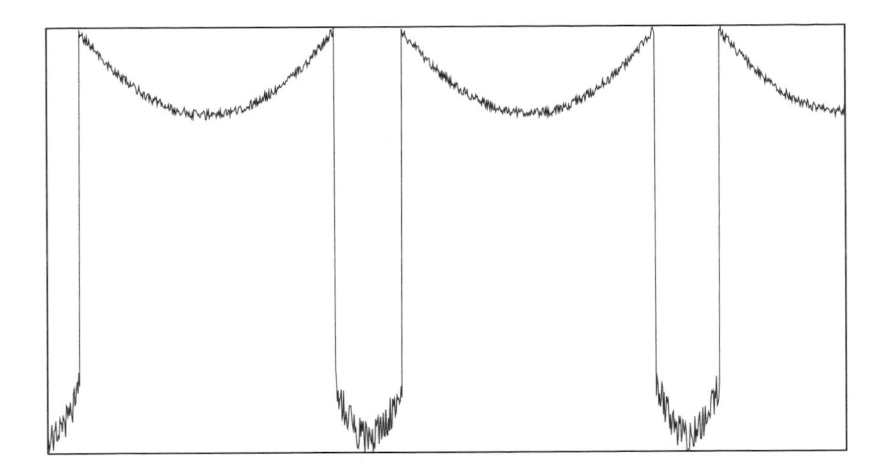

Fig. 1. A minimisation problem

steadily toward the centre. But if the search starts in a shallow valley then it will take a long time to escape to a deeper valley. Random restarts can escape these shallow valleys, but finding the deeper ones requires a lucky guess, and if the deep valleys are very narrow then this is unlikely to occur.

We could use high noise so that the search does not become trapped in a shallow valley, but instead performs a near-random walk until falling into a deep valley. Unfortunately, in the deep valleys high noise is counter-productive, as we need low noise to locate the very deepest minima. Ideally we would like to use dynamic noise that is high when the objective function value is high, and low otherwise. In other examples we might need high noise when the function is low, or some more complex pattern.

This kind of behaviour is largely achieved by Dynamic Local Search methods (see [8] for a survey), in which search stagnation initiates changes to the objective function, so that local minima are transformed into peaks. But these changes must be rediscovered each time stagnation occurs, and there is no attempt to learn from past experience. If a search space contains similar structures repeated many times (as in Figure 1) then there might be an opportunity for machine learning techniques to improve search performance.

1.2 Benefits of Learned Noise

We now show that making noise dependent upon the objective function value can potentially reduce the time complexity of a local search algorithm. Suppose we want to minimise a function of a vector $\mathbf{v}$ of Boolean variables. Let $x(\mathbf{v})$ be the Hamming distance of $\mathbf{v}$ from the fixed vector $\mathbf{0}$, in other words the number of 1s in $\mathbf{v}$, and $y(x)$ be the function in Figure 2: a step function of a single integer variable, changing value at each step. The lowest point corresponds to the global minimum. A local move flips the value of a variable, changing the value of x by 1.

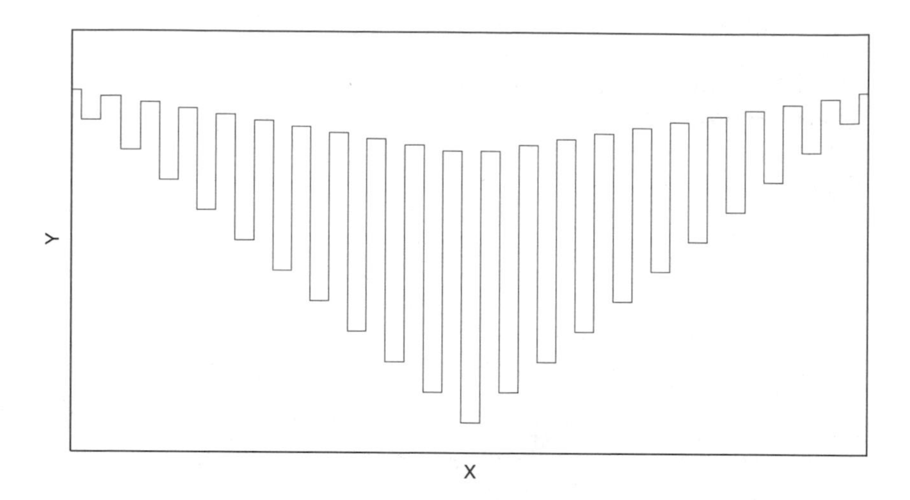

Fig. 2. A motivating example

From now on we will ignore **v** and consider only x, but note that each x value may correspond to an exponential number of states.

Suppose that we would like to find the global minimum using the following simple local search: from a given state x move to a random neighbour $x' = x \pm 1$: if $y(x') \leq y(x) + k$ then accept the move, otherwise reject it and restore the state to x, where k is a fixed runtime parameter chosen by the user that we will refer to as "noise". This is similar to constant-temperature annealing, but instead of accepting bad local moves with a probability that depends on how bad they are, we crudely prohibit moves that are worse than a threshold. The hypothetical algorithm also performs a periodic random restart every r local moves, where r is another parameter specified by the user at runtime. This is not a good or commonly-used local search algorithm but it is easy to analyse and, as we will show, on our example it can be boosted to give very good performance by RL.

Consider the behaviour of this local search algorithm on the problem in Figure 2. Let D_l and D_g denote the difference in y between a deepest local minimum and its lesser and greater neighbouring states, respectively. There are three cases:

- If $k > D_g$ all local moves will be accepted and there is no tendency to improvement. The algorithm will therefore perform a pure random walk among states, and will take an exponential time to find the global minimum.
- Using $k \leq D_l$ causes the algorithm to become trapped in a local minimum (assuming it does not start at the global minimum) until freed by a random restart, so the best restart strategy is to restart at every local move by setting $r = 1$, which is simply *uniform random picking* [8] and will again sample an exponential number of points.

– Using $D_l \leq k < D_g$ causes random walk behaviour everywhere except at the deepest local minima (where it will quickly find the global minimum).

Thus any fixed values of k and r give exponential behaviour. Note that simulated annealing (with any cooling schedule) may also perform poorly on this example. It does have a tendency to improvement because, from a local minimum, there is a greater probability of accepting a move towards the global minimum than away from it. But if the peaks decrease very slowly in height then both left and right moves can be made to have arbitrarily close probabilities, so simulated annealing can be made to behave arbitrarily closely to a random walk and exponential time.

However, suppose that we use a variable value of k that depends purely on the current objective value: $k \equiv k(y)$. Choose k such that at any local minimum x, $k(y)$ always takes a value between $y(x-1)$ and $y(x+1)$, that is the peaks immediately to its left and right. Then any move away from the global minimum will be rejected and any move towards it will be accepted. So the algorithm will reach the global minimum in an expected number of moves that is linear in the Hamming distance between the initial state and the global minimum.

Thus for a very specific form of noise function $k(y)$ we find linear execution time, but for any other form we find exponential time. If we could learn the noise function and remember it when exploring similar valleys (and apply random restarts appropriately) then we could find each central minimum in linear instead of exponential time. This would allow us to sample the central minima rapidly, greatly speeding up the search for the global minimum. We might not obtain such dramatic gains on realistic problems, but this thought experiment shows that the potential gains make the approach worth investigating.

1.3 Our Approach

We use RL to learn appropriate noise levels for different objective function values. Section 2 describes a new combination of standard RL and local search algorithms for SAT problems. Section 3 presents experimental results. Section 4 surveys related work. Section 5 summarises the work and discusses possible extensions.

2 Temporal Difference Learning for Noise

Escaping shallow valleys in order to reach deeper ones requires the sacrifice of short-term rewards in the hope of gaining long-term rewards. The RL technique of *temporal difference learning* (TD) is specifically designed to maximise long-term rewards at the expense of short-term rewards, based on trial-and-error and the propagation of information between reward estimates (*bootstrapping*). Perhaps its most impressive application is to the game of Backgammon [22]. The TD-Gammon system was trained by playing against itself, and reached grandmaster status. Kit Woolsey, rated fifth in the world in 1992, noted that:

> "There is no question in my mind that its positional judgement is far better than mine. Only on small technical areas can I claim a definite advantage over it."

(Cited in [22]). TD-Gammon has even changed how humans select opening moves. This attests to the ability of TD to maximise long-term rewards in the face of randomness and uncertainty. It therefore seems like an appropriate tool for tuning noise in local search.

2.1 Total Reward

When applying RL to optimisation problems it is common to use a *delta cost reward*. Suppose that the objective function is f, the search starts with an initial objective function value f_0, and global minima have value f^*. We could define the reward for each local move to be Δf, the reduction in the value of f as a result of the move. Maximising the reward will solve the problem: this form of optimality is called *total reward optimality* [10] and is used by standard TD algorithms such as Q-learning [25] and SARSA [17]. A drawback of total reward optimality is that, in *continuing* (non-episodic) applications that run indefinitely, the reward is unbounded. A common solution is to use the TD technique of *discounting*, which distorts the rewards. But in our application the total reward cannot exceed $f^* - f_0$ so discounting is unnecessary.

A more serious drawback with total reward optimality for this task is that it does not try to solve the problem *quickly*, which is our real aim. It produces a bias towards states with low f values but might do so by a circuitous route. The total reward is the same for any policy that manages to solve the problem, and there is no distinction between policies that solve the problem quickly and those that solve it slowly. In fact if the underlying local search algorithm is probabilistically approximately complete [8] then it can be solved from any state, and all policies are equally good.

A similar situation is discussed by Sutton & Barto [21] (pp. 56–57). Suppose a robot uses TD to navigate through a maze. We might fix all rewards to 0, except for a reward of 1 on successful navigation. This seems reasonable as the same is done for Backgammon. But then there is no pressure to exit the maze quickly, so the robot will not improve its navigation. A solution is to replace the rewards of 0 by a small negative number such as -0.001, so that better navigation has higher total reward. We could use the same trick when applying TD to optimisation problems: add a small negative reward to each local move. But this ad hoc method introduces an arbitrary number, and Sutton & Barto note the importance of choosing a reward that reflects what we truly want, though it can be tempting to tweak the reward to alter the search heuristics.

2.2 Average Reward

We argue that the natural optimisation criterion for each local move is the *average progress towards a solution*: that is the average difference in the objective function before and after the local move. This form of optimality is called *gain optimality* [10]. Suppose that the objective function is reduced by an average of Δf per local move. Then the problem will be solved in exactly $\Delta f(f_0 - f^*)$ local moves, and the greater the average reward the quicker the problem will be solved. This direct link between average reward and local search performance makes average reward more appropriate than total reward.

Figure 2.2 shows an average-reward TD algorithm called R-learning [18]. It learns an action-value function $Q(s,a)$ which estimates the reward for following each action a from each state s, initially set to arbitrary values. From each state s it selects an action a: usually the one with the best estimated reward, but occasionally a random action. This is typical of TD and is intended to balance the *exploitation* of the current state of learning with *exploration* for further learning. We use a standard behaviour policy called ϵ-*greedy*: choose a random action with probability ϵ, otherwise choose the best action according to current $Q(s,a)$ estimates. Whichever action is taken, a new state s' and a reward r are observed. The average reward is estimated by ρ which is updated in a similar way to Q, and also initialised to an arbitrary value.

initialise ρ and the $Q(s,a)$ arbitrarily
repeat forever
($s \leftarrow$ current state
  choose action a using a behaviour policy such as ϵ-greedy
  take action a and observe r, s'
  $Q(s,a) \leftarrow Q(s,a) + \alpha\,[r - \rho + \max_{a'} Q(s',a') - Q(s,a)]$
  if $Q(s,a) = \max_a Q(s,a)$ then
    $\rho \leftarrow \rho + \beta\,[r - \rho + \max_{a'} Q(s',a') - \max_a Q(s,a)]$
)

Fig. 3. The R-learning algorithm

A drawback of gain optimality that has been pointed out [10] is that it does not distinguish between policies that have the same average reward but incur different "startup costs". That is, some policies might produce a slow but steady decrease in f, while others rapidly reduce f at the start of the search (SAT local search typically behaves in the latter way [6]). To make this distinction we might use an alternative such as *bias optimality* [10]. This objection is relevant in *anytime optimisation*, in which we aim for the best solution in a limited time, but it is irrelevant if we are interested in finding the global minimum in the shortest time possible. In this paper we consider local search applied to SAT problems, so gain optimality is appropriate, but bias optimality might be useful for a problem such as MAX-SAT. Interestingly, standard SAT local search algorithms do not seem to be the best choice for MAX-SAT [23], illustrating the fact that different objectives require different search heuristics.

2.3 SAT Local Search with R-Learning

The SAT problem is to determine whether a Boolean expression has a satisfying labelling (set of truth assignments). The problems are usually expressed in conjunctive normal form: a conjunction of clauses $c_1 \wedge \ldots \wedge c_m$ where each clause c is a disjunction of literals $l_1 \vee \ldots \vee l_n$ and each literal l is either a Boolean variable v or its negation $\bar{v}$. A Boolean variable can be labelled true (T) or false (F).

There has been considerable research on local search for SAT, but we consider just three algorithms: the well-known SKC variant of WalkSAT [11,19]; a recent variant called VW1 [16]; and HWSAT [7], a version of WalkSAT/G [11,19] that breaks ties by preferring the least recently flipped variable. All start from an arbitrary total assignment of truth values to variables, and proceed by flipping the values of single variables with the aim of satisfying all clauses. WalkSAT/SKC is shown in Figure 2.3, where the phrase "break fewest clauses" means that we choose a variable that minimises the number of currently satisfied clauses that would be violated if its truth value were flipped. A *freebie move* is one that breaks no satisfied clause. Noise is controlled by the parameter p which is a probability chosen at runtime by the user. The VW1 algorithm [16] is exactly the same algorithm, except that in the last line it breaks ties between variables by choosing the one that has been flipped fewest times so far, a heuristic that was shown to improve performance on structured problems at the expense of performance on random problems. HWSAT was chosen for variety: like WalkSAT/G it tries to minimise the total number of clause violations, instead of the number of satisfied clauses that would become violated, and it does not use freebie moves. The algorithm is shown in Figure 2.3.

```
initialise all variables to randomly selected truth values
repeat until no clause is violated
(   randomly select a violated clause C
    if C contains freebie variables
        randomly flip one of them
    else with probability p
        flip a variable in C chosen randomly
    else with probability 1 − p
        flip a variable in C that would break fewest clauses
)
```

Fig. 4. The WalkSAT/SKC algorithm

We propose to use R-learning to learn noise functions for these algorithms while solving a problem instance. In our approach a *state* is the number of violated clauses of a given local search state, an *action* is the selection of a new noise level from the set $\{0.05, 0.1, 0.15, \ldots, 0.95, 1.00\}$ followed by a local move, and the *reward* is the reduction (possibly zero or negative) in the number of violated clauses since the last state. The local search algorithm is unchanged by the addition of R-learning, except that its noise level may be reset at each move. We terminate both R-learning and local search on solving the SAT problem.

In this scheme the choice of action only affects the choice of local move in a probabilistic way, via the noise parameter. This is not a problem, as TD learning

```
initialise all variables to randomly selected truth values
repeat until no clause is violated
(   randomly select a violated clause C
    with probability p
        flip a variable in C chosen randomly
    else with probability 1 − p
        flip a variable in C giving fewest violations
)
```

Fig. 5. The HWSAT algorithm

is often applied to problems in which actions have a probabilistic effect. These are *stochastic* as opposed to *deterministic* policies, and in fact stochastic policies can be arbitrarily more efficient than deterministic policies [20].

Note that we have replaced a single noise parameter p by the three R-learning parameters α, β, ϵ. But these are "second-order" parameters that we hope will be robust over a wide range of instances, whereas noise is strongly instance-dependent. This approach has worked well with dynamic local search algorithms, which often have several parameters of their own that can be set to default values [8].

3 Experiments

First we experimented with WalkSAT/SKC and VW1 using selected SAT benchmark problems from the SATLib repository,[2] with results shown in Tables 1 and 2 respectively. In both tables we compare (i) the original algorithm with optimal constant noise parameter ("best") after trying noise values $p \in \{0.1, 0.2, \ldots, 0.9\}$, (ii) a random noise value at each step ("random") as a sanity check in case the instances are insensitive to the noise setting, and (iii) the algorithm with noise guided by R-learning using three values of α. We set $\epsilon = 0.01$ and $\beta = 0.001$ in all cases. The figures are the median number of flips taken to solve the problem over 100 runs, and entries marked "—" denote medians greater than 10^7 flips. Note that different `aim` instances appear in the two tables: 100-variable instances were much harder for SKC so we used 50-variable instances.

These results show the potential of the method. With SKC the best results using R-learning are close to the best results using fixed noise, with the exception of the random 3-SAT problem `f1000`, on which its results are worse than those using random noise. Note that on this instance random noise is roughly equivalent to the best fixed noise level of 0.5, so this is not as strange as it first appears. But R-learning clearly fails on `f1000`. The `aim` problems are random problems that have been slightly modified to make them harder for local search. They are not very noise-sensitive but R-learning does reasonably well on them. The

[2] http://www.cs.ubc.ca/~hoos/SATLIB/

Table 1. The effect of R-learning on WalkSAT/SKC

instance	best (p)	random	with R-learning $\alpha = 0.1$	$\alpha = 0.9$	$\alpha = 0.99$
logistics.a	64,866 (0.3)	159,544	85,252	106,027	91,402
logistics.b	88,186 (0.2)	230,029	141,603	148,182	213,870
logistics.c	104,610 (0.2)	502,150	192,134	241,706	373,994
logistics.d	425,746 (0.4)	718,580	492,059	329,326	503,427
bw_large.a	14,459 (0.5)	16,808	28,107	26,258	15,731
bw_large.b	422,723 (0.3)	972,153	1,249,395	739,082	632,595
ais6	891 (0.4)	838	1,160	832	824
ais8	19,306 (0.4)	25,639	17,190	25,926	29,066
ais10	106,752 (0.3)	315,685	148,630	249,420	314,948
ais12	1,219,293 (0.1)	9,199,786	1,677,166	3,391,396	2,699,864
f600	130,625 (0.5)	118,268	1,707,566	188,158	143,536
f1000	491,262 (0.5)	609,146	—	1,962,880	2,071,272
aim50-2.0-1	151,840 (0.5)	189,208	244,746	179,538	150,210
aim50-2.0-2	15,605 (0.4)	16,542	25,140	18,454	22,965
aim50-2.0-3	123,724 (0.5)	151,556	218,648	190,526	222,628
aim50-2.0-4	92,836 (0.4)	86,094	151,358	114,570	193,079

Table 2. The effect of R-learning on VW1

instance	best (p)	random	with R-learning $\alpha = 0.1$	$\alpha = 0.9$	$\alpha = 0.99$
logistics.a	27,236 (0.4)	37,412	29,936	29,898	29,070
logistics.b	18,302 (0.3)	38,399	19,951	22,700	31,693
logistics.c	29,483 (0.3)	67,874	33,047	36,366	46,764
logistics.d	168,500 (0.3)	797,784	210,562	219,786	338,060
bw_large.a	8,085 (0.4)	8,924	10,620	7,398	8,249
bw_large.b	80,468 (0.3)	147,120	120,894	112,175	125,012
bw_large.c	582,056 (0.3)	—	609,188	616,455	1,290,333
bw_large.d	630,464 (0.3)	—	615,151	5,503,318	—
ais6	984 (0.3)	652	827	1,190	1,068
ais8	18,661 (0.4)	21,757	29,820	15,924	18,721
ais10	101,157 (0.3)	169,546	148,438	142,182	166,760
ais12	816,645 (0.3)	5,705,746	834,994	1,424,670	4,546,006
f600	148,749 (0.4)	237,162	367,262	113,244	154,618
f1000	939,022 (0.4)	3,795,379	4,611,894	879,957	710,552
aim100-2.0-1	259,875 (0.3)	649,063	525,126	414,677	993,611
aim100-2.0-2	221,535 (0.3)	450,447	240,532	281,849	376,057
aim100-2.0-3	64,717 (0.3)	363,624	122,032	170,495	195,893
aim100-2.0-4	175,616 (0.3)	304,225	317,986	292,277	505,835

logistics, bw_large and ais instance are structured problems from planning and music theory, and R-learning does better on these.

The VW1 algorithm results are better, and even on f1000 R-learning equals the performance with best fixed noise. It is encouraging that R-learning works better on the stronger of the algorithms: new techniques often seem promising when applied to weaker algorithms, but have little effect when combined with stronger ones. It is also encouraging that R-learning works well on the larger instances (except for SKC on f1000): again, new techniques often pay off on small instances but fail to scale up to harder problems.

The negative results might be caused simply by suboptimal values for the R-learning parameters. In further experiments we found that reducing β to 0.000001 greatly improved SKC with R-learning on the f1000 instance: with α, ϵ as before it now takes a median of 708,736 flips, which is not much more than SKC with optimal fixed noise. This is gratifying as f1000 was our worst result, and using the best α, ϵ values from above this new β value does not harm SKC or VW1 performance on other instances: in fact it improves VW1 performance on logistics.d to 170,237 flips, again close to optimal fixed noise performance. We now have uniformly good results on this set of benchmarks for SKC and VW1.

Guided by these results, we experimented with HWSAT using $\beta = 0.000001$ (again with $\epsilon = 0.01$) with results shown in Table 3. As with SKC we used 50-variable aim instances. R-learning's worst result is on the very easy ais6 problem, where it takes three times as long as the best fixed noise. Its next worst results are on f600 and bw_large.a where it takes less than twice as long as with the best fixed noise. The R-learning results are slightly worse on the logistics problems but better than random noise; similarly for the larger ais instances. On bw_large.b R-learning matches the best fixed noise result and beats that of random noise. On aim instances 2, 3 and 4 there is little difference between best fixed noise, random noise, and the best R-learning results. In summary, R-learning never fails very badly, and then only on the easiest problems. On the harder problems it matches or is slightly worse than using the best fixed noise value.

It is promising that R-learning seems to work best with stronger local search algorithms and harder SAT problems. We have not seen any dramatic improvements such as those we speculated about in Section 1.2, but this work is still at an early stage. Our next step will be to find a robust setting for the α parameter and to experiment with more instances.

4 Related Work

The first use of RL for combinatorial optimisation was by Zhang & Dietterrich [26] to learn local repair moves on a variant of the TSP. Total reward was optimised using a final reward for solving the problem, and a small negative reward for each step to encourage fast solution. Nareyek [15] also used RL to choose local moves during search and optimised total rewards, but did not use a standard

Table 3. The effect of R-learning on HWSAT

instance	best (p)	random	with R-learning $\alpha = 0.1$	$\alpha = 0.9$	$\alpha = 0.99$
logistics.a	137,332 (0.4)	210,242	215,580	174,430	187,724
logistics.b	151,214 (0.3)	485,512	205,414	249,155	314,829
logistics.c	270,154 (0.3)	1,151,212	318,720	659,552	841,401
logistics.d	281,172 (0.4)	775,631	400,186	346,294	389,250
bw_large.a	12,372 (0.5)	10,834	38,972	21,514	18,946
bw_large.b	484,932 (0.4)	716,006	885,652	475,520	442,115
ais6	1,268 (0.5)	1,110	9,246	3,791	3,430
ais8	27,688 (0.4)	26,978	78,162	34,810	36,462
ais10	165,035 (0.3)	498,898	211,262	387,969	357,534
ais12	1,357,178 (0.2)	—	2,898,064	8,713,594	9,067,544
f600	136,267 (0.6)	147,172	—	440,695	257,892
aim50-2.0-1	139,502 (0.5)	104,493	230,238	149,092	144,672
aim50-2.0-2	14,126 (0.5)	13,562	16,753	24,742	13,243
aim50-2.0-3	99,860 (0.5)	88,126	178,398	94,383	81,532
aim50-2.0-4	23,124 (0.4)	29,215	61,870	27,488	30,726

RL algorithm or exploit TD-style bootstrapping to estimate long-term rewards. Boyan & Moore's STAGE [2] algorithm estimates the quality of a search state according to the quality of local minimum that can be reached from it using a fixed local search algorithm. Another local search is performed on the learned value function. Varrentrapp's Guided Adaptive Iterated Local Search (GAILS) [24] uses Q-learning to learn composite moves in an Iterating Local Search between local optima. This has the effect of smoothing out the effects of single local moves. Again total rewards are optimised. Moll et al. [13] use standard TD(λ) algorithms from RL, with a non-standard use of exploration, and two variations on total reward that penalise local moves in order to encourage fast search. They use a training phase to learn how to solve a class of problems.

Some hybrid algorithms use distributed systems of agents that individually exploit RL, and collectively behave like local or evolutionary search algorithms. Crites & Barto [3] solve an elevator scheduling problem by applying RL to each elevator separately, and the agents learn to cooperate. Gambardella & Dorigo [5] describe Ant-Q, a generalisation of ant systems with similarities to a distributed form of Q-learning, and apply it to the TSP. Miagkikh & Punch [12] apply a distributed RL algorithm to the QAP.

Gagliolo & Schmidhuber [4] treat the problem of algorithm selection as a bandit problem, a form of RL in which actions are learned independently of the current state, and attempts to balance exploitation with exploration. On a sequence of problem instances this approach learns which algorithms work best, giving online performance improvement. Lagoudakis & Littman [9] apply a version of Q-learning to recursive problems, learning which algorithm to apply at each recursion.

In summary, our approach has novel features with respect to other work on RL for combinatorial optimisation: few approaches use RL on single problem instances; few hybridise standard RL algorithms with standard local search algorithms; and none (to the best of our knowledge) optimise the average reward.

5 Conclusion and Future Work

We described an application of Reinforcement Learning to SAT local search: using temporal difference learning to select a noise level at each local move, as a function of the current objective function value. This simple technique allows the easy combination of standard Reinforcement Learning algorithms with standard local search algorithms, so that neither is compromised. We use R-learning instead of the more usual Q-learning or SARSA algorithms, in order to optimise average reward instead of total reward with or without local move penalties, and we argue that this is a more natural approach. Our experimental results show that noise can be successfully automated by using this technique.

We plan to extend this work in several directions. Firstly, generalising states to objective function values is quite drastic, and additional features could be used. These need not be raw features, but could instead be derived features such as a smoothed estimate of local gradients, time spent near the current value, or a measure of objective function variance (which is a useful invariant for noise tuning [11]). Function approximation techniques such as neural networks could be used to generalise search states. We will also generalise the method to tune more parameters such as the random restart period, and the two extra parameters in a more complex version of the VW1 algorithm [16].

A possible drawback with using TD to tune runtime parameters is an assumption made by TD methods: that the problem to be solved is a *Markov Decision Process*, in other words the optimum action depends only on the state. Our states are generalised versions of the search state, and this assumption clearly does not hold. The problem we are trying to solve is actually a *Partially Observable MDP* (POMDP). TD algorithms sometimes give good results on POMDPs, but better results can usually be obtained by more complex algorithms with some form of memory, whereas the policies we find in this work are *memoryless* (or *reactive*). This is an interesting direction for future work.

Acknowledgment

This material is based in part upon works supported by the Science Foundation Ireland under Grant No. 00/PI.1/C075.

References

1. Boese, K.D.: Cost Versus Distance in the Travelling Salesman Problem. Technical report CSD-950018, UCLA Computer Science Department
2. Boyan, J.A., Moore, A.W.: Learning Evaluation Functions for Global Optimization and Boolean Satisfiability. In: 15th National Conference on Artificial Intelligence and 10th Innovative Applications of Artificial Intelligence Conference, pp. 3–10. AAAI Press / MIT Press (1998)

3. Crites, R., Barto, A.: Improving Elevator Performance Using Reinforcement Learning. In: Conference on Advance in Neural Information Processing Systems, pp. 1017–1023. MIT Press, Cambridge (1999)
4. Gagliolo, M., Schmidhuber, J.: Gambling in a Computationally Expensive Casino: Algorithm Selection as a Bandit Problem. In: Online Trading of Exploration and Exploitation, NIPS 2006 Workshop, Whistler, BC, Canada (2006)
5. Gambardella, L.M., Dorigo, M.: Ant-Q: A Reinforcement Learning Approach to the Traveling Salesman Problem. In: 12th International Conference on Machine Learning, pp. 252–260. Morgan Kaufmann, San Francisco (1995)
6. Gent, I.P., Walsh, T.: An Empirical Analysis of Search in GSAT. Journal of Artificial Intelligence Research 1, 47–59 (1993)
7. Gent, I.P., Walsh, T.: Unsatisfied Variables in Local Search. In: Hallam, J. (ed.) Hybrid Problems, Hybrid Solutions, pp. 73–85. IOS Press, Amsterdam (1995)
8. Hoos, H.H., Stützle, T.: Stochastic Local Search: Foundations and Applications. Morgan Kaufmann, San Francisco (2004)
9. Lagoudakis, M.G., Littman, M.L.: Algorithm Selection Using Reinforcement Learning. In: 17th International Conference on Machine Learning, pp. 511–518. Morgan Kaufmann, San Francisco (2000)
10. Mahadevan, S.: Average Reward Reinforcement Learning: Foundations, Algorithms, and Empirical Results. Machine Learning 22, 159–196 (1996)
11. McAllester, D.A., Selman, B., Kautz, H.A.: Evidence for Invariants in Local Search. In: 14th National Conference on Artificial Intelligence and Ninth Innovative Applications of Artificial Intelligence Conference, pp. 321–326. AAAI Press / MIT Press (1997)
12. Miagkikh, V., Punch, W.: Global Search in Combinatorial Optimization using Reinforcement Learning Algorithms. In: Congress on Evolutionary Computation, vol. 1, pp. 189–196. IEEE, Los Alamitos (1999)
13. Moll, R., Barto, A., Perkins, T., Sutton, R.: Learning Instance-Independent Value Functions to Enhance Local Search. In: Advances in Neural Information Processing Systems 11, pp. 1017–1023. MIT Press, Cambridge (1999)
14. Morris, P.: The Breakout Method for Escaping from Local Minima. In: 11th National Conference on Artificial Intelligence, pp. 40–45. AAAI Press / MIT Press (1993)
15. Nareyek, A.: Choosing Search Heuristics by Non-Stationary Reinforcement Learning. Metaheuristics: Computer Decision-Making, pp. 523–544. Kluwer, Dordrecht (2004)
16. Prestwich, S.D.: Random Walk With Continuously Smoothed Variable Weights. In: Bacchus, F., Walsh, T. (eds.) SAT 2005. LNCS, vol. 3569, pp. 203–215. Springer, Heidelberg (2005)
17. Rummery, G.A., Niranjan, M.: On-line Q-learning Using Connectionist Systems. Technical report CUED/F-INFENG/TR 166, Engineering Dept., Cambridge University, UK (1994)
18. Schwartz, A.: A Reinforcement Learning Method for Maximizing Undiscounted Rewards. In: 10th International Conference on Machine Learning, pp. 298–305. Morgan Kaufmann, San Francisco (1993)
19. Selman, B., Kautz, H.A., Cohen, B.: Noise Strategies for Improving Local Search. In: 12th National Conference on Artificial Intelligence, pp. 337–343. AAAI Press, Menlo Park (1994)

20. Singh, S., Jaakkola, T., Jordan, M., Cohen, W.W., Hirsh, H. (eds.): Learning Without State-Estimation in Partially Observable Markovian Decision Processes. Eleventh International Conference on Machine Learning, pp. 284–292. Morgan Kaufmann, San Francisco (1994)
21. Sutton, R.S., Barto, A.G.: Reinforcement Learning: An Introduction. MIT Press, Cambridge (1998)
22. Tesauro, G.: Temporal Difference Learning and TD-Gammon. Communications of the ACM 38(3), 58–67 (1995)
23. Tompkins, D.A.D., Hoos, H.H.: Scaling and Probabilistic Smoothing: Dynamic Local Search for Unweighted MAX-SAT. In: Xiang, Y., Chaib-draa, B. (eds.) Canadian AI 2003. LNCS, vol. 2671, pp. 145–159. Springer, Heidelberg (2003)
24. Varrentrapp, K.E.: A Practical Framework for Adaptive Metaheuristics. PhD thesis, Fachgebiet Intellektik, Fachbereich Informatik, Technische Universität Darmstadt, Darmstadt, Germany (2005)
25. Watkins, C.J.C.H.: Learning From Delayed Rewards. PhD thesis. Cambridge University (1989)
26. Zhang, W., Dietterrich, T.D.: A Reinforcement Learning Approach to Job-Shop Scheduling. In: 14th International Joint Conference on Artificial Intelligence, pp. 1114–1120. Morgan Kaufmann, San Francisco (1995)

Evolution of Fitness Functions to Improve Heuristic Performance*

Stephen Remde, Peter Cowling, Keshav Dahal, and Nic Colledge

MOSAIC Research Group, University of Bradford, Bradford, BD7 1DP, United Kingdom
{s.m.remde,p.i.cowling,k.p.dahal,n.j.colledge}@bradford.ac.uk
http://mosaic.ac/

Abstract. In this paper we introduce the variable fitness function which can be used to control the search direction of any search based optimisation heuristic where more than one objective can be defined, to improve heuristic performance. The method is applied to a multi-objective travelling salesman problem and the performance of heuristics enhanced with the variable fitness function is compared to the original heuristics, yielding significant improvements. The structure of the variable fitness functions is analysed and the search is visualised to better understand the process.

Keywords: Fitness Function, Evolution, Heuristic, Local Search, VFF.

1 Introduction

Optimisation heuristics are used when an optimal solution cannot be found in a reasonable amount of time. When the problem is just too complex to solve exactly, a heuristic method is used to find a sufficiently good or near optimal solution. One such type of heuristic is local search, which takes an initial solution and tries to improve it through a series of local perturbations. Local search has the problem of getting stuck in local optima. These are solutions from which a local change cannot improve the solution, however the solution is not globally optimal. Hence, a local move may not be the best globally, and the global fitness function may not be adequate for assessing local moves. Many ways to escape local optima have been introduced, including [1, 2, 3]. However, all these methods require modification of the local search. The method we introduce changes the search direction to avoid and escape local optima and can be applied to any search based optimisation heuristic without modification so long as two conditions are satisfied: (i) The objective function can be expressed in terms of two or more sub objectives (which is almost always the case in our experience of practical problems); and (ii) We have enough CPU time to run the heuristic many times. The search direction is determined by a Variable Fitness Function (VFF), which is evolved using a genetic algorithm. We apply this enhancement method to various heuristic methods for the TSP [4]. We show that given an optimisation

* This work was funded by EPSRC and @Road Ltd, a Trimble Company under an EPSRC CASE studentship, which was made available through and facilitated by the Smith Institute for Industrial Mathematics and System Engineering.

V. Maniezzo, R. Battiti, and J.-P. Watson (Eds.): LION 2007 II, LNCS 5313, pp. 206–219, 2008.

heuristic H, we can create a better heuristic *VFF(H)* and that the extra time used by the VFF is far better than using random perturbations with the original heuristic.

In this paper we introduce the Variable Fitness Function and show how it can be used to enhance local searches for a multi objective problem, by varying the fitness function over the course of the search. Throughout this paper, the term *global fitness function* will be used where we mean the global fitness function that is used to assess the overall quality of solution so as to make a clear distinction between it and the *variable fitness function*.

The next section reviews literature related to the variable fitness function and the TSP test case. Section 3 describes how a variable fitness function is represented and evolved. Section 4 details the implementation of the variable fitness function and section 5 describes computational experiments and results. Section 6 draws final conclusions and analysis from the work and identifies future work.

2 Related Work

Local search (including constructive heuristics, which are a special case of local search) may fall well short of a global optimum when the search converges to a local optimum or a basin of attraction. A local optimum is a solution which has no better neighbouring solution, but such a solution may be far worse than the global optimum. Several methods have been described in literature. Here we describe three that are effective and have the practical advantage that they are easy to implement for complex, practical problems.

Simulated Annealing is a local search method that was inspired by the physical annealing process [3]. Simulated annealing works by changing the acceptance criteria of a local search operator. It will always accept moves which lead to a better solution, however it also has a chance to accept moves that make the solution worse. This probability of accepting a worse move is controlled by a cooling scheme and is inversely proportional to how bad the move is and how far into the search the process is. This helps diversity at the beginning and helps intensify the search toward the end. In the survey done by Kolisch and Hartmann [5] the heuristic was shown to be competitive and performed well ranking about midway of the tested heuristics for a the Resource Constrained Project Scheduling Problem (RCPSP).

Guided Local Search [2] attempts to modify the fitness function to change the direction the search heads when a local optimum has been found. Features of a solution are identified and penalties for solutions exhibiting these features are increased when the solution is stuck in a local optimum. It redefines the objective function thus:

$$f'(s) = f(s) + \lambda \sum_{i=1}^{F} p_i I_i(s)$$

where λ is the weighting for the guided local search, F is the number of features, p_i is the penalty value for the i-th feature and $I_i(s)=1$ when s exhibits feature i, 0 otherwise. When the search settles on a local optimum s^* the utility of penalising a feature is defined by:

$$util_i(s^*) = I_i(s^*) \times \frac{c_i}{1+p_i}$$

The feature or features with the largest utility will be penalised by increasing their penalty values. This has the effect of changing the fitness function and forces the search to move in another direction.

Variable Neighbourhood Search (VNS) [1] is based on the idea of systematically changing the neighbourhood of a local search algorithm. Variable Neighbourhood Search enhances local search using a variety of neighbourhoods to "shake" the search into a new position after it reaches a local optimum. Several variants of VNS exist as extensions to the VNS framework [6] which have been shown to work well on various optimisation problems.

These local search metaheuristics all require modification of the local search. In the case of simulated annealing, only a minor change to the criteria of accepting a neighbouring solution is needed, however in guided local search and variable neighbourhood search, much larger changes are needed. The methods we introduce require no modification of the underlying local search and hence can easily be used to enhance any local search method. This becomes particularly important when trying to solve complex, real-world problems with a wide range of objectives and a detailed model, where the VFF approach provides a straightforward way to further enhance an existing approach.

The single objective travelling salesman problem [4] is a well known, well studied problem, often used to test single objective metaheuristics (for example [6] for VNS, [7] for GLS). [8] studies a multi-objective version of this which they call the J-objective TSP. Here, J different objectives are associated with travelling between each city. In practical applications these could represent factors such as distance, cost, travel time or some measure of traffic although uncorrelated objectives are used in [8]. Several greedy heuristics for initial tour construction exist for the TSP. Nearest neighbour and Multiple fragment are two we look at in this paper [9]. 2-opt is a local search heuristic often used to find good solutions quickly for the TSP. The basic move consists of choosing two edges and seeing if swapping them improves the tour. The result for Euclidean TSPs is that crossed edges get uncrossed and the tour is shortened. The TSP is a simple-to-express but difficult-to-solve problem which has been studied extensively and proven to be a good benchmark problem. It provides a very useful case study to investigate our VFF approach.

3 The Variable Fitness Function

The Variable Fitness Function describes how the weights of a weighted sum fitness function change over the iterations of a search process. Here we use a weighted sum linear objective, but the approach is immediately applicable to non-linear parameterised objectives. The variable fitness function is piecewise linear, describing the relative importance of objectives at each iteration. We consider two alternatives, the standard variable fitness function fixes the number of discontinuities and the number of iterations between them. The adaptive variable fitness function allows the points of discontinuity to evolve along with the variable fitness function objective weights. These two methods will be described in detail in the next sections.

3.1 Standard Variable Fitness Function

We define a set of weights $\{W_{a,b}\}$ where a indexes the weight set ($a=0...A-1$) and b indexes the objective ($b=1...B$). We define I, the number of iterations between the weight sets. The variable fitness function is now defined as:

$$f(s,i) = \sum_{b=1}^{B} O_b(s) W_b(i)$$

where s is the solution to be evaluated and i is the iteration, $O_b(s)$ is the value of objective b for solution s, and

$$W_b(i) = W_{b,\lfloor i/I \rfloor} + \frac{i \bmod I}{I}(W_{b,\lfloor i/I \rfloor+1} - W_{b,\lfloor i/I \rfloor})$$

(i.e. the linear interpolation of the weight of objective b for iteration i)

Figure 1 shows how the weights of an example variable fitness function change over iterations. In this example, B=4, A=3 and I=100.

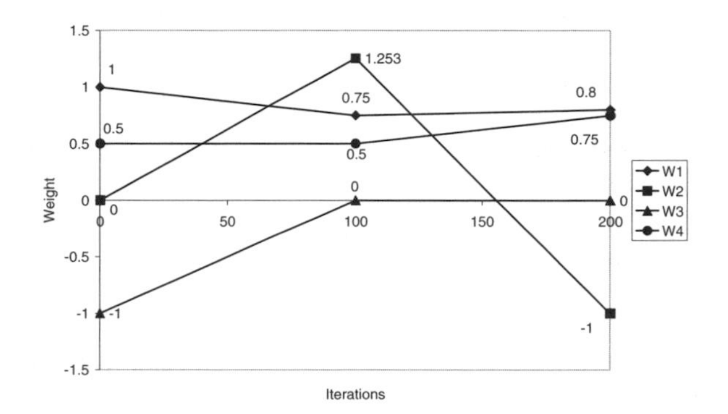

Fig. 1. An example standard variable fitness function. The number of weight sets (3) and the number of iterations between them (100) are fixed.

3.2 Adaptive Variable Fitness Function

Initial experiments with the standard variable fitness function quickly showed its weakness. If the number of iterations was too small, the solution quality would suffer. If the number of iterations was to large, CPU time would be wasted. The adaptive variable fitness function does not require the optimal number of weight sets and iterations between them to be known. These are evolved along with the weight data to find appropriate values. This can lead to more complex variable fitness functions.

Figure 2 shows an example adaptive variable fitness function. This describes how the weights change over the iterations, for example, that the weight of objective 1 (W1) starts off at 2 and then after iteration 200 its importance starts to decrease and objective 3 (that has weight W3) is to be minimized, and its importance is high at the start and end of the search process.

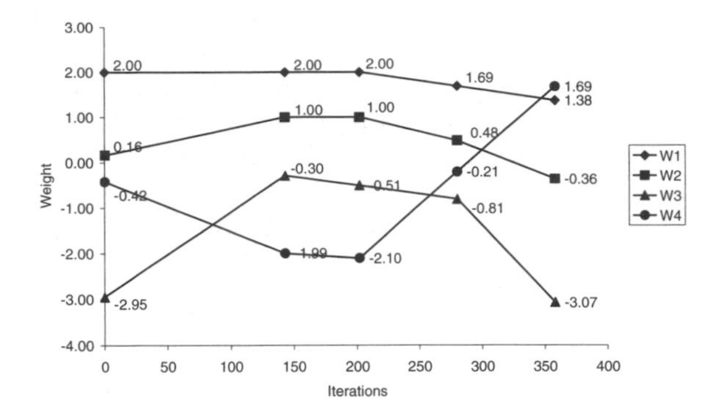

Fig. 2. An example adaptive variable fitness function. In this example, the number of iterations between the weight sets and the number of weight sets may vary.

3.3 Evolution

Little work has been done in encoding piecewise linear functions such as these into chromosomes. [10] uses a complex encoding for polynomial expressions. The encoding is used to optimise a curve to fit a function described by a set of data points and is not an appropriate method in this case. The evolution here is similar to work done on tuning of parameters for another algorithm using genetic algorithms [11].

When optimizing the weights of the variable fitness function, each weight in the variable fitness function appears as a gene in a GA chromosome. When the adaptive variable fitness function is used, the iterations between the weight sets is also included. Figure 3 shows how the weight sets are mapped to the genes of a chromosome.

A modified version of 1 point crossover [12] will be used. It works the same way as normal 1-point crossover but the crossover point may only be on a weight set boundary. This method will keep mutually compatible weight sets together. The thick lines in Figure 3 show these crossover points. Each gene will have a chance to be mutated with a probability of p_{mut}, the mutation rate. Mutation will simply mutate the value of the gene by a random variable normally distributed around 0 and with the standard deviation defined for that weight. Hence $W_{a,b}$ is deviated by a value from the normal distribution $N(0, V_b)$ with probability p_{mut}. Where V_b is the standard deviation of mutation associated with objective b and p_{mut} is the probability of mutation for all alleles. This is similar to work done on mutation of artificial neural network weights evolved using GAs [13] where the network weight is mutated by a random number selected from a normal distribution.

The initial population of variable fitness functions is generated at random. We may also seed the initial population with the global fitness function. These seeds are variable fitness functions where the weights are constant over all iterations and equivalent to the global fitness function. This may give the genetic algorithm a good individual to work from or provide good genetic material to create other individuals. For the random individuals $W_{a,b}$ is picked uniformly at random out of the interval $[L_b, U_b]$.

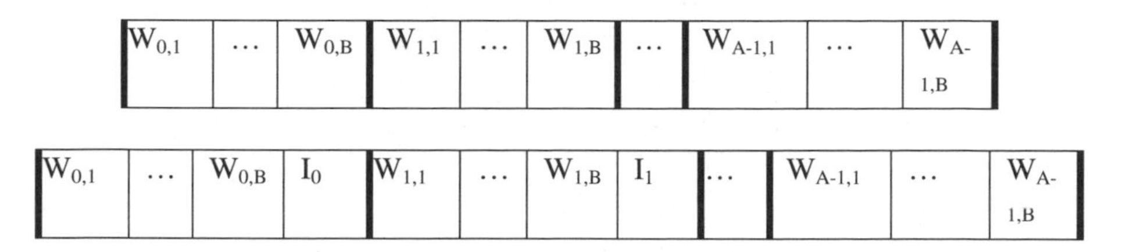

Fig. 3. Mapping the weights to a chromosome for a standard variable fitness function (top) and an adaptive variable fitness function (bottom)

Seeding the initial population with the global fitness function and using an elitist replacement scheme would ensure that in a worst case scenario, the best individual of the final population is a variable fitness function representing the global fitness function.

In the adaptive version there is also a p_{adapt} probability that the chromosome will change length. If a chromosome is to change length there is an equal probability it will either shrink or grow by one weight set. If it is to shrink, a random weight set is chosen and removed from the chromosome. If it is to grow, a new weight set is inserted between two randomly chosen adjacent weight sets. The inserted weight set does not change the shape of the variable fitness function as it is inserted exactly half way between the two adjacent weight sets and has weight values that are the mean of the bordering weight sets. The new weight set is then mutated. Lastly, the I_a genes also have a p_{mut} probability of being mutated. This gives the chromosomes a chance to get more and less complex and to also expand to more or less iterations.

We have introduced a lot of parameters in this section but our experiments using the TSP and other problems show that the performance of VFF is not sensitive to these parameters. L_b and U_b can be set to -1 and 1 respectively without losing any information as any weighted sum objective function can be normalised and the individual weights will lie within this range. We have found that V_b set to 5% of the range ($V_b = 0.05(U_b - L_b)$) also works well. All the experiments here have used these default values. In the experiments carried out in this paper, $p_{mut} = p_{adapt} = 0.05$.

4 Application to the Traveling Salesman Problem

The Travelling Salesman Problem (TSP) is a well studied optimisation problem which is often solved using a basic local search operator 2-opt [14]. We study a multi-objective variant of the TSP and use the variable fitness function to guide a 2-opt local search to find better solutions than 2-opt alone.

The TSP consists of a set of n cities, and a cost matrix c_{ij} such that $1 \leq i,j \leq n$ that defines the cost of travelling from city i to city j. The aim of the TSP is to determine a tour of minimum length visiting each city only once and returning to the starting city. The Symmetric TSP (STSP) add a further constraint that $c_{ij} = c_{ji}$ for all i,j. We study a variant of this such that each journey has B uncorrelated objectives associated with it. The global objective is a weighted sum of the B objectives. It should be noted that

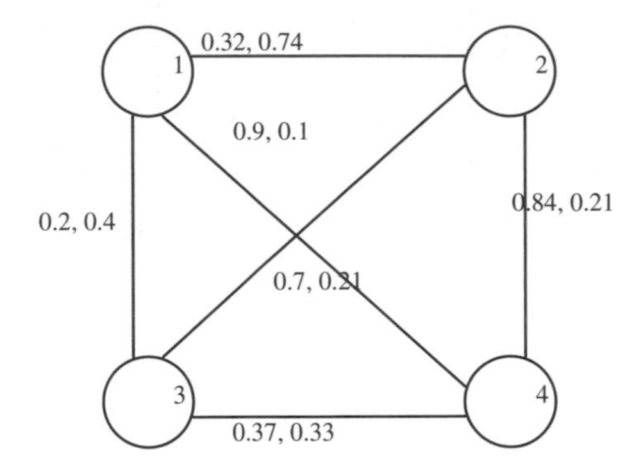

Fig. 4. Example MO-TSP with 4 cities and 2 objectives

these problems can easily be converted into an STSP (and in fact we do to solve them exactly). This way of creating multi objective TSP problems is related to the work of Jaszkiewicz [8] who use uncorrelated objectives to make the symmetric TSP into a multi-objective problem.

Figure 4 shows an example problem. The edge costs are uncorrelated. In the problems we look at, all costs are generated uniformly at random between 0 and 1. "Splitting" each edge in this way provides a way to convert a single-objective TSP to a multiple objective problem, which is readily applicable to other problems. In most of the real-world problems which we study, there are already many (often too many) objectives.

4.1 Variable Fitness Function Usage

We will use 3 different initial solution generation methods and improve them using 2-opt. We will then use the variable fitness function to enhance these methods in two ways. Firstly, just enhancing the 2-opt part and secondly enhancing the initial solution generation method as well. The methods are shown in Table 1. All of these heuristics for the TSP are well known – see [4] for details.

The three initial solution generation methods we will use are an arbitrary solution (*AR*), nearest neighbour (*NN*) and multiple fragment (*MF*). The arbitrary solution is simply the tour where all the nodes are visited in numeric order. Nearest neighbor starts the tour at a given point and repeatedly adds the nearest unvisited city to the city at a fixed end of the current partial tour until a complete tour is found. Multiple fragment is similar to nearest neighbor as at each iteration it adds an edge between the two closest unconnected cities whose connection does not form a cycle (unless it is the last edge to be added).

2-opt is a simple local search heuristic which improves a TSP solution by finding edges (i, j) and (k, l) in the current tour such that $c_{ij} + c_{kl} > c_{ik} + c_{kl}$ and replacing edges (i, j) and (k, l) with (i, k) and (j, l). For each of the NN, MF and 2-opt we consider stochastic versions where instead of greedily choosing the best at each iteration, we choose peckishly [15], with equal probability, one of the best two possibilities at each iteration, so allowing us to use extended CPU time sensibly.

Table 1. Hueristics and variable fitness function enhanced versions used in our experiments

Heuristic	Description
AR	Solution is an arbitrary solution
NN	Solutions are generated using a stochastic nearest neighbor algorithm
MF	Solutions are generated using a stochastic multiple fragment algorithm
AR + 2opt	Solutions are generated using AR and improved using a stochastic 2-opt
NN + 2opt	Solutions are generated using NN and improved using a stochastic 2-opt
MF + 2opt	Solutions are generated using MF and improved using a stochastic 2-opt
AR + VFF(2opt)	Solutions are generated using AR then improved using 2-opt where the fitness function for 2-opt is evolved
NN + VFF(2opt)	Solutions are generated using NN then improved using 2-opt where the fitness function for 2-opt is evolved
MF + VFF(2opt)	Solutions are generated using MF then improved using 2-opt where the fitness function for 2-opt is evolved
VFF(NN + 2opt)	Solutions are generated using NN and improved using 2-opt where the fitness function for both the NN and the 2-opt algorithm is evolved
VFF(MF + 2opt)	Solutions are generated using MF and improved using 2-opt where the fitness function for both the MF and the 2-opt algorithm is evolved

Table 2 shows the parameters used in the evolution. Picking the weights between -1 and 1 gives us the possibility to start the search in every direction, including those negatively correlated with the global fitness function.

Table 2. Parameters used to evolve the variable fitness functions for the MOTSP problems

Objective b	Initial Value $L_b \ldots U_b$	Standard Deviation V_b
all	-1…1	0.1

5 Computational Experiments

Each of our ten methods was run 5 times for each of 5 problem instances. They were given the same CPU time (15 minutes) in which to find a solution and were restarted if they completed before the allotted time. The optimal solution of each of the 5 instances was found using CONCORDE [16]. The five instances have 100 cities and 2 objectives, equally weighted in the global fitness function. The quality of a method will be assessed by the average deviation from the optimal tour, measured by this global objective, over the 25 runs.

To tune the genetic algorithm parameters for the variable fitness function evolution, the genetic algorithm was run for 1000 fitness evaluations with different population sizes of 10, 20 and 40 in an attempt to find the best parameters. The population was seeded with *0*, *1* and *All* global fitness functions to see the difference. The parameter tuning experiments show that the genetic algorithm was not very sensitive to the parameters. A Population size of 20, and seeding with no global fitness functions were among the best set of parameters and were used for the rest of the experiments.

The comparative results are shown in Figure 5, showing each method's average deviation from the global optimum over 5 runs of 5 problems instances together with 90% confidence intervals. We can see that NN provides the weakest result. In fact, the stochastic version of *NN* is worse on average than normal *NN* (not shown here). 2-opt can be seen to improve the NN and MF heuristics considerably as expected. When we enhance the 2-opt with the variable fitness function, we can see significant further improvements. When also enhancing the NN or MF as well as the 2-opt we see different results. In the case of *VFF(NN + 2opt)* the solution quality improves again, however, for *VFF(MF + 2opt)* the solution gets worse when compared to *MF + VFF(2opt)*. This decrease in solution is, however, statistically insignificant at the 90% confidence level and could be because *MF + VFF(2opt)* is already producing very good solutions. Overall the results demonstrate that the *MF + VFF(2opt)* and *VFF(MF + 2opt)* perform the best. For every heuristic H in our experiments, VFF(H) performs significantly better than H (at the 90% confidence level). These results provide good evidence that the variable fitness function can be used to enhance a simple local search without needing knowledge of the problem or modification of the search technique.

Figure 6 show a sample of the best variable fitness functions evolved for different MO-TSP problem instances. They are quite different and have very little in common, ranging from really simple to quite complex. This implies, not surprisingly, that there is not a single good variable fitness function for this set of uncorrelated MO-TSP problems instances. This can be seen where the weights of variable fitness function change priority (for example at approximately iteration 75 of the top left graph). This could be because the search has reached a local optimum and changing the direction toward the other objective avoids or escapes it. We will investigate this in more detail below. Each one is quite different because the problem instances have nothing in common. This is as expected because the objectives were generated randomly and uncorrelated.

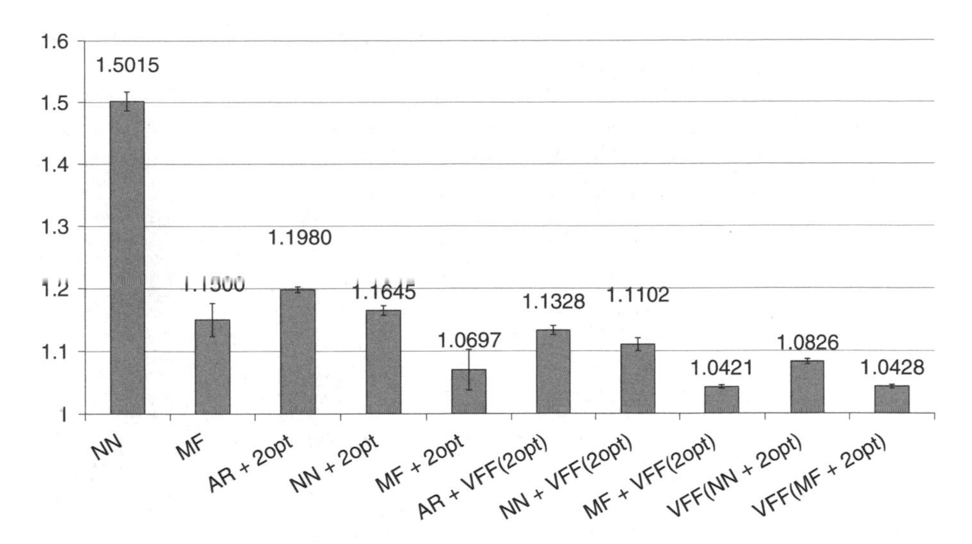

Fig. 5. Average Deviation of the six tested methods from the optimal solution. Error bars show 90% confidence intervals.

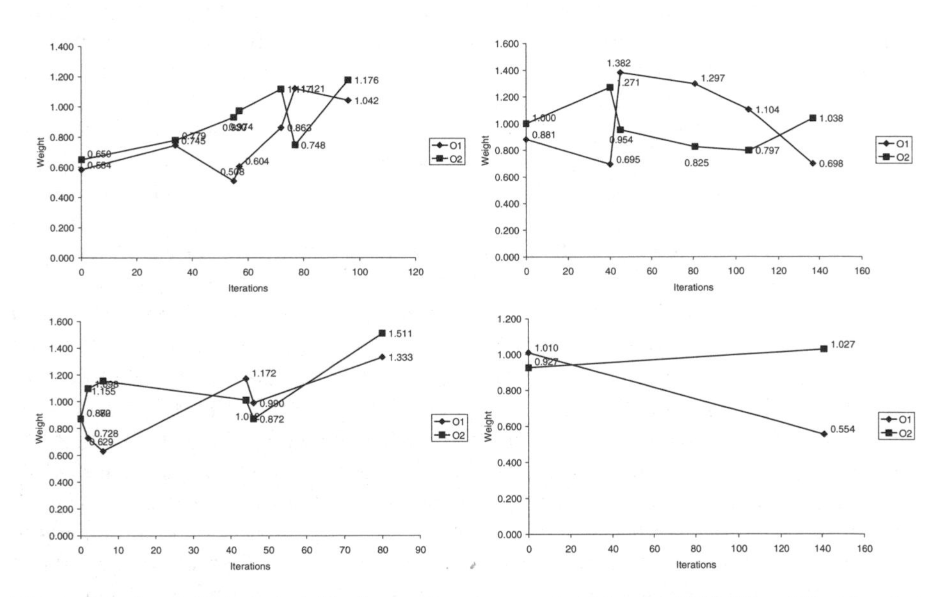

Fig. 6. Sample of the best variable fitness functions evolved for the *VFF(NN + 2Opt)* heuristic for different MO-TSP problems

To show that the evolved variable fitness functions exploit individual problem instances characteristics rather than the characteristics of the TSP itself, we used each of the evolved variable fitness functions for *VFF(NN + 2opt)* on the other problem instances for which it was not evolved. Figure 7 shows this comparison and indicates that using an incorrect variable fitness function is worse than using the global fitness function (comparing *Mismatch VFF(NN + 2 opt)* to *NN + 2 opt*).

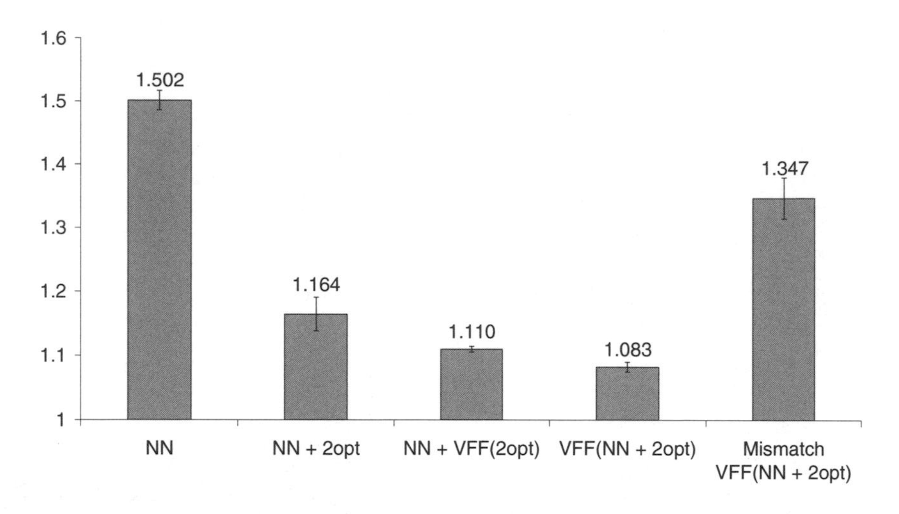

Fig. 8. *Mismatch VFF(NN + 2 opt)* represents the average deviation from the optimal solution when using variable fitness function evolved for other problem instance

For problems where there is significant correlation between the objectives in different instances, a VFF generated using historical problem instances is likely to show good performance on future problem instances. Our initial work on real-world scheduling problems (forthcoming) show that this is indeed the case.

Figure 9 shows a visualisation of the search process of a variable fitness function for a *AR + VFF(2opt)* search. The top plot shows the evolved variable fitness function and the plot below it attempts to visualise the search. *Current Solution Fitness* shows the fitness of the solution at each iteration. *Moves to Local Optima* and *Local Optima Fitness* show the number of 2-opt moves the current solution is to the 2-opt local optimum that would be found if the weights were fixed (at the VFF values) and the fitness of that local optimum. This shows us many things. When the *Moves to Local Optima* reaches zero the search is at a local optimum. When it increases after being zero it has changed search direction and escaped a local optimum. When *Local Optima Fitness* stays constant, the search is heading for the same local optima, and when it changes it has changed direction. When the *Local Optima Fitness* is worse than the *Current Solution Fitness* the search is heading in a non intuitive way (in terms of the global fitness function), away from a globally unpromising region.

From Figure 8 we see that the search is both escaping local optima and changing direction to avoid local optima throughout the search process. Until a good local optimum is reached the *Moves to Local Optimum* stays high, keeping the search away from local optima. After a good local optima is reached, the *Moves to Local Optimum* is increased by more radical changes in the objective weights. During the beginning of the search the *Current Solution Fitness* does not improve much and the *Local Optimum Fitness* varies a lot. This could indicate the search is moving to a different area of the search space. We then see a period of intensification where there is a large improvement in the solution quality and the *Local Optimum Fitness* varies less and the *Moves to Local Optimum* steadily decreases indicating it is heading toward the

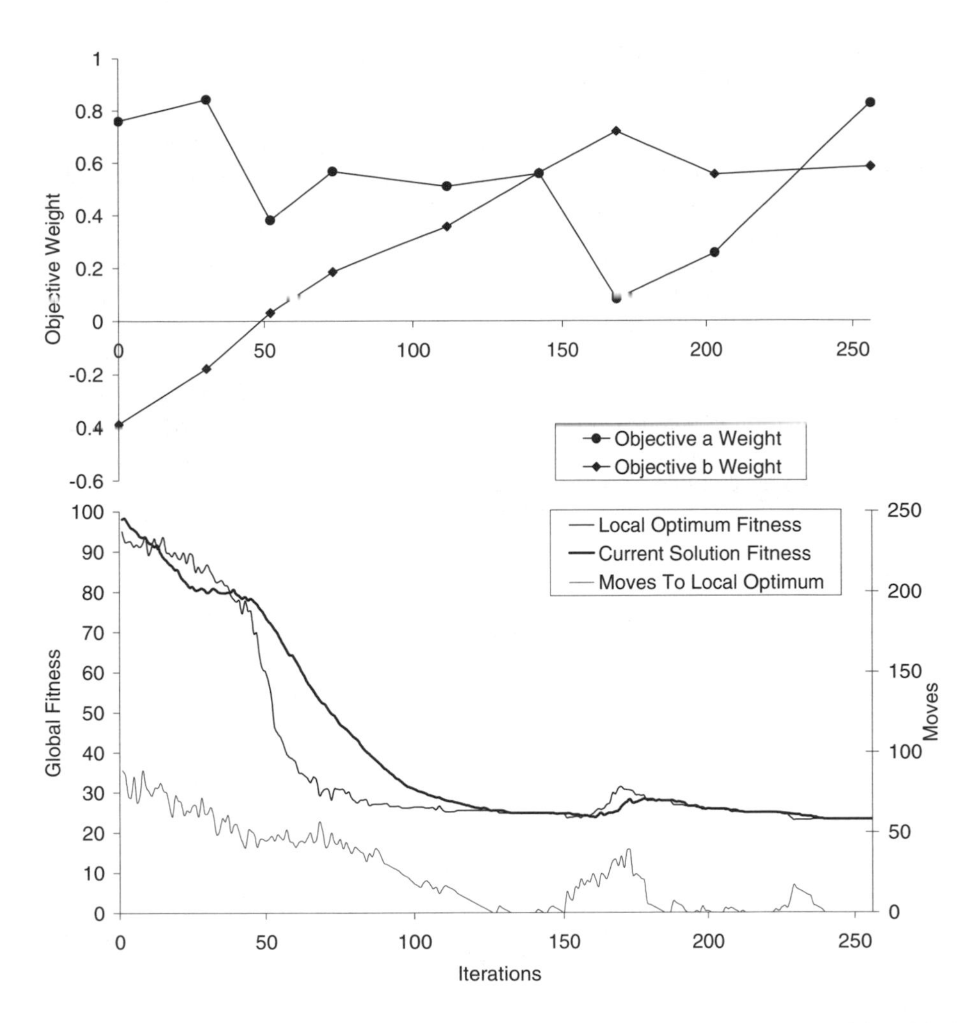

Fig. 8. Visualising the search. Fitness use the right axis and are measured in term of the global fitness function.

same local optima. At around iteration 140, the search has reached a local optima after which, we see a change in the priority of weights in the variable fitness function which leads the search to another local optima at around iteration 200. During this period of "diversification" we can see that the *Local Optimum Fitness* is worse than the *Current Solution Fitness.* This is because the *Objective a Weight* has become very small and the search is probably pushing toward the other objective at the expense of this one.

6 Conclusions

The variable fitness function has demonstrated its ability to enhance local search methods to provide better solutions. We have shown that given a optimization heuristic H and a sufficiently large quantity of CPU time we can produce a better heuristic *VFF(H)*. We have also shown evidence that the variable fitness functions are learning to move the search to different parts of the search space when local optimum are encountered

The generation of a solution using VFF takes longer due to the evolutionary process. However, if the time is available, this method appears to be better than random perturbations to a local search and requires no modification of the local search unlike most common meta-heuristics. Although we have not empirically tested this method against and in conjunction with other common meta-heuristics, we feel that from limited testing, the variable fitness function would result in improvements and we will try this in future work.

Preliminary experiments with more complex, real world problems indicate that when using variable fitness functions on problems with correlated objectives and rather unusual, non-linear, real-world objectives, variable fitness functions can be evolved to work on a range of problem instances and hence the VFF can be evolved offline and then used online with no increase in CPU time.

References

1. Mladenovic, N., Hansen, P.: Variable neighborhood search. Computers & Operational Research 24(11), 1097–1100 (1997)
2. Tsang, E., Voudouris, C.: Fast local search and guided local search and their application to British Telecom's workforce scheduling problem. Operations Research Letters 20(3), 119–127 (1997)
3. Aarts, E.H.L., Korst, J.H.M.: Simulated Annealing and Boltzmann Machines: A Stochastic Approach to Computing. Whiley, Chichester (1989)
4. Lawler, E.L., Lenstra, J.K., Rinnooy Kan, A.H.G., Shmoys, D.B.: The Traveling Salesman Problem: A Guided Tour of Combinatorial Optimization. Wiley, Chichester (1985)
5. Kolisch, R., Hartmann, S.: Experimental Investigations of Heuristics for RCPSP: An Update. European Journal of Oper. Res. 174(1), 23–37 (2006)
6. Hansen, P., Mladenovic, N.: Variable neighborhood search: Principles and applications. European Journal of Oper. Res. 130(3), 449–467 (2001)
7. Burke, E.K., Cowling, P.I., Keuthen, R.: Effective Local And Guided Variable Neighbourhood Search Methods for the Asymetric Travelling Salesman Problem. Applications Of Evolutionary Computing. In: Boers, E.J.W., Gottlieb, J., Lanzi, P.L., Smith, R.E., Cagnoni, S., Hart, E., Raidl, G.R., Tijink, H. (eds.) EvoIASP 2001, EvoWorkshops 2001, EvoFlight 2001, EvoSTIM 2001, EvoCOP 2001, and EvoLearn 2001. LNCS, vol. 2037, pp. 203–212. Springer, Heidelberg (2001)
8. Jaszkiewicz, A.: Genetic Local Search for Multi-Objective Combinatorial Optimisation. European Journal of Operational Research 137(1), 50–71 (2002)
9. Johnston, D.S., McGeoch, L.A.: The Travelling Salesman Problem: a case study in local optimisation. In: Aarts, E.J.L., Lenstra, J.k. (eds.) Local Search in Combinatorial Optimisation. Wiley, London (1997)
10. Potgieter, G., Engelbrecht, A.P.: Genetic Algorithms for the Structure Optimisation of learned Polynomial Expressions. Applied Mathematics and Computation 186(2), 1441–1466 (2007)
11. Shimojika, K., Fukuda, T., Hasehawa, Y.: Self-Tuning Fuzzy Modeling with adaptive membership function, rules, and hierarchical structure-based on Genetic Algorithm. Fuzzy Sets And Systems 71(3), 295–309 (1995)

12. Reeves, C.R.: Genetic Algorithms and Combinatorial Optimization. In: Rayward-Smith, V.J. (ed.) Applications of Modern Heuristic Methods, Alfred Waller, Henley-on-Thames, pp. 111–125 (1995)
13. Yao, X.: Evolving artificial neural networks. Proceedings Of The IEEE 87(9), 1423–1447 (1999)
14. Croes, G.A.: A Method for Solving Traveling Salesman Problems. Operations Res. 6, 791–812 (1958)
15. Corne, D., Ross, P.: Peckish Initialisation Strategies for Evolutionary Timetabling. In: Burke, E.K., Ross, P. (eds.) PATAT 1995. LNCS, vol. 1153, pp. 227–240. Springer, Heidelberg (1996)
16. Applegate, D., Bixby, R., Chvatal, V., Cook, W.: Concorde TSP Solver, `http://www.tsp.gatech.edu/concorde.html`

A Continuous Characterization of Maximal Cliques in k-Uniform Hypergraphs

Samuel Rota Bulò and Marcello Pelillo

Dipartimento di Informatica.
Università Ca' Foscari di Venezia
{srotabul,pelillo}@dsi.unive.it

Abstract. In 1965 Motzkin and Straus established a remarkable connection between the local/global maximizers of the Lagrangian of a graph G over the standard simplex Δ and the maximal/maximum cliques of G. In this work we generalize the Motzkin-Straus theorem to k-uniform hypergraphs, establishing an isomorphism between local/global minimizers of a particular function over Δ and the maximal/maximum cliques of a k-uniform hypergraph. This theoretical result opens the door to a wide range of further both practical and theoretical applications, concerning continuous-based heuristics for the maximum clique problem on hypergraphs, as well as the discover of new bounds on the clique number of hypergraphs. Moreover we show how the continuous optimization task related to our theorem, can be easily locally solved by mean of a dynamical system.

1 Introduction

Many problems of practical interest are inherently intractable, in the sense that it is not possible to find fast (i.e., polynomial time) algorithms to solve them exactly, unless the classes P and NP coincide. The Maximum Clique Problem (MCP) is one of the most famous intractable combinatorial optimization problems, that asks for the largest complete subgraph of a given graph. This problem is even hard to approximate within a factor of $n/2^{(\log n)^{1-\epsilon}}$ for any $\epsilon > 0$ where n is the number of nodes in the graph [15]. Although this pessimistic state of affairs and because of its important applications in different fields such as computer vision, experimental design, information retrieval and fault tolerance, much attention has gone into developing efficient heuristics for the MCP, even if no formal guarantee of performance may be provided, but are nevertheless useful in practical applications. Moreover many important problems can be easily reduced to maximum clique problem e.g. boolean satisfiability problem, subgraph isomorphism problem, vertex cover problem etc.

Plenty of heuristics have been proposed over the last 50 years and we refer to [7] for a complete survey about complexity issues and applications of the MCP. In this introduction, we will focus our attention in particular on the continuous-based class of heuristics, since they are strongly related to the topics addressed in this paper. The heuristics of this class are mostly based on a result

V. Maniezzo, R. Battiti, and J.-P. Watson (Eds.): LION 2007 II, LNCS 5313, pp. 220–233, 2008.

of Motzkin and Straus [17] that establishes a remarkable connection between the maximum clique problem and the extrema of the Lagrangian of a graph (1). In Section 2 we will see in deeper details the Motzkin-Straus theorem, but briefly it states, especially in its regularized version, an isomorphism between the set of maximal/maximum cliques of an undirected graph G and the set of local/global maximizers of the Lagrangian of G. This continuous formulation of the MCP suggests a fundamental new way of solving this problem, by allowing a shift from the discrete domain to the continuous one in an elegant manner. As pointed out in [20] continuous formulations of discrete problems are partic ularly attractive, because they not only allow us to exploit the full arsenal of continuous optimization techniques, thereby leading to the development of new algorithms, but may also reveal unexpected theoretical properties.

From an applicative point of view the Motzkin-Straus result led to the development of several MCP heuristics [6,8,13,21,23], but very interesting are also its theoretical implications. This result in fact was originally achieved by Motzkin and Straus to support an alternative proof of a slightly weaker version of the fundamental Turán theorem [27], moreover it was successfully used to achieve several bounds for the clique number of graphs [9,28,29]. The Motzkin-Straus theorem was also successfully generalized to vertex- weighted graphs [14] and edge-weighted graphs [22].

Recently the interest of researchers in many fields is focusing on hypergraphs, i.e. generalizations of graphs where edges are subsets of vertices, because of their greater expressiveness in representing higher-order relations. Just as graphs naturally represent many kinds of information in mathematical and computer science problems, hypergraphs also arise naturally in important practical problems [10,19,30]. Moreover, many theorems involving graphs, as for example the Ramsey's theorem or the Szemerédi lemma, also hold for hypergraphs, in particular for k-uniform hypergraphs (or more simply k-graphs), i.e. hypergraphs whose edges have all cardinality k. Nevertheless, all known intractable problems on graphs can be reformulated on hypergraphs and in particular the maximum clique problem.

Even if clique problems on hypergraphs are gaining increasing popularity in several scientific communities, a bridge from these discrete structures to the continuous domain is still missing. With our work we will fill up this gap, in the same way as the Motzkin-Straus theorem filled it up in the context of graphs. Hence the contribution of this paper is purely theoretical and basically consists in a generalization of the Motzkin-Straus theorem to k-uniform hypergraphs. However, as happened for the Motzkin-Straus theorem, our hope is to open the door to a wide range of further both practical and theoretical applications. First of all, we furnish a continuous characterization of maximal cliques in k-graphs, allowing the development of continuous-based heuristics for the maximum clique problem over hypergraphs based on it. Thereby, in Section 5 we provide a discrete dynamical system to elegantly find maximal cliques in k-graphs, that turns out to include the heuristic for MCP developed by Pelillo [23] on graphs as a special case (in fact graphs are 2-uniform hypergraphs). Moreover our theorem can be used to achieve new bounds for the clique number on k-graphs, a very popular problem in the extremal graph theory field, however we leave this topic as a future development of this work.

2 The Motzkin-Straus Theorem

Let $G = (V, E)$ be an (undirected) graph, where $V = \{1, \dots, n\}$ is the vertex set and $E \subseteq \binom{V}{2}$ is the edge set, with $\binom{V}{k}$ denoting the set of all k-element subsets of V. A *clique* of G is a subset of mutually adjacent vertices in V. A clique is called *maximal* if it is not contained in any other clique. A clique is called *maximum* if it has maximum cardinality. The maximum size of a clique in G is called the *clique number* of G and is denoted by $\omega(G)$.

Consider the following function $L_G : \Delta \mapsto \mathbb{R}$, sometimes called the *Lagrangian* of graph G

$$L_G(\boldsymbol{x}) = \sum_{\{i,j\} \in E} x_i x_j \tag{1}$$

where

$$\Delta = \{\boldsymbol{x} \in \mathbb{R}^n : \boldsymbol{x} \geq 0, \ \sum_{i=1}^{n} x_i = 1\}$$

is the *standard simplex*.

In 1965, Motzkin and Straus [17] established a remarkable connection between the maxima of the Lagrangian of a graph and its clique number.

Theorem 1 (Motzkin-Straus). *Let G be a graph with clique number $\omega(G)$, and $\boldsymbol{x}^*$ a maximizer of L_G then*

$$L_G(\boldsymbol{x}^*) = \frac{1}{2}\left[1 - \frac{1}{\omega(G)}\right]$$

Additionally Motzkin and Straus proved that a subset of vertices S is a maximum clique of G if and only if its *characteristic vector* $\boldsymbol{x}^S$ is a global maximizer of L_G.[1] The characteristic vector of a set S is the vector in Δ defined as:

$$\boldsymbol{x}_i^S = \frac{1_{i \in S}}{|S|}$$

where $|S|$ indicates the cardinality of the set S and 1_P is an indicator function giving 1 if property P is satisfied and 0 otherwise. With $\sigma(\boldsymbol{x})$ we will denote the *support* of a vector $\boldsymbol{x} \in \Delta$, i.e. the set of positive components in $\boldsymbol{x}$. For example, the support of the characteristic vector of a set S is S.

Gibbons, Hearn, Pardalos and Ramana [14], and Pelillo and Jagota [24], extended the theorem of Motzkin and Straus, providing a characterization of maximal cliques in terms of local maximizers of L_G, however not all local maximizers

[1] Actually, Motzkin and Straus provided just the "only if" part of this theorem, even if the converse direction is a direct consequence of their results [24].

were in the form of a characteristic vector. Finally Bomze et al. [6] introduced a regularizing term in the graph Lagrangian obtaining $L_G^\tau : \Delta \mapsto \mathbb{R}$ defined as

$$L_G^\tau(\boldsymbol{x}) = L_G(\boldsymbol{x}) + \tau \sum_{i \in V} x_i^2$$

and proved that all local maximizers of L_G^τ are strict and in one-to-one relation with the characteristic vector of the maximal cliques of G, provided that $0 < \tau < \frac{1}{2}$.

Theorem 2 (Bomze). *Let G be a graph and $0 < \tau < \frac{1}{2}$. A vector $\boldsymbol{x} \in \Delta$ is a global/local maximizer of L_G^τ over Δ if and only if it is the characteristic vector of a maximum/maximal clique of G.*

The Motzkin Straus theorem was successfully extended also to vertex-weighted graphs by Gibbons et al. [14] and edge-weighted graphs by Pavan and Pelillo [22].

In this paper we provide a further generalization of the Motzkin-Straus as well as the Bomze theorems to k-uniform hypergraphs, but firstly, we will introduce hypergraphs and review another generalization of the Motzkin-Straus theorem due to Sós and Straus [26].

3 k-Uniform Hypergraphs

Let $\mathcal{P}(A)$ be the power set of A. A *hypergraph* G is a pair (V, E) where $V = \{1, \ldots, n\}$ is a set of vertices and $E \subseteq \mathcal{P}(V)$ is a set of hyperedges. If all hyperedges have cardinality k, then the hypergraph is *k-uniform* (or more easily it is called *k-graph*). A *clique* C of G is a set of vertices such that every subset of C of order k forms an hyperedge of G. A clique is *maximal* if it is not contained in any other clique. It is *maximum* if it has maximum cardinality. The *clique number* $\omega(G)$ of a k-graph G is the cardinality of a maximum clique.

The Lagrangian of a k-graph $G = (V, E)$ is denoted by $L_G : \Delta \mapsto \mathbb{R}$ and defined as

$$L_G(\boldsymbol{x}) = \sum_{e \in E} \prod_{j \in e} x_j. \tag{2}$$

Unfortunately L_G cannot be directly used to extend the Motzkin-Straus theorem to k-graphs. Frankl and Rödl [12] proved that by taking a maximizer $\boldsymbol{x}^*$ of L_G with support as small as possible, the subhypergraph induced by the support of $\boldsymbol{x}^*$ is a 2-cover, i.e. a hypergraph such that every pair of vertices is contained in some hyperedge. Since 2-covers in graphs are basically cliques, we could expect a possible generalization of the Motzkin-Straus theorem where the clique number is replaced by the size l of the maximum 2-cover in the hypergraph. However $\boldsymbol{x}^*$ is not necessarily in the form of a characteristic vector, and it is not in general possible to express l as a function of $L_G(\boldsymbol{x}^*)$. Nevertheless, this result was used by Mubay [18] to achieve a bound for $L_G(\boldsymbol{x}^*)$ in terms of l on k-graphs, obtaining

$$L_G(\boldsymbol{x}^*) \leq \binom{l}{k} l^{-k}.$$

and he used it to provide an hypergraph extension of the Turán's theorem.

The only real case of generalization of the Motzkin-Straus theorem to k-uniform hypergraphs, although not explicit and not general, is due to Sós and Straus [26]. They attach a nonnegative weight $x(H_l)$ to every complete l-subgraph H_l of a graph G, normalized by the condition

$$\sum_{H_l \subseteq G} x(H_l)^l = 1$$

and to every complete $(l+1)$-subgraph H_{l+1} of G they attach the weight

$$x(H_{l+1}) = \prod_{H_l \subset H_{l+1}} x(H_l)$$

and they define

$$f_G(x) = \sum_{H_{l+1} \subseteq G} x(H_{l+1}).$$

Then they get the following.

Theorem 3. $\max_x f_G(x) = \binom{k}{l+1} / \binom{k}{l}^{(l+1)/l}$, *where k is the order of a maximum clique K of G. This maximum is attained by attaching weights $\binom{k}{l}^{-1/l}$ to the l-subgraphs of K and weight 0 to all other complete l-subgraphs.*

Note that, the case $l = 1$ is exactly the Motzkin-Straus theorem.

Even if this result does not explicitly apply to hypergraphs, actually this theorem could be extended to k-graphs by attaching weights to subsets of hyperedges. However in order this theorem to work, the k-graph should satisfy a strong property that it to be a complete-subgraph graph of an ordinary graph (also said to be *conformal* [4]). This restricts the applicability of this theorem to a class of hypergraphs isomorphic to a subclass of 2-graphs having cliques of cardinality $\geq k$. We will see in the subsequent sections that our generalization applies to all k-uniform hypergraphs.

4 Characterization of Maximal Cliques on k-Graphs

Let $G = (V, E)$ be a k-graph. The *complement* of G is given by $\bar{G} = (V, \bar{E})$ where $\bar{E} = \binom{V}{k} \setminus E$.

Consider the following function $h_G(\boldsymbol{x}) : \Delta \mapsto \mathbb{R}$ defined as

$$h_G(\boldsymbol{x}) = L_G(\boldsymbol{x}) + \tau \sum_{i=1}^{n} x_i^k \tag{3}$$

where $\tau \in \mathbb{R}$ and L_G is the Lagrangian of the hypergraph G defined in (2).

We will indicate with $h_G^j(\boldsymbol{x})$ the partial derivative of h_G with respect to x_j, i.e.

$$h_G^j(\boldsymbol{x}) = \frac{\partial h_G(\boldsymbol{x})}{\partial x_j} = \sum_{e \in E} 1_{j \in e} \prod_{i \in e \setminus \{j\}} x_i + \tau k x_j^{k-1}.$$

Lemma 1. *Let G be a k-graph and let $\boldsymbol{x}$ be a global/local minimizer of $h_{\bar{G}}$ with $\tau > 0$. If $C = \sigma(\boldsymbol{x})$ is a clique of G then it is a maximum/maximal clique and $\boldsymbol{x}$ is the characteristic vector of C.*

Proof. Since $\boldsymbol{x}$ is a local minimizer of $h_{\bar{G}}$ over the simplex, it satisfies the first order necessary Karush-Kuhn-Tucker (KKT) condition. Therefore for all $j \in C$ we have that $\lambda = \tau k x_j^{k-1}$ and this implies that $\boldsymbol{x}$ is the characteristic vector of C. Moreover if there exists a larger clique D that contains C, then there exists a vertex $j \in D \setminus C$ such that $h_{\bar{G}}^j(\boldsymbol{x}) = 0 < \lambda$. This contradicts the KKT condition and hence C is a maximal clique of G.

Finally, $h_{\bar{G}}(\boldsymbol{x}) = \tau|\sigma(\boldsymbol{x})|^{1-k}$ attains its global minimum when $\boldsymbol{x}$ is the characteristic vector of a maximum clique. □

Lemma 2. *Let G be a k-graph and $\boldsymbol{x}^C$ the characteristic vector of a maximum/maximal clique C of G. Then $\boldsymbol{x}^C$ is a strict global/local minimizer of $h_{\bar{G}}$, provided that $0 < \tau < \frac{1}{k}$.*

Proof (Sketch). For simplicity let $\boldsymbol{x} = \boldsymbol{x}^C$. We can prove [25] that $\boldsymbol{x}$ is a strict local minimizer of $h_{\bar{G}}$ by checking that the second order sufficiency conditions for local minimum are satisfied.

Finally, $h_{\bar{G}}(\boldsymbol{x}^C) = \tau|C|^{1-k}$ attains its global minimum where C is as large as possible, i.e. a maximum clique. □

Lemma 3. *Let G be a k-graph. If $\boldsymbol{x}$ is a global/local minimizer of $h_{\bar{G}}$ then it is the characteristic vector of a maximum/maximal clique of G, provided that $\tau < \frac{1}{2^k-2}$.*

Proof (Sketch). Let $\boldsymbol{x}$ be a local minimizer of $h_{\bar{G}}$. We claim that its support forms a clique of G. Otherwise suppose that an edge $\tilde{e} \subseteq \sigma(\boldsymbol{x})$ is missing. Let $w, j \in \tilde{e}$ and take $\boldsymbol{y} = \boldsymbol{x} + \varepsilon(\boldsymbol{e}^j - \boldsymbol{e}^w)$, where $\boldsymbol{e}^j$ denotes a zero vector except for the j-th element set to 1 and where $0 < \varepsilon < x_w$ and assume without loss of generality that $x_w \leq x_j \leq \min_{z \in \tilde{e} \setminus \{j,w\}} x_z$.

We can prove [25] that $h_{\bar{G}}(\boldsymbol{y}) - h_{\bar{G}}(\boldsymbol{x}) < 0$, contradicting the minimality of $\boldsymbol{x}$ and hence $C = \sigma(\boldsymbol{x})$ is a clique of G. Finally by applying Lemma 1 we conclude the proof. □

The following theorem generalizes the Bomze's Theorem (2) on k-graphs.

Theorem 4. *Let G be a k-graph and $0 < \tau < \frac{1}{2^k-2}$. A vector $\boldsymbol{x} \in \Delta$ is a global/local minimizer of $h_{\bar{G}}(\boldsymbol{x})$ if and only if it is the characteristic vector of a maximum/maximal clique of G.*

Proof. It follows from Lemmas 2 and 3. □

Note that if we take $k = 2$ and $0 < \tau < \frac{1}{2}$ then global/local minimizers of h correspond to global/local maximizers of $L_G^{\frac{1}{2}-\tau}$. In fact

$$h(\boldsymbol{x}) = \sum_{\{i,j\}\in\bar{E}} x_i x_j + \tau \sum_{i=1}^{n} x_i^2 = \frac{1}{2} - \sum_{\{i,j\}\in E} x_i x_j + \left(\tau - \frac{1}{2}\right) \sum_{i=1}^{n} x_i^2 =$$

$$= \frac{1}{2} - \left[\sum_{\{i,j\}\in E} x_i x_j + \left(\frac{1}{2} - \tau\right) \sum_{i=1}^{n} x_i^2 \right] = \frac{1}{2} - L_G^{\frac{1}{2}-\tau}(\boldsymbol{x})$$

Since $0 < \frac{1}{2} - \tau < \frac{1}{2}$, what we obtain is an equivalent formulation of the Bomze Theorem on graphs in terms of a minimization task.

The following corollary is our generalization of the Motzkin-Straus Theorem (1) to k-graphs, with the only difference that we deal for convenience with minimizers of a function instead of maximizers.

Corollary 1. *Let G be a k-graph with clique number $\omega(G)$. Then $h_{\bar{G}}$ attains its minimum at $\tau\,\omega(G)^{1-k}$ provided that $0 < \tau \leq \frac{1}{2^k - 2}$.*

Proof (Sketch). Let $\boldsymbol{x}$ be a global minimizer of $h_{\bar{G}}$ with support as small as possible. We claim that its support forms a clique of G. Otherwise suppose that an edge $\tilde{e} \subseteq \sigma(\boldsymbol{x})$ is missing. Let $w, j \in \tilde{e}$ and take $\boldsymbol{y} = \boldsymbol{x} + x_w(\boldsymbol{e}^j - \boldsymbol{e}^w)$ and assume without loss of generality that $x_w \leq x_j \leq \min_{z\in\tilde{e}\setminus\{j,w\}} x_z$. Then we can prove [25] that $h_{\bar{G}}(\boldsymbol{y}) \leq h_{\bar{G}}(\boldsymbol{x})$. If the inequality is strict, it clearly contradicts the minimality of $\boldsymbol{x}$, and if $h_{\bar{G}}(\boldsymbol{y}) = h_{\bar{G}}(\boldsymbol{x})$ it contradicts the minimality of the support size of $\boldsymbol{x}$.

Hence $\sigma(\boldsymbol{x})$ is a clique of G. By Lemma 1 follows that $\boldsymbol{x}$ is the characteristic vector of a maximum clique of G and thereby $h_{\bar{G}}(\boldsymbol{x}) = \tau|C|^{1-k} = \tau\omega(G)^{1-k}$. □

Note that this result is equivalent to the original Motzkin-Straus Theorem (1) for graphs, if we take $k = 2$ and $\tau = \frac{1}{2}$. In fact, in this case we obtain

$$L_G(\boldsymbol{x}) = \sum_{\{i,j\}\in E} x_i x_j = \frac{1}{2} - \sum_{\{i,j\}\in\bar{E}} x_i x_j - \frac{1}{2}\sum_{i=1}^{n} x_i^2 = \frac{1}{2} - h(\boldsymbol{x})$$

and it follows that

$$\max_{\boldsymbol{x}\in\Delta} L_G(\boldsymbol{x}) = \frac{1}{2} - \min_{\boldsymbol{x}\in\Delta} h(\boldsymbol{x}) = \frac{1}{2} - \frac{1}{2\omega(G)} = \frac{1}{2}\left[1 - \frac{1}{\omega(G)}\right].$$

5 Finding Maximal Cliques of k-Graphs

Summarizing our results, we propose a generalization of a well-known theorem in the extremal graph theory field to k-graphs that turns out to provide a continuous characterization of a purely discrete problem, i.e. finding maximal cliques in k-graphs. More precisely, we implicitly provide an isomorphism between the set of maximal/maximum cliques of a k-graph G and the set of local/global minimizers of a particular function $h_{\bar{G}}$ over Δ, that permits to perform local optimization on $h_{\bar{G}}$ in order to extract, through the isomorphism, a maximal clique of the k-graph G. In this section we will see that the optimization of $h_{\bar{G}}$ may be easily carried out thanks to a theorem due to Baum and Eagon [1].

In the late 1960s, Baum and Eagon [1] introduced a class of nonlinear transformations in probability domain and proved a fundamental result which turns out to be very useful for the optimization task at hand. Their result generalizes an earlier one by Blakley [5] who discovered similar properties for certain homogeneous quadratic transformations. The next theorem introduces what is known as the Baum-Eagon inequality.

Theorem 5 (Baum-Eagon). *Let $P(\boldsymbol{x})$ be a homogeneous polynomial in the variables x_i with nonnegative coefficients, and let $\boldsymbol{x} \in \Delta$. Define the mapping $\boldsymbol{z} = \mathcal{M}(\boldsymbol{x})$ as follows:*

$$z_i = x_i \frac{\partial P(\boldsymbol{x})}{\partial x_i} \Big/ \sum_{j=1}^{n} x_j \frac{\partial P(\boldsymbol{x})}{\partial x_j}, \qquad i = 1, \ldots, n. \tag{4}$$

Then $P(\mathcal{M}(\boldsymbol{x})) > P(\boldsymbol{x})$, unless $\mathcal{M}(\boldsymbol{x}) = \boldsymbol{x}$. In other words $\mathcal{M}$ is a growth transformation for the polynomial P.

This result applies to homogeneous polynomials, however in a subsequent paper, Baum and Sell [3] proved that Theorem 5 still holds in the case of arbitrary polynomials with nonnegative coefficients, and further extended the result by proving that $\mathcal{M}$ increases P homotopically, which means that

$$P(\eta\mathcal{M}(\boldsymbol{x}) + (1-\eta)\boldsymbol{x}) \geq P(\boldsymbol{x}), \qquad 0 \leq \eta \leq 1$$

with equality if and only if $\mathcal{M}(\boldsymbol{x}) = \boldsymbol{x}$.

The Baum-Eagon inequality provides an effective iterative means for maximizing polynomial functions in probability domains, and in fact it has served as the basis for various statistical estimation techniques developed within the theory of probabilistic functions of Markov chains [2]. As noted in [3], the mapping $\mathcal{M}$ defined in Theorem 5 makes use of the first derivative only and yet is able to take finite steps while increasing P. This contrasts sharply with classical gradient methods, for which an increase in the objective function is guaranteed only when infinitesimal steps are taken, and determining the optimal step size entails computing higher-order derivatives. Additionally, performing gradient ascent in Δ requires some projection operator to ensure that the constraints not be violated, and this causes some problems for points lying on the boundary [11,16]. In (4), instead, a computationally simple vector normalization is required.

It is worth noting that not all stationary points of the mapping $\mathcal{M}$ correspond to local maxima of the polynomial P; consider the vertices of Δ as an example. However all local maxima are the only stationary states that are stable, or even asymptotically stable if they are strict. Therefore if the dynamics gets trapped in non optimal stationary states, it suffices a small perturbation to get rid of the problem. We will see an example of this fact in Section 6.

Moving a step back to our function $h_{\bar{G}}$, it satisfies the hypothesis of Theorem 5 since it is a homogeneous polynomial of degree k with nonnegative coefficients in the variables x_i with $\boldsymbol{x} \in \Delta$. However our targets are not local maxima but local minima. Fortunately, it turns out that we can transform our minimization problem into an equivalent maximization one, by keeping the conditions of Theorem 5 still satisfied.

Note that all coefficients of $h_{\bar{G}}(\boldsymbol{x})$ are positive and upper bounded by 1. Furthermore, let $\xi = \max\left[\tau, \frac{1}{k!}\right]$ and note that ξ can be expressed as a complete homogeneous polynomial $\pi(\boldsymbol{x})$ of degree k in the variables x_i as follows

$$\xi = \xi \left(\sum_{i=1}^{n} x_i \right)^k = \pi(\boldsymbol{x}), \qquad \forall \boldsymbol{x} \in \Delta.$$

It is trivial to verify that the polynomial $\pi(\boldsymbol{x}) - h_{\bar{G}}(\boldsymbol{x})$ is a homogeneous polynomial of degree k with nonnegative coefficients. Moreover

$$\arg\min_{\boldsymbol{x} \in \Delta} h_{\bar{G}}(\boldsymbol{x}) = \arg\max_{\boldsymbol{x} \in \Delta} \left[\xi - h_{\bar{G}}(\boldsymbol{x})\right] = \arg\max_{\boldsymbol{x} \in \Delta} \left[\pi(\boldsymbol{x}) - h_{\bar{G}}(\boldsymbol{x})\right].$$

Therefore, in order to minimize $h_{\bar{G}}$ we can apply Theorem 5 considering $P(\boldsymbol{x}) = \pi(\boldsymbol{x}) - h_{\bar{G}}(\boldsymbol{x})$, and since

$$\frac{\partial \pi(\boldsymbol{x})}{\partial x_i} = k\xi \left(\sum_{i=1}^{n} x_i \right)^{k-1} = k\xi$$

we end up with the following dynamics for the minimization of $h_{\bar{G}}$ over Δ

$$x_i^{(t+1)} = \frac{x_i^{(t)} \left[k\xi - h_{\bar{G}}^{i}(\boldsymbol{x}^{(t)})\right]}{k\xi - \sum_{j=1}^{n} x_j^{(t)} h_{\bar{G}}^{j}(\boldsymbol{x}^{(t)})}, \tag{5}$$

that will converge to a local minima of $h_{\bar{G}}$ starting from any state $\boldsymbol{x}$ in the interior of Δ, which corresponds by Theorem 4 to a maximal clique of the k-graph G.

6 A Toy Example

This section is not intended to provide experimental evidence that the dynamics (5) works, since we have a proof that guarantees it. Indeed, we provide a very simple toy example.

Figure 1 represents a 3-graph T, and the two sets that seem to be 4-edges are actually complete 3-subgraphs on the respective vertex sets. Hence T contains all possible 3-edges on the 5 vertices except for $\{0, 3, 4\}$ and $\{1, 3, 4\}$. T is a small example of a non conformal graph, i.e. it is not a complete-subgraphs graph of an ordinary graph. In other words, there exists no ordinary graph that has the same maximal cliques and therefore the generalization of the Motzkin-Straus theorem due to Sós and Straus [26] does not hold on this 3-graph. The set of maximal cliques of T is $\{\{0, 1, 2, 3\}, \{0, 1, 2, 4\}, \{2, 3, 4\}\}$.

We illustrate the behaviour of the dynamics (5) when applied to our toy example. Our parameters choice in this test is $\tau = \frac{1}{12}$, but no matter what

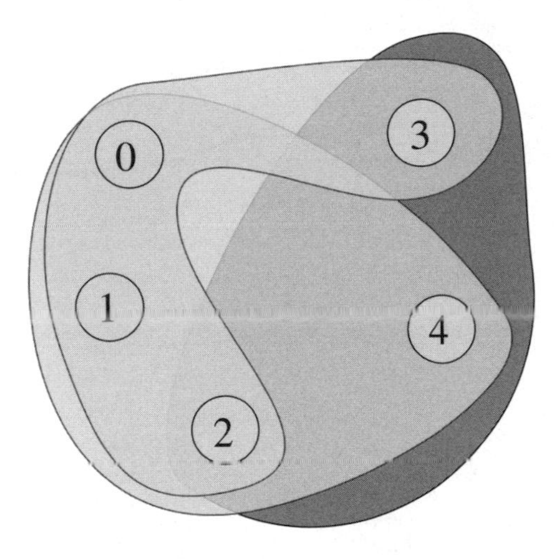

Fig. 1. A non conformal 3-graph T. Note that the sets $\{0, 1, 2, 3\}$ and $\{0, 1, 2, 4\}$ should be interpreted as complete 3-subgraphs on the respective vertex set. We draw them as 4-edges only for graphical clarity. In other words T contains all possible 3-edges on the 5 vertices except for $\{0, 3, 4\}$ and $\{1, 3, 4\}$.

is chosen as long as $0 < \tau < \frac{1}{6}$ as stated in Theorem 4, and therefore in the dynamics we have $k\xi = \frac{1}{2}$. The initial state encodes the hypothesis we make about the likelihood of a vertex to be part of a maximal clique, in fact if we set for example the i-th component of the initial state vector to zero then the i-th vertex will never be considered in a solution. Figure 2 presents three plots of the evolution of the state vector of the dynamics (5) for the 3-graph T over time. The initial states are respectively set to the simplex barycenter in the first two plots in order to have full uncertainty, and to $\boldsymbol{x}^{(0)} = (0.1, 0.1, 0.1, 0.35, 0.35)'$ in the last one in order to provide an initial stronger preference on the vertices 3 and 4.

Analysing our toy graph, we see that vertex 2 belongs to every maximal clique of T, while vertices 0 and 1 are shared between the two maximal 4-cliques and finally vertices 3 and 4 belong individually to a different maximal 4-clique, but together to the maximal 3-clique. Considering the first 114 iterations of the first two plots in Figure 2, we see that without advancing preferences of vertices, i.e. we start from the barycenter of the simplex, the dynamics converges to a stationary state, that is not optimal and hence not stable, but very informative. In fact, vertex 2 that certainly belongs to a maximal clique, has the highest likelihood, followed by vertices 0 and 1, that are shared between the two biggest maximal cliques in T and finally we find vertices 3 and 4. By inducing a small perturbation at that point, we introduce some random preference on vertices that leads the dynamics to a certain solution; in the first case we end up with the maximal 4-clique $\{0, 1, 2, 3\}$, while in the second one we end up with the maximal 4-clique $\{0, 1, 2, 4\}$. Let us move now our attention on the last plot in Figure 2. In order to extract the smallest maximal clique, that has a smaller basin

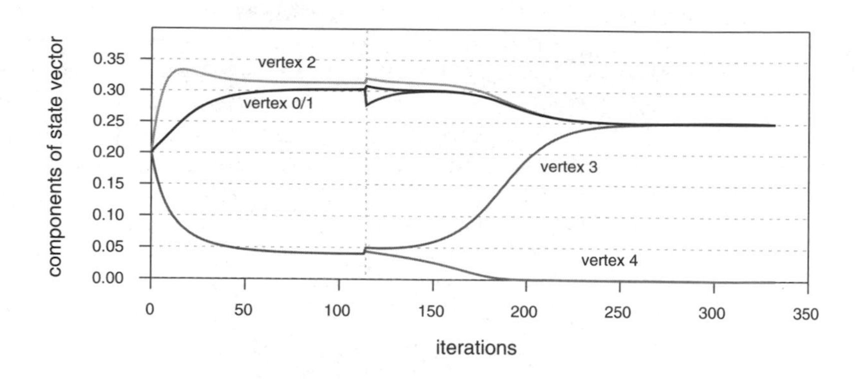

(a) Extraction of the maximal clique $\{0, 1, 2, 3\}$

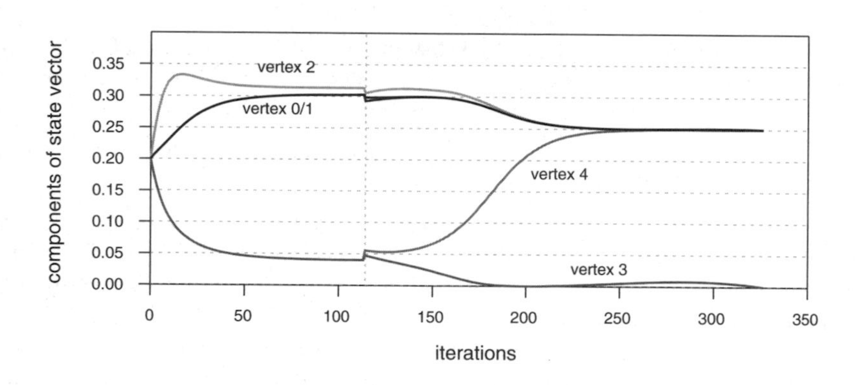

(b) Extraction of the maximal clique $\{0, 1, 2, 4\}$

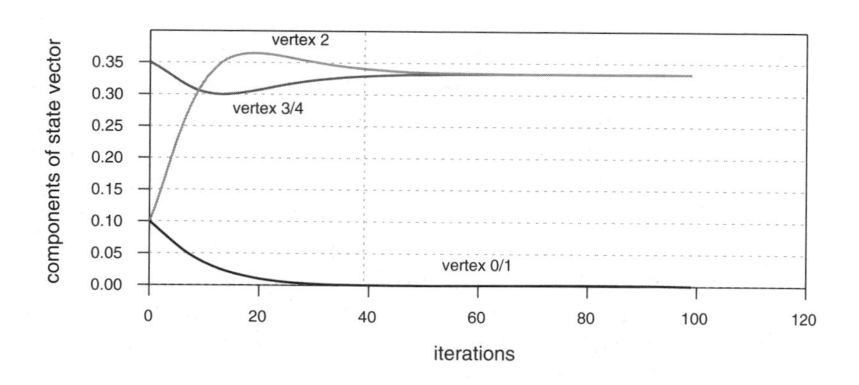

(c) Extraction of the maximal clique $\{2, 3, 4\}$

Fig. 2. Evolution of the components of the state vector $\boldsymbol{x}^{(t)}$ for the k-graph in Figure 1, using the dynamics (5)

of attraction we have to put stronger preferences on some vertices; for example we put stronger hypothesis on the vertices 3 and 4, since the only maximal clique they share is the smallest one. As we can see, the dynamics is able to extract also the maximal clique $\{2, 3, 4\}$. The solutions we found for this small example are the only stable ones for the dynamics (5) when applied to T. Hence randomly choosing the initial state we will certainly end up with a maximal clique, but clearly the maximal cliques with a larger basin of attraction are more likely to be extracted.

7 Conclusions and Future Work

In this paper we provide a generalization of a well-known extremal graph theory result, i.e. the Motzkin-Straus theorem, to k-uniform hypergraphs, and through it, we are able to provide a bridge between the purely discrete problem of finding maximal cliques in k-graphs and a minimization task of a continuous function. More precisely we introduce an isomorphism from the set of maximal/maximum cliques of a k-graph G and the set of local/global minima of the function $h_{\bar{G}}$. In this way we can focus our attention on minimizing $h_{\bar{G}}$ in order to find maximal cliques in G. Nevertheless, in the last section we provide also a dynamical system, derived from a result due to Baum and Eagon, to easily solve the optimization problem at hand. This basically furnishes an heuristic for the maximum clique problem on k-graphs.

This result opens a wide range of possible future works. First of all, we may conduct experiments on the effectiveness of our heuristic for the maximum clique problem on k-graphs, however we expect in general performances on hypergraphs comparable with those obtained by Pelillo [23] on simple graphs. Even more interesting could be the theoretical applications carrying on with our work, such as finding new bounds on the clique number of k-uniform hypergraphs, or further generalizing the Motzkin-Straus theorem to vertex-weighted and edge-weighted hypergraphs.

References

1. Baum, L.E., Eagon, J.A.: An inequality with applications to statistical estimation for probabilistic functions of Markov processes and to a model for ecology. Bull. Amer. Math. Soc. 73, 360–363 (1967)
2. Baum, L.E., Petrie, T., Soules, G., Weiss, N.: A maximization technique occurring in the statistical analysis of probabilistic functions of Markov chains. Ann. Math. Statist. 41, 164–171 (1970)
3. Baum, L.E., G.R.: Sell Growth transformations for functions on manifolds. Pacific J. Math. 27, 211–227 (1968)
4. Berge, C.: Hypergraphs. Combinatorics of Finite Sets. Ed. North-Holland, Amsterdam (1989)
5. Blakley, G.R.: Homogeneous nonnegative symmetric quadratic transformations. Bull. Amer. Math. Soc. 70, 712–715 (1964)

6. Bomze, I.M., Pelillo, M., Giacomini, R.: Evolutionary approach to the maximum clique problem: empirical evidence on a larger scale. Developments in Global Optimization, 95–108 (1997)
7. Bomze, I.M., Budinich, M., Pardalos, P.M., Pelillo, M.: The maximum clique problem. In: Handbook of Combinatorial Optimization (Supplement Volume A), pp. 1–74 (1999)
8. Bomze, I.M., Budinich, M., Pelillo, M., Rossi, C.: Annealed replication: a new heuristic for the maximum clique problem. Discr. Appl. Math. 121(1-3), 27–49 (2002)
9. Budinich, M.: Exact bounds on the order of the maximum clique of a graph. Discr. Appl. Math. 127, 535–543 (2003)
10. Bunke, H., Dickinson, P.J., Kraetzl, M.: Theoretical and Algorithmic Framework for Hypergraph Matching. In: ICIAP, pp. 463–470 (2005)
11. Faugeras, O.D., Berthod, M.: Improving consistency and reducing ambiguity in stochastic labeling: an optimization approach. IEEE Trans. Pattern Anal. Machine Intell. 3, 412–424 (1981)
12. Frankl, P., Rödl, V.: Hypergraphs do not jump. J. Combinatorica 4, 149–159 (1984)
13. Gibbons, L.E., Hearn, D.W., Pardalos, P.M.: A continuous based heuristic for the maximum clique problem. In: Cliques, Coloring and Satisfiability: 2nd DIMACS Impl. Chall., vol. 26, pp. 103–124 (1996)
14. Gibbons, L.E., Hearn, D.W., Pardalos, P.M., Ramana, M.V.: Continuous characterizations of the maximum clique problem. Math. Oper. Res. 22, 754–768 (1997)
15. Khot, S.: Improved inapproximability results for maxclique, chromatic number and approximate graph coloring. In: Proc. of 42nd Ann. IEEE Symp. on Found. of Comp. Sc., pp. 600–609 (2001)
16. Mohammed, J.L., Hummel, R.A., Zucker, S.W.: A gradient projection algorithm for relaxation labeling methods. IEEE Trans. Pattern Anal. Machine Intell. 5, 330–332 (1983)
17. Motzkin, T.S., Straus, E.G.: Maxima for graphs and a new proof of a theorem of Turán. Canad. J. Math. 17, 533–540 (1965)
18. Mubay, D.: A hypergraph extension of Turán's theorem. J. Combin. Theory B 96, 122–134 (2006)
19. Papa, D.A., Markov, I.: Hypergraph Partitioning and Clustering. In: Approximation Algorithms and Metaheuristics, pp. 61.1– 61.19 (2007)
20. Pardalos, P.M.: Continuous approaches to discrete optimization problems. In: Nonlinear Optimization and Applications, pp. 313–328 (1996)
21. Pardalos, P.M., Phillips, A.T.: A global optimization approach for solving the maximum clique problem. Int. J. Comput. Math. 33, 209–216 (1990)
22. Pavan, M., Pelillo, M.: Generalizing the Motzkin-Straus theorem to edge-weighted graphs, with applications to image segmentation. In: Rangarajan, A., Figueiredo, M.A.T., Zerubia, J. (eds.) EMMCVPR 2003. LNCS, vol. 2683, pp. 485–500. Springer, Heidelberg (2003)
23. Pelillo, M.: Relaxation labeling networks for the maximum clique problem. J. Artif. Neural Networks 2, 313–328 (1995)
24. Pelillo, M., Jagota, A.: Feasible and infeasible maxima in a quadratic program for maximum clique. J. Artif. Neural Networks 2, 411–420 (1995)
25. Rota Bulò, S., Pelillo, M.: A Continuous Characterization of Maximal Cliques in k-uniform Hypergraphs Tech. Report CS-2007-4, "Ca' Foscari" University of Venice (2007)
26. Sos, V.T., Straus, E.G.: Extremal of functions on graphs with applications to graphs and hypergraphs. J. Combin. Theory B 63, 189–207 (1982)

27. Turán, P.: On an extremal problem in graph theory (in Hungarian). Mat. ès Fiz. Lapok 48, 436–452 (1941)
28. Wilf, H.S.: The eigenvalues of a graph and its chromatic number. J. London Math. Soc. 42, 330–332 (1967)
29. Wilf, H.S.: Spectral bounds for the clique and independence numbers of graphs. J. Combin. Theory Ser. B 40, 113–117 (1986)
30. Zhou, D., Huang, J., Schölkopf, B.: Learning with hypergraphs: clustering, classification, embedding. Neural Inform. Proc. Systems 19 (2006)

Hybrid Heuristics for Multi-mode Resource-Constrained Project Scheduling

Celso Tchao[1,⋆] and Simone L. Martins[2,⋆⋆]

[1] Instituto Aerus de Seguridade Social, Rio de Janeiro, RJ, Brazil
[2] Universidade Federal Fluminense, Departamento de Ciência de Computação, Niterói, RJ, Brazil

Abstract. This paper describes some tabu search based heuristics with path relinking for the multi-mode resource-constrained project scheduling problem. Path relinking is used as a post optimization strategy, so that it explores paths that connect elite solutions found by the tabu search based heuristics. Computational results show that path relinking is able to improve the tabu search based heuristics, and that these hybrid heuristics are able to find good quality solutions in quite short computational times.

1 Introduction

The resource constrained project scheduling problem (RCPSP) consists in minimizing the time to execute a project composed by activities, which are interrelated by precedence relations and demand scare resources. This is an NP-complete problem [1].

This paper deals with its multiple mode version (MRCPSP), which is a generalization of the RCPSP and is a strong NP-hard problem. In this problem, each activity may be performed in one of several modes, where each mode specifies the duration of the activity, the number of needed resources and their types. There are two types of resources: renewable and non-renewable. The renewable resources such as machines and workers have a limited period availability, and the non-renewable resources such as raw materials are limited for the entire project.

The MRCPSP consists in minimizing the makespan of a project, composed by a set of activities J, subject to a set of precedence constraints P and to a set of resource constraints. A set of resources R consists of renewable R^{re} and non-renewable R^{non} resources. The total amount of a k renewable resource $r_k^{re} \in R^{re}$, required simultaneously by more than one activity in each time t, should not exceed R_k^{re}. For each k resource $r_k^{non} \in R^{non}$, the total amount required by all activities should not exceed R_k^{non} within the whole execution period. Each activity $j \in J$ can execute in one mode m_j chosen from a set M_j. Each mode

⋆ Celso Tchao developed this work as a M.Sc. student at Universidade Federal Fluminense and was sponsored by FAPERJ research grant E-26/180.478/2006.
⋆⋆ Work sponsored by FAPERJ research grant E-26/180.478/2006.

V. Maniezzo, R. Battiti, and J.-P. Watson (Eds.): LION 2007 II, LNCS 5313, pp. 234–242, 2008.

$m_j \in M_j$ specifies, for an activity j, its duration p_{m_j} and the number of required resources of each type, r^{re}_{k,m_j} and r^{non}_{k,m_j}.

We can find some exact procedures developed to solve this problem [2,3,4]. Due to the NP-hardness of the problem, it is difficult to optimally solve large projects using exact methods. Some heuristic methods have been developed in order to find near optimal solutions in reduced computation times. Simulated annealing algorithms can be found in [5,6,7,8] and genetic algorithms were proposed in [9,10,11]. Nonobe and Ibaraki [12] developed a tabu search algorithm.

In this paper, we describe some heuristics developed for the MRCPSP based on tabu search, and a path relinking method applied to the pool of solutions obtained by these heuristic procedures. We show that path relinking can improve the quality of these solutions.

The paper is organized as follows. The next section describes the developed heuristics and Section 3 reports the computational results. The last section presents conclusions and future work.

2 Proposed Heuristics

Some algorithms using tabu search for RCPSP and MRCPSP achieved very good results [12,13,14]. Therefore, we decided to develop some heuristics based on tabu search and tried to improve the obtained results using a path relinking intensification procedure.

We use the same solution representation described in [12]. Given the set of activities J, the set of resources R, the set of modes M_j of each activity j, a solution is represented by a scheme (m, s) consisting of a vector of activity modes $m = (m_j | j \in J)$ and a list of activity starting time $s = (s_j | j \in J)$. Each mode m_j specifies a duration p_{m_j} for each activity j, the units of each k renewable resource r^{re}_k required by the mode in each period and the units of each k non-renewable resource r^{non}_k consumed. The conclusion time of an activity j is $c_j = s_j + p_{m_j}$.

2.1 Heuristics Based on Tabu Search

The underlying ideas of tabu search were proposed by Glover [15] and it was later developed in [16,17,18]. It may be viewed as a dynamic neighborhood method that makes use of memory to drive the search by escaping from local optima and avoiding cycling [19]. Contrarily to memoryless heuristics such as simulated annealing, and to methods that use rigid memory structures such as branch-and-bound, tabu search makes use of flexible and adaptive memory designs.

The developed heuristics are based on [12]. In order to improve the efficiency of the algorithm, a solution is represented by a scheme (m, l), instead of (m, s), where the list l represents an ordered relation among the activities. There is a special procedure to construct feasible solutions (m, s) from a scheme (m, l), which respects the precedence and resource constraints, and generates start and end times for all activities found in the ordered list l using the established modes

procedure TabuAlgorithm($J, P, M, R^{re}, R^{non}, T_{max}, maxcbt, Max\_Tabu\_Iter, Tabu\_Tenure$)
1. $S \leftarrow$GenerateInitialSolution();
2. $S^* \leftarrow S$;
3. $cbt \leftarrow 0$;
4. **while** $cbt < maxcbt$ **do**
5. $S \leftarrow TS\_neigh\_activity(S, Max\_Tabu\_Iter, Tabu\_Tenure)$;
6. **if** $c(S) < c(S^*)$ **then**
7. $S^* \leftarrow S$
8. **end_if**
9. $S \leftarrow TS\_neigh\_mode(S, Max\_Tabu\_Iter, Tabu\_Tenure)$;
10. **if** $c(S) < c(S^*)$ **then**
11. $S^* \leftarrow S$
12. **end_if**
13. $cbt \leftarrow cbt + 1$;
14. **end_while**;
15. **return** S;

end.

Fig. 1. Tabu search for MRCPSP

in m. A solution is considered not feasible only if its makespan is larger than a parameter T_{max}.

Figure 1 shows the developed algorithm based on tabu search. The input parameters are the activity list J, the precedence relations among activities P, the set of modes allowable for each activity M, the set of renewable R^{re} and nonrenewable R^{non} resources, the maximum time allowable for the makespan so that a solution is considered feasible T_{max}, the maximum number of iterations $maxcbt$, the maximum number of iterations for each Tabu Search procedure Max_Tabu_Iter and the number of iterations to consider a move as tabu, $Tabu\_Tenure$.

We try to find an initial solution $S = (m^{(0)}, l^{(0)})$, in line 1, Figure 1, which simultaneously does not violate the constraints related to maximum makespan, precedence and nonrenewable resources. First, an incumbent mode list $m^{(0)}$ is generated by choosing a mode for each activity with the smallest relative nonrenewable resource consumption. Then a tabu search is applied on the neighborhood of this solution which is obtained by changing a mode of an activity j to another mode. The aim of this tabu search is to choose minimum durations for p_{m_j} and for the number of non-renewable resources r^{non}_{k,m_j}. After obtaining $m^{(0)}$, the list $l^{(0)}$ is generated using the Most Total Successors Rule, where the first activity to be scheduled presents more direct or indirect successors.

After generating an initial solution, two algorithms $TS\_neigh\_activity$ and TS_neigh_mode based on tabu search algorithm are applied $maxcbt$ times, as shown in line 4, Figure 1. The neighborhoods used by each algorithm are very distinct among themselves.

The *TS_neigh_activity* procedure looks for the best feasible non-tabu neighbor solution $S' \in N(S)$ from all neighbors $n \in N(S)$. Each neighbor n is obtained by taking an activity pair (i, j) in l, where j is scheduled after i, and reschedule j immediately before i. The incumbent solution is replaced by S' and the related pair (i', j') is inserted into the tabu list remaining in this list for *Tabu_Tenure* iterations. This procedure is repeated *Max_Tabu_Iter* iterations. The best known solution is updated in line 7.

The *TS_neigh_mode* procedure looks for the best feasible non-tabu neighbor solution $S' \in N(S)$. But this procedure uses a different neighborhood. Each neighbor n is obtained by choosing an activity j from J and changing its current mode m_j to another m'_j. There are two ways used to consider $N(S)$: greedy and non-greedy. In the greedy mode, an activity j is randomly selected, all mode changes related to it are evaluated, and S' is set to the best among them. In the non-greedy mode, we evaluate mode changes for all activities, and S' is set to the best among them. For both cases, the incumbent solution is replaced by S' and the related pair (m_j, m'_j) is inserted into the tabu list remaining in this list for *Tabu_Tenure* iterations. This procedure is repeated *Max_Tabu_Iter* iterations. The best known solution is updated in line 11.

2.2 Path Relinking

Path-relinking was originally proposed by Glover [20] as an intensification strategy exploring trajectories connecting elite solutions obtained by tabu search or scatter search. Starting from one or more elite solutions, paths in the solution space leading toward other elite solutions are generated and explored in the search for better solutions. To generate paths, moves are selected to introduce attributes in the current solution that are present in the elite guiding solution.

Several alternatives have been considered and combined in recent successful implementations of path-relinking in conjunction with tabu search, GRASP, and genetic algorithms [21,22,23,24,25,26,27].

We implemented a post-optimization path-relinking applied to a pool of solutions generated by the algorithms described in previous section. Each time a better solution is found by the tabu search procedure, it is inserted in the pool N_{el}, which is a vector composed by the best solutions generated by the algorithm, ordered by their makespan values (from worst to best). The path-relinking is applied to each adjacent pair in N_{el} in order to reduce the path-relinking execution time. The solution $N_{el}[i]$ is the initial solution and $N_{el}[i+1]$ is the guiding solution. We implemented the path-relinking by adjusting only the modes of the initial solution to the guiding solution. Therefore, we verify the activities in $N_{el}[i]$ that are associated to a different mode in $N_{el}[i+1]$. For all these activities, we generate a new feasible solution (if possible) by changing the mode of an activity in the initial solution to the mode of the same activity in the guiding solution. The best generated solution is selected as the new initial solution and the procedure is repeated until all activities in the initial solution have the same modes of the guiding solution.

3 Computational Results

We implemented eight versions of the algorithms previously described, as shown in Table 1. The first column is the name of the algorithm. The second column indicates if *TS_neigh_activity* procedure is executed. The third column shows the mode that *TS_neigh_mode* procedure (greedy or non-greedy) is implemented and the fourth column presents the number of iterations that a move is considered tabu. For example, algorithm V1 uses the tabu search executing only the *TS_neigh_mode* procedure, its greedy version and infinite tabu tenure, i.e, once a move is considered tabu, it will never be selected again.

For all algorithms the parameter *maxcbt* is equal to 5 and *Max_tabu_Iter* is set to 1000. We chose these parameters for the tabu search procedure to execute the same number of iterations performed in [12], as we based our procedures in this work.

The path-relinking is applied in all versions and the pool has a maximum size of 100. We chose this maximum size in order to allow all better solutions to be inserted in the pool. So, each instance of the problem may present different number of solutions in the pool, because all better solutions found by the tabu search procedure are inserted in the pool.

Table 1. Algorithm versions

Version	TS_neigh_activity	TS_neigh_mode	Tabu_tenure
V1	NO	Greedy	∞
V1.a	NO	Greedy	30
V1.b	NO	Greedy	1500
V2	NO	Non-Greedy	∞
V2.a	NO	Non-Greedy	30
V2.b	NO	Non-Greedy	1500
V3	YES	Non-Greedy	∞
V4	YES	Greedy	∞

We evaluated the algorithms by using some benchmark problems found in http:129.187.106.231/psplib.The available instances in this library present distinct characteristics related to number of activities, activity modes, use of resources, etc. We used the j30mm instances which consists of multi-mode problems with 30 activities. They are considered the most difficult instances of this type in this library. There are 640 instances in this group.

All codes were written in C++ and compiled using the Integrated Development Environment DEV C++ and run on an IBM compatible PC with a 1 G HZ Pentium Processor.

For the 640 instances, there are 506 known optimal solutions and 46 known feasible solutions [4]. The results obtained are compared to the optimal or best known solutions available. In Table 2, we present the average and maximum relative deviation from the optimal or the best known solution, the percentage

Table 2. Results for all algorithms

Algorithm	Aver. Dev.(%)	Max. Dev.(%)	Feas. (%)	Best (%)
V1	5.4	44.7	96.4	44.4
V1.a	5.5	31.0	96.4	42.1
V1.b	9.1	36.8	96.4	32.9
V2	11.7	61.0	96.4	33.8
V2.a	11.7	61.0	96.4	33.8
V2.b	11.7	61.0	96.4	33.8
V3	10.2	61.0	96.2	38.4
V4	4.5	29.8	96.2	46.0

of feasible solutions and the percentage of best solutions (optimal or best know solution) found by each algorithm.

We can see that V4 presents the best results, V1 and its variations present the second best results and V2 and its variations and V3 present worse results. These results show that the greedy version of the *TS_neigh_mode* procedure performs better that the non-greedy version, and that the *TS_neigh_activity* procedure improves the results when executed together with *TS_neigh_mode* procedure.

In order to show the influence of path relinking for all procedures, Table 3 shows the number of solutions improved by the path-relinking technique for each algorithm and the number of best solutions achieved. The use of this technique brings some benefits by improving 45% of the solutions generated by the tabu search for V1 and V1.a, 34% for V4 and 33% for V1.b. We can also see that the number of optimal or equal to the best known solutions was increased when using path-relinking. There were no improvements for V2, V2.a, V2.b and V3.

Table 3. Number of solutions improved by path-relinking

	V1	V1.a	V1.b	V2	V2.a	V2.b	V3	V4
Number of solutions improved	239	239	174	0	0	0	2	182
Number of best solutions achieved	56	45	40	0	0	0	0	42

Table 4 shows the number of initial solutions improved by the algorithms. We can see that 59.6% of the solutions were improved for V1, 60.7% for V1.a, 48.5% for V1.b, 53% for V2, 57.6% for V3 and 68% for V4, showing that all versions were able to improve the initial solutions.

Table 4. Number of initial solutions improved by tabu search

V1	V1.a	V1.b	V2	V2.a	V2.b	V3	V4
317	323	258	282	282	282	306	361

Table 5. Computational times (seconds)

	V1	V1.a	V1.b	V2	V2.a	V2.b	V3	V4
mean time	14	15	14	13	40	40	262	264
max. time	18	33	28	17	50	58	660	556
std dev.	1	2	1	1	3	4	79	102

Table 5 shows statistics for times obtained by processing all instances for each algorithm. The mean times for algorithms V3 and V4 are larger than those obtained for V1 and V2. It follows because V3 and V4 executes both procedures *TS_neigh_activity* and *TS_neigh_mode*. We observed that the time needed to generate the initial solution and to perform path relinking is much shorter than the processing time demanded by the procedures based on tabu search.

4 Concluding Remarks

This paper presented some versions of hybrid heuristics to solve the multi-mode resource constrained project scheduling problem (MRCPSP). Some heuristics based on tabu search were hybridized with a path-relinking strategy. Experimental results showed that we were able to find good quality solutions in quite short computational times.

An important contribution of this work is to show that the use of path relinking significantly improves the quality of the solutions generated by the tabu search based heuristics and does not cause a significant increase in computational time.

As described in [27], there are three components that are critical in the design of a path-relinking procedure: rules for building the elite set, rules for choosing the initial and guiding solution and the neighborhood structure for moving from the initial to the guiding solution. In this paper, we implemented path-relinking as a post-optimization strategy and adopted just one option for each of these components. As we obtained good results using this strategy, we intend to explore other options and also to study the use of path-relinking as an intensification strategy applied during the execution of the tabu search based heuristics.

References

1. Blazewicz, J., Lenstra, J., Kan, A.R.: Scheduling projects subject to resource constraints: classification and complexity. Discrete Applied Mathematics (5), 11–24 (1983)
2. Patterson, J., Slowinski, R., Talbot, F., Weglarz, J.: An algorithm for a general class of precedence and resource constrained scheduling problem. In: Slowinski, R., Weglarz, J. (eds.) Advances in Project Scheduling, pp. 3–28. Elsevier, Amsterdam (1989)
3. Hartmann, S., Drexl, A.: Project scheduling with multiple modes: A comparison of exact algorithms. Networks (32), 283–297 (1998)

4. Zhu, G., Bard, J., Yu, G.: A branch-and-cut procedure for the multimode resource-constrained project-scheduling problem. INFORMS Journal on Computing 18(3), 377–390 (2006)
5. Slowinski, R., Soniewicki, B., Weglarz, J.: DSS for multiobjective project scheduling. Eur. J. Opl. Res. 18(79), 220–229 (1994)
6. Boctor, F.: Resource-constrained project scheduling by simulated annealing. International Journal of Production Research 8(34), 2335–2351 (1996)
7. Bouleimen, K., Lecocq, H.: A new efficient simulated annealing algorithm for the resource-constrained project scheduling problem and its multiple mode version. European Journal of Operational Research (149), 268–281 (2003)
8. Jozefowska, J., Mika, M., Rozycki, R., Waligora, G., Weglarz, J.: Simulated annealing for multi-mode resource-constrained project scheduling. Annals of Operations Research (102), 137–155 (2001)
9. Alcaraz, J., Maroto, C., Ruiz, R.: Solving the multi-mode resource constrained project scheduling problem with genetic algorithms. Journal of the Operational Research Society (54) (54), 614–626 (2003)
10. Hartmann, S.: Project scheduling with multiple modes: a genetic algorithm. Ann. Opns. Res (102), 111–135 (2001)
11. Ozdamar, L.: A genetic algorithm approach to a general category project scheduling problem. IEEE Trans. Syst. Man Cybern (29), 44–59 (1999)
12. Nonobe, K., Ibaraki, T.: Formulation and tabu search algorithm for the resource constrained project scheduling problem. In: Ribeiro, C., Hansen, P. (eds.) Essays and Surveys in Metaheuristics, pp. 557–588. Kluwer Academic Publishers, Dordrecht (2002)
13. Baar, T., Brucker, P., Knust, S.: Tabu search algorithms for resource-constrained project scheduling problems. In: Voss, S., Martello, S., Osman, I., Roucairol, C. (eds.) Metaheuristics: Advances and Trends in Local Search Paradigms for Optimisation, pp. 1–18. Kluwer Academic Publishers, Dordrecht (1997)
14. Cavalcante, C., Cavalcante, V., Ribeiro, C., de Souza, C.: Parallel cooperative approaches for the labor constrained scheduling problem. In: Ribeiro, C., Hansen, P. (eds.) Essays and Surveys in Metaheuristics, pp. 201–225. Kluwer Academic Publishers, Dordrecht (2002)
15. Glover, F.: Heuristics for integer programming using surrogate constraints. Decision Sciences 8(1), 156–166 (1977)
16. Glover, F.: Future paths for integer programming and links to artificial intelligence. Computers and Operations Research 13, 533–549 (1986)
17. Glover, F.: Tabu search - Part I. ORSA Journal on Computing 1, 190–206 (1989)
18. Glover, F.: Tabu search - Part II. ORSA Journal on Computing 2, 4–32 (1990)
19. Glover, F., Laguna, M.: Tabu Search. Kluwer, Dordrecht (1997)
20. Glover, K.: Tabu search and adaptive memory programming - advances, applications and challenges. In: Barr, R., Helgason, R., Kennington, J. (eds.) Interfaces in Computer Science and Operations Research, pp. 1–75. Kluwer Academic Publishers, Dordrecht (1996)
21. Festa, P., Pardalos, P., Resende, M., Ribeiro, C.: Randomized heuristics for the max-cut problem. Optimization Methods and Software 7, 1033–1058 (2002)
22. Martins, S., Ribeiro, C., Rosseti, I.: Applications and parallel implementations of metaheuristics in network design and routing. In: Manandhar, S., Austin, J., Desai, U., Oyanagi, Y., Talukder, A.K. (eds.) AACC 2004. LNCS, vol. 3285, pp. 205–213. Springer, Heidelberg (2004)

23. Resende, M., Ribeiro, C.: GRASP with path-relinking: Recent advances and applications. In: Ibaraki, T., Nonobe, K., Yagiura, M. (eds.) Metaheuristics: Progress as Real Problem Solvers, pp. 29–63. Springer, Heidelberg (2005)
24. Ribeiro, C., Uchoa, E., Werneck, R.: A hybrid GRASP with perturbations for the Steiner problem in graphs. INFORMS Journal on Computing 14, 228–246 (2002)
25. Silva, G., de Andrade, M., Ochi, L., Martins, S., Plastino, A.: New heuristics for the maximum diversity problem. J. Heuristics 13, 315–336 (2007)
26. Ribeiro, C., Vianna, D.: A genetic algorithm for the phylogeny problem using an optimized crossover strategy based on path-relinking. Revista Tecnologia da Informação 3, 67–70 (2003)
27. Ho, S.C., Gendreau, M.: Path relinking for the vehicle routing problem. J. Heuristics 12, 55–72 (2006)

Author Index

Printing: Mercedes-Druck, Berlin
Binding: Stein + Lehmann, Berlin